Studies in Fuzziness and Soft Computing

Volume 412

The series "Studies in Fuzziness and Soft Computing" contains publications on various topics in the area of soft computing, which include fuzzy sets, rough sets, neural networks, evolutionary computation, probabilistic and evidential reasoning, multi-valued logic, and related fields. The publications within "Studies in Fuzziness and Soft Computing" are primarily monographs and edited volumes. They cover significant recent developments in the field, both of a foundational and applicable character. An important feature of the series is its short publication time and world-wide distribution. This permits a rapid and broad dissemination of research results.

Indexed by SCOPUS, DBLP, WTI Frankfurt eG, zbMATH, SCImago.

All books published in the series are submitted for consideration in Web of Science.

More information about this series at http://www.springer.com/series/2941

Tofigh Allahviranloo · Soheil Salahshour
Editors

Advances in Fuzzy Integral and Differential Equations

Springer

Editors
Tofigh Allahviranloo
Faculty of Engineering and Natural Sciences
Bahçeşehir University
Istanbul, Turkey

Soheil Salahshour
Faculty of Engineering and Natural Sciences
Bahçeşehir University
Istanbul, Turkey

ISSN 1434-9922 ISSN 1860-0808 (electronic)
Studies in Fuzziness and Soft Computing
ISBN 978-3-030-73710-8 ISBN 978-3-030-73711-5 (eBook)
https://doi.org/10.1007/978-3-030-73711-5

This Springer imprint is published by the registered company Springer Nature Switzerland AG
The registered company address is: Gewerbestrasse 11, 6330 Cham, Switzerland

Contents

Fuzzy Differential Equations

Differential and Integral Calculus for Fuzzy Number-Valued Functions with Interactivity

Laécio Carvalho de Barros, Francielle Santo Pedro, and Estevão Esmi

Abstract This chapter presents the theory of differential and integral calculus for autoregressive fuzzy processes, that is, fuzzy number-valued functions F such that $F(t)$ and $F(t+h)$ are interactive fuzzy numbers for every $|h|$ sufficiently small. Recall that interactivity between fuzzy variables are described in terms of joint possibility distributions and plays a similar role to that of dependence between random variables. We present a theory of differential calculus for certain autoregressive fuzzy processes based on a special type of interactivity called $\mathcal{F}$-interactivity. Next, we introduce a Banach space, denoted by $\mathbb{R}_{\mathsf{F}(A)}$, which is composed by fuzzy numbers that are interactive with each other, such that the vector addition corresponds to an interactive operation. The concepts of differential and integral for mappings from $\mathbb{R}$ to $\mathbb{R}_{\mathsf{F}(A)}$ correspond to the classical notions of Fréchet derivative and Riemann integral in Banach spaces. Thus, the theory of differential and integral calculus for functions taking values in $\mathbb{R}_{\mathsf{F}(A)}$ is different from those found in fuzzy literature that are based in (generalized) Hukuhara differentiability and fuzzy Aumann (or Riemann) integral, which involve non-interactive arithmetic. In contrast to the usual fuzzy Aumann integral, the diameter of a fuzzy function defined in terms of the Riemann integral in $\mathbb{R}_{\mathsf{F}(A)}$ may decrease. We also study fuzzy ordinary and partial differential equations in $\mathbb{R}_{\mathsf{F}(A)}$.

L. C. de Barros · E. Esmi (✉)
Universidade Estadual de Campinas, Campinas, SP 13081-970, Brazil
e-mail: eesmi@unicamp.br

L. C. de Barros
e-mail: laeciocb@unicamp.br

F. S. Pedro
Universidade Federal de São Paulo, Osasco, SP 06110-295, Brazil
e-mail: fsimoes@unifesp.br

T. Allahviranloo and S. Salahshour (eds.), *Advances in Fuzzy Integral and Differential Equations*, Studies in Fuzziness and Soft Computing 412,
https://doi.org/10.1007/978-3-030-73711-5_1

1 Introduction

The first approach to calculus theory for fuzzy number-valued function was proposed by Puri and Ralescu where they introduced the notion of fuzzy Hukuhara differentiability or, for short, H-differentiability [34]. The concept of fuzzy Aumann integrability was introduced by [35]. Subsequently, Kaleva established a fundamental theorem of calculus using these two concepts [27]. However, from a practical point of view, the application of such a notion of differentiability is quite limited. For example, let A be a fuzzy number and $F(t) = g(t)A$ be such that g is a differentiable in $\mathbb{R}$. The fuzzy function F is not H-differentiable at t_0 if $|g|$ is a strictly decreasing function at t_0. Stefanini and Bede proposed the concept of generalized Hukuhara differentiability (or, for short, gH-differentiability) which extends the H-differentiability and is the most common and well-known [47]. The gH-derivative of the function $F(t) = g(t)A$ at t_0 is given by $g'(t_0)A$ which is an intuitive result.

One the other hand, when we are dealing with a fuzzy function F that describes an uncertain process, it is reasonable to assume that the value of this process in a near future $F(t+h)$ is close (or related) to the value of the present $F(t)$, even though the quantities are inaccurate and/or uncertain. In addition, depending on the phenomenon described by the fuzzy process F, one can expect that the level of uncertainty at instant $t+h$ should be reduced or increased in comparison to the instant t. In fuzzy set theory, interactivity (or dependence) between $F(t)$ and $F(t+h)$ can be described by the notion of joint possibility distributions. The concept of interactivity for fuzzy variables is analogous to concept of dependence for random variables that arises from joint probability distributions. Intuitively, according Zadeh, two fuzzy quantities "are interactive if the assignment of a value to one affects the fuzzy restrictions placed on the other" [52]. Once the interactivity between two fuzzy quantities is known, we can perform arithmetic operations taking into account it. This approach leads to the so-called interactive arithmetic (see Sect. 3). The gH-derivative is given in terms of the gH-difference of $F(t+h)$ and $F(t)$ which, in its turn, implicitly takes into account the existence of a type of interactivity described by a certain joint possibility distribution [49]. Thus, the gH-difference can be viewed as an interactive subtraction. If we assume other types of interactivity between $F(t)$ and $F(t+h)$, then we obtain other interactive differences of $F(t)$ and $F(t+h)$ and, consequently, other notions of differentiability for fuzzy functions F [6]. The theory of calculus based on interactive arithmetic operations is called of interactive fuzzy calculus. In this chapter, we focus on a type of interactivity between fuzzy numbers, namely, the $\mathcal{F}$-interactivity, which leads to the notion of $\mathcal{F}$-interactive differentiability.

Despite the fact that gH-difference is a particular case of interactive fuzzy subtraction, many researchers consider the other arithmetic operations on fuzzy numbers, namely, addition, multiplication, and division, as being the standard operations which is an extension of interval arithmetic. Recall that, in this case, the sum of two uncertain quantities given by fuzzy numbers never results in a real number. However, in many situations and phenomena, we may have the sum of two fuzzy numbers (or uncertain quantities) is a real number. For example, the weight of a pregnant woman is a real number but consist of the sum of the weights of the woman and her fetus that are uncertain approximations.

In the context of fuzzy dynamical system, interactivity also arises naturally. A damped harmonic oscillator may be governed by an equation of the form $aX'' + bX' + cX = 0$ where X, X', and X'' are uncertain quantities. Salgado et al. studied these types of systems with fuzzy interactivity [38, 39]. Another interesting dynamic system where the usual arithmetic can not be properly applied is the compartmental epidemiological model without vital dynamics whose the initial conditions and/or parameters are uncertain [11]. For example, in the susceptible-infected (SI) model without vital dynamics, we have that the sum of the infected individuals with the successive ones is equal to the total population, that is, $S + I = N$. Note that, from a practical point of view, it is reasonable to assume that the intial conditions are uncertain, since it may be impossible to determine the number of infected due to various factors such as false positives, the existence asymptomatic individuous, impossibility to test all population at same time, etc. Let S and I be fuzzy numbers, we have that the sum of S and I is equal to the real number N. To ensure that the total population is a real number, the initial conditions must be interactive fuzzy numbers. A similar situation appears in chemical reactions where at each stage of the reaction we have that the mass of intermediate reagents is uncertain but the total is known [50]. In [41], authors investigate some boundary value problems with boundary conditions given by interactive fuzzy numbers. These examples suggest the use of interactive arithmetic operations and, consequently, the interactive calculus to model and study fuzzy process described by such fuzzy dynamical equations.

Given a fuzzy process F, determining gH-derivative of F at a point t may be a hard task. Most of this difficulty lies in verifying the hypothesis of the Stacking theorem. The calculation of the $\mathcal{F}$-interactive derivative assumes that $F(t)$ and $F(t+h)$ are $\mathcal{F}$-interactive which may be difficult to verify in practice too. In a further step, obtaining solutions for fuzzy differential equations given in terms of gH-differentiability may also be a hard and costly task from both practical and analytical point of views. Recall that the calculation of such solutions may require the analysis of types of gH-differentiability, existence of switch points [13], application of Stacking theorem [32], and to solve two differential equations for each $\alpha \in [0, 1]$. Again, in general, the verification of the hypothesis of Stacking theorem may be difficult in practical problems. Similarly, solving fuzzy differential equations using $\mathcal{F}$-interactive differentiability is not a easy task as well. In contrast, the concept A-linear interactivity, which is a special type of $\mathcal{F}$-interactive relation with respect to a fixed fuzzy number A, can be used to develop a simple and intuitive theory of calculus for functions taking values in a certain subclass of fuzzy numbers denoted by $\mathbb{R}_{\mathsf{F}(A)}$. A fuzzy function from $\mathbb{R}$ to $\mathbb{R}_{\mathsf{F}(A)}$ is also called an A-linearly interactive fuzzy process. The simplicity of such a theory of calculus is due to the fact the space $\mathbb{R}_{\mathsf{F}(A)}$ forms a Banach space that isomorphic to $\mathbb{R}^2$ [21]. Roughly speaking, the calculation of the derivative (or integral) of a A-linearly interactive fuzzy process is given in terms of the classical derivatives (or integrals) of a certain pair of two real functions. For example, the function $F(t) = g(t)A$, that we considered at the beginning of Introduction, corresponds to an A-linearly interactive fuzzy process.

Using this theory of calculus, we have that for every t_0 such that $g'(t_0)$ exists, the function F has derivative equal to $g'(t_0)A$.

This chapter is organized as follows. Section 2 provides some mathematical foundations used throughout this chapter. Section 3 presents the concept of interactivity for fuzzy numbers and the arithmetic operations with interactive operands which are the basis for the development of the interactive fuzzy calculus. A particular case of interactive differential calculus for certain fuzzy functions, namely, $\mathcal{F}$-interactive fuzzy processes, is presented in Sect. 4. Section 5 deals with another particular case of interactive calculus that is simpler to understand and use. Thus, beginner readers can choose to read this section first, before the previous ones which deal with more general cases.

2 Mathematical Background

This section reviews some basic concepts and notations from fuzzy set theory that are used throughout this chapter.

A fuzzy (sub)set A of the universe $X \neq \emptyset$ is described by its membership function $\mu_A : X \longrightarrow [0, 1]$, where $\mu_A(x)$ means the degree to which x belongs to A. For notational convenience, we write $A(x)$ instead of $\mu_A(x)$. We use the symbol $\mathbb{F}(X)$ to denote the set of all fuzzy subsets of X. The α-levels of the fuzzy subset A are classical subsets defined as follows:

$$[A]_\alpha = \{x \in X : A(x) \geq \alpha\}$$

for $0 < \alpha \leq 1$ and, when X is a topological space, such as $\mathbb{R}^n$, $[A]_0 = \overline{\{x \in U : A(x) > 0\}}$, where $\bar{Y}$ denotes the closure of $Y \subseteq X$. Here, we focus on a particular subclass of fuzzy sets of $\mathbb{R}$ called fuzzy numbers.

Definition 1 *(Fuzzy number)* [4, 20] The fuzzy subset A of $\mathbb{R}$ is a fuzzy number if

i. all their α-levels are closed and nonempty intervals of $\mathbb{R}$, and
ii. the support of A, $supp(A) = \{x \in \mathbb{R} : A(x) > 0\}$, is bounded.

A well-known example of fuzzy number is a triangular fuzzy number, denoted by a triple $(a; b; c)$ with $a \leq b \leq c$, that can be defined levelwise by

$$[a + \alpha(b - a), c - \alpha(c - b)], \ \forall \alpha \in [0, 1].$$

The family of fuzzy numbers is denoted by $\mathbb{R}_\mathcal{F}$.

For every fuzzy number A, we denote its α-levels by $[A]_\alpha = [a_\alpha^-, a_\alpha^+], \alpha \in [0, 1]$. The length of the α-level of A is defined by

$$\text{len}([A]_\alpha) = a_\alpha^+ - a_\alpha^-, \text{ for all } \alpha \in [0, 1].$$

The diameter of A is defined as being $\text{diam}(A) = \text{len}([A]_0)$.

Definition 2 [14] The metric d_∞ on $\mathbb{R}_\mathsf{F}$ is defined for every $A \in \mathbb{R}_\mathsf{F}$ and $B \in \mathbb{R}_\mathsf{F}$ as follows:

$$d_\infty(A, B) = \sup_{0 \le \alpha \le 1} \max\{|a_\alpha^- - b_\alpha^-|, |a_\alpha^+ - b_\alpha^+|\}.$$

Proposition 1 *[14] $(\mathbb{R}_\mathsf{F}, d_\infty)$ is a complete metric space and moreover, the following properties hold true:*

i. $d_\infty(U + W, V + W) = d_\infty(U, V), \forall U, V, W \in \mathbb{R}_\mathsf{F}$ i.e., d_∞ is translation invariant.
ii. $d_\infty(k \cdot U, k \cdot V) = |k| d_\infty(U, V), \forall U, V \in \mathbb{R}_\mathsf{F}, k \in \mathbb{R}$;
iii. $d_\infty(U + V, W + E) \le d_\infty(U, W) + d_\infty(V, E), \forall U, V, W, E \in \mathbb{R}_\mathsf{F}$.

A norm of a fuzzy number A is defined by

$$\|A\| = d_\infty(A, 0).$$

The Zadeh extension Principle can be viewed as a mathematical tool to extend classical functions to the fuzzy domain. Let $f : X \to Y$ and $A \in \mathbb{F}(X)$. The Zadeh extension of f at A is the fuzzy set $\hat{f}(A) \in \mathbb{F}(Y)$ given by [51]

$$\hat{f}(A)(y) = \sup_{x \in f^{-1}(y)} A(x) \tag{1}$$

where $f^{-1}(y)$ stands for the preimage of y under f, that is, $f^{-1}(y) = \{x \in X \mid f(x) = y\}$. By definition, $\sup \emptyset = 0$. The Zadeh extension principle can also be defined for functions with multiple arguments as follows. Let $f : X_1 \times \ldots \times X_n \to Y$ and $A_i \in \mathbb{F}(X_i)$, $i = 1, \ldots, n$. The Zadeh extension of f at $(A_1, \ldots, A_n)$ is the fuzzy set $\hat{f}(A_1, \ldots, A_n) \in \mathbb{F}(Y)$ given by

$$\hat{f}(A_1, \ldots, A_n)(y) = \sup_{(x_1, \ldots, x_n) \in f^{-1}(y)} \min\{A_1(x_1), \ldots, A_n(x_n)\}. \tag{2}$$

The next theorem characterizes the α-levels of the Zadeh extension of a continuous functions in terms of the α-levels of the corresponding arguments.

Theorem 1 *[5, 33]*

(a) If $f : \mathbb{R} \to \mathbb{R}$ is continuous and $A \in \mathbb{R}_\mathsf{F}$, then $\hat{f}(A) \in \mathbb{R}_\mathsf{F}$, and

$$[\hat{f}(A)]_\alpha = f([A]_\alpha), \quad \forall \alpha \in [0, 1].$$

(b) If $f : \mathbb{R}^n \to \mathbb{R}$ is continuous and $A_i \in \mathbb{R}_\mathsf{F}$, $i = 1, \ldots, n$, then $\hat{f}(A_1, \ldots, A_n) \in \mathbb{R}_\mathsf{F}$, and

$$[\hat{f}(A_1, A_2, \ldots, A_n)]_\alpha = f([A_1]_\alpha, [A_2]_\alpha, \ldots, [A_n]_\alpha), \quad \forall \alpha \in [0, 1].$$

The Zadeh extension principle can be used to define the standard arithmetic operations for fuzzy numbers as follows.

Definition 3 *(Standard Arithmetic)* Let $A, B \in \mathbb{R}_{\mathsf{F}}$.

1. The sum of A and B is given by $A + B = \hat{+}(A, B)$, where $+(a, b) = a + b$ for all $a, b \in \mathbb{R}$.
2. The difference of A and B is given by $A - B = \hat{-}(A, B)$, where $-(a, b) = a - b$ for all $a, b \in \mathbb{R}$.
3. The multiplication of A and B is given by $A \cdot B = \hat{\cdot}(A, B)$, where $\cdot(a, b) = ab$ for all $a, b \in \mathbb{R}$. If $A = \lambda \in \mathbb{R}$, then we speak of scalar multiplication and denote by λB.
4. The division of A by $B, 0 \notin [B]_0$, is given by $A \div B = \hat{\div}(A, B)$, where $\div(a, b) = a \div b$ for all $a, b \in \mathbb{R}$, $b \neq 0$.

The next proposition is a consequence of Theorem 1 and characterizes the standard arithmetic operations in terms of the α-levels of the corresponding operands.

Proposition 2 *[4, 29] Let $A, B \in \mathbb{R}_{\mathsf{F}}$ with $[A]_\alpha = [a_\alpha^-, a_\alpha^+]$ and $[B]_\alpha = [b_\alpha^-, b_\alpha^+]$. For every $\alpha \in [0, 1]$. We have*

1. $[A + B]_\alpha = [a_\alpha^- + b_\alpha^-, a_\alpha^+ + b_\alpha^+]$.
2. $[A - B]_\alpha = [a_\alpha^- - b_\alpha^+, a_\alpha^+ - b_\alpha^-]$.
3. $[A \cdot B]_\alpha = [\min Y_\alpha, \max Y_\alpha]$, where $Y_\alpha = \{a_\alpha^- b_\alpha^-, a_\alpha^- b_\alpha^+, a_\alpha^+ b_\alpha^-, a_\alpha^+ b_\alpha^+\}$.
4. If $A = \lambda \in \mathbb{R}$, then

$$[\lambda B]_\alpha = \lambda [B]_\alpha = \begin{cases} \left[\lambda b_\alpha^-, \lambda b_\alpha^+\right] & \text{if } \lambda \geq 0 \\ \left[\lambda b_\alpha^+, \lambda b_\alpha^-\right] & \text{if } \lambda < 0 \end{cases}.$$

5. $[A \div B]_\alpha = [\min Z_\alpha, \max Z_\alpha]$, where $Z_\alpha = \left\{\frac{a_\alpha^-}{b_\alpha^-}, \frac{a_\alpha^-}{b_\alpha^+}, \frac{a_\alpha^+}{b_\alpha^-}, \frac{a_\alpha^+}{b_\alpha^+}\right\}$ if $0 \notin [B]_0$.

The next proposition reveals the space of fuzzy numbers with the standard arithmetic and the scalar product does not satisfy the algebraic structure of a vector space.

Proposition 3 *[8, 16] Let $A, B, C \in \mathbb{R}_{\mathsf{F}}$ and $p, q \in \mathbb{R}$.*

1. $A + B = A + B$.
2. $A + (B + C) = (A + B) + C$.
3. $A + 0 = A$.
4. there exists $D \in \mathbb{R}_{\mathsf{F}}$ such that $A + D = 0$ iff $A \in \mathbb{R}$.
5. $(p + q)A = pA + qA$ if $pq \geq 0$.
6. $p(A + B) = pA + pB$.
7. $(pq)A = p(qA)$.

The standard subtraction is not useful to develop an interesting theory of differential calculus for fuzzy number-valued functions since $A - A \neq 0$ for every

$A \in \mathbb{R}_F \setminus \mathbb{R}$. To this end, other definitions of subtraction were proposed in the literature such as Hukuhara difference and its extensions. Given $A, B \in \mathbb{R}_F$, a fuzzy number D is said to be the Hukuhara difference of A and B, denoted by $D = B -_H A$, if [34]

$$D = A -_H B \Leftrightarrow A = B + D.$$

Note that if $A -_H B$ exits, then $len[A]_\alpha > len[B]_\alpha$, for all $\alpha \in [0, 1]$. This necessary condition brings a considerable restriction for the use of this fuzzy difference in practical problems. The notion of generalized Hukuhara difference extends the Hukuhara difference and does not require such conditions.

Definition 4 [47] Let $A, B \in \mathbb{R}_F$, the generalized Hukuhara difference (or, for short, gH-difference) is the fuzzy number D, if it exists, such that

$$A -_{gH} B = D \Longleftrightarrow \begin{cases} A = B + D \\ \text{or } B = A - D \end{cases},$$

where "+" and "−" represent, respectively, the standard addition and subtraction of fuzzy numbers given by of the Zadeh's extension principle.

If the gH-difference of two fuzzy number A and B exists, then [8]

$$[B -_{gH} A]_\alpha = [\min\{b_\alpha^- - a_\alpha^-, b_\alpha^+ - a_\alpha^+\}, \max\{b_\alpha^- - a_\alpha^-, b_\alpha^+ - a_\alpha^+\}]. \qquad (3)$$

In order to compare the method described in Sect. 4 to solve FIVPs with the method based on gH, we recall this notion of gH-differentiability.

Definition 5 [47] Let $F : [a, b] \to \mathbb{R}_\mathcal{F}$. We say that F is generalized Hukuhara differentiable (gH- differentiable for short) at $t_0 \in [a, b]$ if there exists an element $F'_{gH}(t_0) \in \mathbb{R}_\mathcal{F}$, such that

$$F'_{gH}(t_0) = \lim_{h \to 0} \frac{1}{h} \left[F(t_0 + h) -_{gH} F(t_0) \right],$$

where the limit is taken in the metric d_∞. At the end points, we consider only the one-sided derivative.

Definition 6 [9] Let $F : [a, b] \to \mathbb{R}_F$ be a fuzzy function such that $[F(t)]_\alpha = \left[f_\alpha^-(t), f_\alpha^+(t) \right]$ with f_α^- and f_α^+ differentiable at t_0. We say that

i. F is gH_1-differentiable at t_0 if

$$\left[F'_{gH_1}(t_0) \right]_\alpha = [(f_\alpha^-)'(t_0), (f_\alpha^+)'(t_0)], \quad \forall \alpha \in [0, 1].$$

ii. F is gH_2-differentiable at t_0, if

$$\left[F'_{gH_2}(t_0) \right]_\alpha = [(f_\alpha^+)'(t_0), (f_\alpha^-)'(t_0)], \quad \forall \alpha \in [0, 1].$$

The notion of fuzzy Aumann integral was introduced by Puri and Ralescu [35]. Fundamental theorems of calculus considering this notion of integrability and the H-differentiability is given in [27].

Definition 7 *(Aumann integral)* [27, 35] Let $F : [a, b] \to \mathbb{R}_{\mathsf{F}}$ be a fuzzy function. The fuzzy Aumann integral of F over $[a, b]$, denoted by $(FA) \int_a^b F(t)dt$, is defined by α-levels:

$$\left[(FA)\int_a^b F(t)dt\right]_\alpha = (A)\int_a^b [F(t)]_\alpha dt = (A)\int_a^b [f_\alpha^-(t), f_\alpha^+(t)]dt$$
$$= \left\{\int_a^b y(t)dt \mid y : [a, b] \to \mathbb{R} \text{ is a measurable selection of } [F(\cdot)]_\alpha\right\}, \tag{4}$$

for all $\alpha \in [0, 1]$, provided that (4) defines a fuzzy numbers.

According to [27, 35], a function $F : [a, b] \to \mathbb{R}_{\mathsf{F}}$ is said to be strongly measurable if, for every $\alpha \in [0, 1]$, the set-valued mapping $F_\alpha(t) = [F(t)]_\alpha$ is (Lebesgue) measurable on $[a, b]$ with the metric d_∞. Moreover, $F : [a, b] \to \mathbb{R}_{\mathsf{F}}$ is said to be integrable bounded if there exists an integrable function $h : [a, b] \to \mathbb{R}$ such that $\| F(t) \| \leq h(t)$ for all $t \in [a, b]$.

Proposition 4 *[27] If $F : [a, b] \to \mathbb{R}_{\mathsf{F}}$ is strongly measurable and integrably bounded function, then F is fuzzy Aumann integrable over $[a, b]$.*

By Remark 4.2 of [27], if the fuzzy number-valued function F is fuzzy Aumann integrable over $[a, b]$, then the integrals of the endpoint functions f_α^- and f_r^+ over $[a, b]$ exist and the following equality holds:

$$\left[(FA)\int_a^b F(t)dt\right]_\alpha = \left[\int_a^b f_\alpha^-(t)dt, \int_a^b f_\alpha^+(t)dt\right], \tag{5}$$

for all $\alpha \in [0, 1]$.

Besides the notion of fuzzy Aumann integral, there are other concepts of fuzzy integrability. Below we provide the definition of the fuzzy Riemann integral of a fuzzy number-valued function [24].

Definition 8 A function $F : [a, b] \longrightarrow \mathbb{R}_{\mathsf{F}}$ is said to be fuzzy Riemann integrable if there exists $S \in \mathbb{R}_{\mathsf{F}}$ such that for every $\epsilon > 0$ there exists $\delta > 0$ such that for all partition $a = x_0 < x_1 < \cdots < x_n = b$ with $x_k - x_{k-1} < \delta$, $k = 1, \ldots, n$, we have

$$d_\infty\left(\sum_{k=1}^n F(t_k)(x_k - x_{k-1}),\ S\right) < \epsilon,$$

where $t_k \in [x_{k-1}, x_k]$, $k = 1, \ldots, n$, and the summation is given in terms of the standard addition. In this case, S is said to be the fuzzy Riemann integral of F over $[a, b]$ and is also denote by $(FR) \int_a^b F(s)ds$.

The next proposition states that the notion of fuzzy Aumann integrability and of fuzzy Riemann integrability coincide if the integrand function is continuous.

Proposition 5 *[8] If $F : [a, b] \longrightarrow \mathbb{R}_F$ is a continuous function, then the fuzzy Aumann and Riemann integrals of F exist and are equal, that is,*

$$(FA)\int_a^b F(s)ds = (FR)\int_a^b F(s)ds.$$

Example 1 Consider $F(t) = tA$ for all $t \in \mathbb{R}$ and $A \in \mathbb{R}_F \setminus \mathbb{R}$. We have that

$$F'_{gH}(t) = A, \qquad \forall t \in \mathbb{R}$$

and

$$\begin{aligned}\left[\int_{-1}^{1} F(t)dt\right]_\alpha &= \left[\int_{-1}^{1} f_\alpha^-(t)dt, \int_{-1}^{1} f_\alpha^+(t)dt\right]\\ &= \left[\int_{-1}^{0} ta_\alpha^+ dt + \int_0^1 ta_\alpha^- dt, \int_{-1}^{0} ta_\alpha^- dt + \int_0^1 ta_\alpha^+ dt\right]\\ &= \frac{1}{2}\left[a_\alpha^- - a_\alpha^+, a_\alpha^+ - a_\alpha^-\right].\end{aligned}$$

for all $\alpha \in [0, 1]$. Thus, we obtain

$$\int_{-1}^{1} F(t)dt = \frac{1}{2}(A - A) \neq 0.$$

Based on the concept of fuzzy Riemann integrability, Allahviranloo et al. [2] proposed the fuzzy Laplace transform that have been used to solve many first and second order fuzzy differential equations given in terms of gH-derivative [2, 36].

Definition 9 [2] The fuzzy Laplace transform of a fuzzy valued function $F : [0, +\infty) \longrightarrow \mathbb{R}_F$ is defined as follows:

$$\mathbf{L}_{(FR)}\{F(t)\} = (FR)\int_0^{+\infty} F(t)e^{-st}dt = \lim_{b\to+\infty}(FR)\int_0^b F(t)e^{-st}dt, \quad (6)$$

whenever the limit exists.

If $[F(t)]_\alpha = \left[f_\alpha^-(t), f_\alpha^+(t)\right]$, for $t \in [0, \infty)$ and $\alpha \in [0, 1]$ such that $\mathbf{L}_{(FR)}\{F(t)\}$ exist (for all $s \in [0, \infty)$), then one can show that [2]

$$\left[\mathbf{L}_{(FR)}\{F(t)\}\right]_\alpha = \left[L\left\{f_\alpha^-(t)\right\}, L\left\{f_\alpha^+(t)\right\}\right] \quad (7)$$

where L denotes the usual Laplace transform, that is, if $g : [0, \infty) \longrightarrow \mathbb{R}$, then $L\{g(t)\} = \int_0^\infty g(t)e^{-st}dt$.

3 Interactive Fuzzy Numbers

The concept of interactivity between two fuzzy sets was proposed by Zadeh [52]. This concept plays an analogous role to the concept of dependence and independence for random variables, where the joint probability distribution reveals whether there is dependence or independence between the random variables. Similarly, the notion of interactivity between two fuzzy variables arises from a given joint possibility (or joint membership) distribution, where the uncertainty involved is given by means of membership functions instead of probability density functions such as in the case of random variables [15].

The operations between fuzzy numbers given by the Zadeh extension principle, which is defined in terms of the minimum t-norm, can be generalized by considering a general t-norms in Eq. (2), which, in its turn, is a particular case of a joint possibility distribution. The concept of joint possibility distributions is inspired by the one of joint probability distributions for random variables.

According to [12, 23], a fuzzy set $J \in \mathbb{F}(\mathbb{R}^n)$ is said to be a joint possibility distribution (JPD) for $A_1, \ldots, A_n \in \mathbb{R}_\mathsf{F}$ if each A_i is the projections of J, that is,

$$P_i(J) = A_i, \qquad \forall i = 1, \ldots, n$$

where

$$P_i(J)(y) = \sup_{x \in \mathbb{R}^n, x_i = y} J(x), \quad \forall y \in \mathbb{R}.$$

If $J(x_1, x_2, \ldots, x_n) = A_1(x_1) \wedge \ldots \wedge A_n(x_n)$, then $A_1, \ldots, A_n$ are said to be non-interactive and, in this case, $[J]_\alpha = [A_1]_\alpha \times \cdots \times [A_n]_\alpha$, $\forall \alpha \in [0, 1]$. Otherwise, $A_1, \ldots, A_n$ are said to be J-interactive or, simply, interactive and, in this case, $[J]_\alpha \subset [A_1]_\alpha \times \cdots \times [A_n]_\alpha$, $\forall \alpha \in [0, 1]$. Here, we denote the family of joint possibility distributions of $\mathbb{R}^n$ by the symbol $\mathbb{F}_J(\mathbb{R}^n)$.

Let $J \in \mathbb{F}_J(\mathbb{R}^2)$. We have that J is a joint possibility distribution of $A \in \mathbb{R}_\mathsf{F}$ and $B \in \mathbb{R}_\mathsf{F}$ if

$$\max_{x_2} J(x_1, x_2) = A(x_1) \text{ and } \max_{x_1} J(x_1, x_2) = B(x_2), \text{ for any } x_1,\ x_2 \in \mathbb{R}. \quad (8)$$

In this case, A and B are also called marginal possibility distributions of J. Figure 1 illustrates an arbitrary joint possibility distribution J with marginals possibility distributions A and B.

Remark 1 In stochastic theory, the expression to the marginal probability distributions of a joint probability distribution resembles Eq. (8) and is given by

$$P_A(x_1) = \int f(x_1, x_2)dx_2 \text{ and } P_B(x_2) = \int f(x_1, x_2)dx_1,$$

where f is the joint probability density function of the random variables A and B.

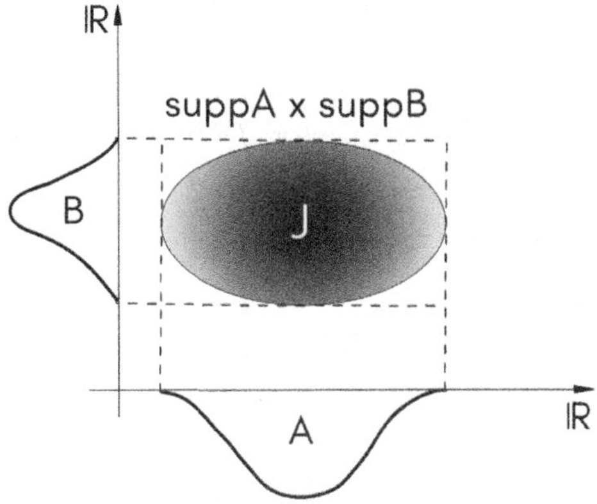

Fig. 1 A joint possibility distribution of $A \in \mathbb{R}_F$ and $B \in \mathbb{R}_F$

Remark 2 For the interval case, a relation $J \subseteq \mathbb{R} \times \mathbb{R}$ is a joint possibility distribution of intervals $A, B \subseteq \mathbb{R}$ if

$$\{x_1 \in \mathbb{R} \mid (x_1, x_2) \in J, x_2 \in \mathbb{R}\} = A \text{ and } \{x_2 \in \mathbb{R} \mid (x_1, x_2) \in J, x_1 \in \mathbb{R}\} = B.$$

If $J = A \times B$, then the intervals A and B are said to be non-interactive. Otherwise, A and B are said to be J-interactive or, simply, interactive.

Let A and B be J-interactive fuzzy numbers and let $f : \mathbb{R}^2 \to \mathbb{R}$. The sup-$J$ extension of f at (A, B) is the fuzzy set $f_J(A, B)$ of $\mathbb{R}$ such that

$$f_J(A, B)(z) = \hat{f}(J)(z) = \sup_{(x,y)\in f^{-1}(z)} J(x, y), \forall z \in \mathbb{R}, \tag{9}$$

where $f^{-1}(z) = \{(x, y) \in \mathbb{R}^2 \mid z = f(x, y)\}$ [12, 23].

Theorem 2 *[5, 22, 33] Let $A, B \in \mathbb{R}_F$, J be a joint possibility distribution whose marginal possibility distributions are A and B such that every $[J]_\alpha$ is a non-empty, connected, and compact set. If $f : \mathbb{R}^2 \longrightarrow \mathbb{R}$ is a continuous function, then $f_J : \mathbb{R}_F \times \mathbb{R}_F \longrightarrow \mathbb{R}_F$ is well-defined and*

$$[f_J(A, B)]_\alpha = f([J]_\alpha) \textit{ for all } \alpha \in [0, 1]. \tag{10}$$

Arithmetic operations for interactive fuzzy numbers can be defined in a similar way that the standard arithmetic operations. To this end, the Zadeh extension principle (Eq. (2)) is replaced by the sup-J extension principle (Eq. (9)). Klir studied a close related approach through the concept of constraints of variables modeled by a fuzzy relation [28].

Definition 10 *(Interactive Arithmetic)* Let A, B be J-interactive fuzzy numbers.

1. The J-sum (or, the J-interactive sum) of A and B is given by

$$A +_J B = +_J(A, B),$$

where $+(a, b) = a + b$ for all $a, b \in \mathbb{R}$.

2. The J-difference (or, J-interactive difference) of A and B is given by

$$A -_J B = -_J(A, B),$$

where $-(a, b) = a - b$ for all $a, b \in \mathbb{R}$.

3. A J-multiplication (or, J-interactive multiplication) of A and B is given by

$$A \cdot_J B = \cdot_J(A, B),$$

where $\cdot(a, b) = ab$ for all $a, b \in \mathbb{R}$.

4. If $A = \lambda \in \mathbb{R}$, λB is given as in the standard case.
5. A J-division (or, J-interactive division) of A by B, $0 \notin [B]_0$, is given by

$$A \div_J B = \div_J(A, B),$$

where $\div(a, b) = a \div b$ for all $a, b \in \mathbb{R}$, $b \neq 0$.

In this chapter we focus on a special type of interactivity called $\mathcal{F}$-interactivity [11]. More precisely, two J-interactive fuzzy numbers A and B are said to be $\mathcal{F}$-interactive if there exists a function $\mathcal{F} : \mathbb{R} \to \mathbb{R}$ such that the corresponding joint possibility distribution J is given by

$$J(x, y) = A(x)\chi_{\{(u,v);\ v=\mathcal{F}(u)\}}(x, y) = B(y)\chi_{\{(u,v);\ v=\mathcal{F}(u)\}}(x, y), \tag{11}$$

where $\chi_{\{(u,v):v=\mathcal{F}(u)\}}$ denotes the characteristic function of the graph of the function $\mathcal{F}$. Since J satisfies $P_2(J) = B$, one can show that (11) is equivalent to $B = \hat{\mathcal{F}}(A)$, that is

$$B(y) = \sup_{z \in \mathcal{F}^{-1}(y)} \mu_A(z) \tag{12}$$

for all $z \in \mathbb{R}$. Thus, if $\mathcal{F}$ is a continuous function, from Theorem 1, the α-levels are given by

$$[B]_\alpha = \mathcal{F}([A]_\alpha), \quad \forall \alpha \in [0, 1].$$

If $\mathcal{F}$ invertible, then $A = \mathcal{F}^{-1}(B)$ and, in this case,

$$[J]_\alpha = \{(x, \mathcal{F}(x)) \in \mathbb{R}^2 | x \in [A]_\alpha\} = \{(\mathcal{F}^{-1}(y), y) \in \mathbb{R}^2 | y \in [B]_\alpha\}. \tag{13}$$

Let A and B be $\mathcal{F}$-interactive fuzzy numbers. In view of the definition of interactive arithmetic operations and the comments above, the arithmetic operations $B \otimes_{\mathcal{F}} A$ are defined by

$$(B \otimes_{\mathcal{F}} A)(y) = \sup_{x \in \Phi_{\otimes}^{-1}(y)} A(x) \tag{14}$$

where $\Phi_{\otimes}^{-1}(y) = \{x | y = x \otimes z,\ z = \mathcal{F}(x)\}$, and $\otimes \in \{+, -, \times, \div\}$. By Theorem 2, if $\mathcal{F}$ is a continuous function, then the four arithmetic operations of $\mathcal{F}$-interactive fuzzy numbers are given, for all $\alpha \in [0, 1]$, by

- $[B +_{\mathcal{F}} A]_\alpha = \{\mathcal{F}(x) + x \in \mathbb{R} | x \in [A]_\alpha\}$;
- $[B -_{\mathcal{F}} A]_\alpha = \{\mathcal{F}(x) - x \in \mathbb{R} | x \in [A]_\alpha\}$;
- $[B \cdot_{\mathcal{F}} A]_\alpha = \{x\mathcal{F}(x) \in \mathbb{R} | x \in [A]_\alpha\}$;
- $[B \div_{\mathcal{F}} A]_\alpha = \{\mathcal{F}(x) \div x \in \mathbb{R} | x \in [A]_\alpha\}$, $0 \notin [A]_0$.

The scalar multiplication λB with $B = \hat{\mathcal{F}}(A)$ is given by

$$[\lambda B]_\alpha = [\lambda \hat{\mathcal{F}}(A)]_\alpha = \lambda \mathcal{F}([A]_\alpha) = \{\lambda \mathcal{F}(x) \in \mathbb{R} | x \in [A]_\alpha\}.$$

In addition, if $\mathcal{F}$ is differentiable in $[A]_0$, then

$$[B +_{\mathcal{F}} A]_\alpha = \begin{cases} \left[b_\alpha^- + a_\alpha^-, b_\alpha^+ + a_\alpha^+\right], & \text{if } \mathcal{F}'(x) \geq 0 \\ \left[b_\alpha^+ + a_\alpha^-, b_\alpha^- + a_\alpha^+\right], & \text{if } -1 \leq \mathcal{F}'(x) < 0 \\ \left[b_\alpha^- + a_\alpha^+, b_\alpha^+ + a_\alpha^-\right], & \text{if } \mathcal{F}'(x) < -1 \end{cases}, \ \forall x \in [A]_\alpha \tag{15}$$

and

$$[B -_{\mathcal{F}} A]_\alpha = \begin{cases} \left[b_\alpha^- - a_\alpha^-, b_\alpha^+ - a_\alpha^+\right], & \text{if } \mathcal{F}'(x) \geq 1 \\ \left[b_\alpha^+ - a_\alpha^+, b_\alpha^- - a_\alpha^-\right], & \text{if } 0 \leq \mathcal{F}'(x) < 1 \\ \left[b_\alpha^- - a_\alpha^+, b_\alpha^+ - a_\alpha^-\right], & \text{if } \mathcal{F}'(x) < 0 \end{cases}, \ \forall z \in [A]_\alpha. \tag{16}$$

From Eq. (3), we can observe that the $\mathcal{F}$-interactive difference $B -_{\mathcal{F}} A$ coincides to gH-difference $B -_{gH} A$ if $\mathcal{F}'(x) \geq 0$ for all $x \in [A]_0$. This fact corroborates to the fact that the gH-difference can be viewed as particular case of interactive subtraction [49]. On the other hand, if $\mathcal{F}'(x) < 0$ for all $x \in [A]_0$, we have that $B -_{\mathcal{F}} A$ coincides with the standard difference. Similarly, from Proposition 2, if $\mathcal{F}'(x) \geq 0$ for all $x \in [A]_0$, then $B +_{\mathcal{F}} A$ coincides with the standard sum $B + A$. However, as we can note, $B \cdot_{\mathcal{F}} A$ and $B \div_{\mathcal{F}} A$ are different from standard operations in the all cases.

An interesting subtype of $\mathcal{F}$-interactivity arise if we only consider functions $\mathcal{F}$ of the form $\mathcal{F}(x) = qx + r$. This restriction leads us to linear interactivity relation among fuzzy numbers. This concept is formalized in the next definition.

Definition 11 [12] The A, B be fuzzy numbers are linearly interactive (or completely correlated) if there exists $q, r \in \mathbb{R}$ such that the corresponding joint possibility distribution J of A and B is given by

$$J(x, y) = A(x)\chi_{\{(u,v);\ v=\mathcal{L}(u)\}}(x, y) = B(y)\chi_{\{(u,v);\ v=\mathcal{L}(u)\}}(x, y),$$

for all $x, y \in \mathbb{R}$, where $\mathcal{L}(u) = qu + r$. Moreover, if $q > 0$ ($q < 0$) then we speak of positive (negative) correlation.

From Eq. (13), α-levels of the joint possibility distribution J in Definition 11 can be rewritten as follows:

$$[J]_\alpha = \{(x, qx + r) \mid x \in [A]_\alpha\}.$$

Since every $[J]_\alpha$ is a connected and compact set of $\mathbb{R}^2$, we obtain the following immediate corollary.

Corollary 1 *[6] If the fuzzy numbers A and B are $\mathcal{L}$-interactive with $\mathcal{L}(u) = qu + r$, then $B = qA + r$, that is, B is the Zadeh extension of the function $\mathcal{L}$ applied to the fuzzy number A. In addition, $[B]_\alpha = q[A]_\alpha + r$ for all $\alpha \in [0, 1]$.*

From Corollary 1, if $q = 0$ then we have that $B = r \in \mathbb{R}$ is linearly interactive with the fuzzy number A, since $B = r = 0 \cdot A + r$. As a consequence of Theorem 2, the α-levels of arithmetic operations of $\mathcal{L}$-interactive fuzzy numbers A and B, with $\mathcal{L}(u) = qu + r$, are given by

- $[B +_\mathcal{L} A]_\alpha = (q + 1)[A]_\alpha + r$;
- $[B -_\mathcal{L} A]_\alpha = (q - 1)[A]_\alpha + r$;
- $[B \cdot_\mathcal{L} A]_\alpha = \{qx^2 + rx_1 \in \mathbb{R} | x \in [A]_\alpha\}$;
- $[B \div_\mathcal{L} A]_\alpha = \{q + \frac{r}{x} \in \mathbb{R} | x \in [A]_\alpha\}, 0 \notin [A]_0$.

Note that $B +_\mathcal{L} A = r$ if $q = -1$ [12].

The next proposition follows immediate from Eqs. (15) and (16) when the operands are linearly interactive fuzzy numbers.

Proposition 6 *[6] Let A and B be $\mathcal{L}$-interactive fuzzy numbers with $\mathcal{L}(u) = qu + r$. We have*

$$[B +_\mathcal{L} A]_\alpha = \begin{cases} \left[b_\alpha^- + a_\alpha^-, b_\alpha^+ + a_\alpha^+\right], & \text{if } q \geq 0 \\ \left[b_\alpha^+ + a_\alpha^-, b_\alpha^- + a_\alpha^+\right], & \text{if } -1 \leq q < 0 \\ \left[b_\alpha^- + a_\alpha^+, b_\alpha^+ + a_\alpha^-\right], & \text{if } q < -1 \end{cases}$$

and

$$[B -_\mathcal{L} A]_\alpha = \begin{cases} \left[b_\alpha^- - a_\alpha^-, b_\alpha^+ - a_\alpha^+\right], & \text{if } q \geq 1 \\ \left[b_\alpha^+ - a_\alpha^+, b_\alpha^- - a_\alpha^-\right], & \text{if } 0 \leq q < 1 \\ \left[b_\alpha^- - a_\alpha^+, b_\alpha^+ - a_\alpha^-\right], & \text{if } q < 0 \end{cases}.$$

4 Differential Calculus for $\mathcal{F}$-Interactive Fuzzy Processes

A fuzzy process F is noting more than a fuzzy number-valued function $F : [a, b] \longrightarrow \mathbb{R}_\mathcal{F}$. For every $t \in [a, b]$, we denote the α-level of $F(t)$ by $[F(t)]_\alpha = [f_\alpha^-(t), f_\alpha^+(t)]$ for all $\alpha \in [0, 1]$. We say that F is expansive at $t \in [a, b]$ if $diam[F(t)]_0 \leq diam[F(t + h)]_0$ for $0 < h < \delta$ and if $diam[F(t + h)]_0 \leq diam[F(t)]_0$ for $-\delta < h < 0, \delta > 0$. Similarly, F is contractive at $t \in [a, b]$ if $diam(F(t)) \geq diam(F(t + h))$ for $0 < h < \delta$ and if $diam(F(t + h)) \geq diam(F(t))$ for $-\delta < h < 0, \delta > 0$.

Definition 12 [42] A fuzzy process F is δ-locally $\mathcal{F}$-autoregressive or, simply, $\mathcal{F}$-autoregressive at $t \in [a, b]$ if there exists a family of continuous real functions $\mathcal{F}_{t,h}$, for all $0 < |h| < \delta$, such that

$$[F(t+h)]_\alpha = \mathcal{F}_{t,h}([F(t)]_\alpha), \forall \alpha \in [0,1].$$

In addition, F is said to be linearly autoregressive (or $\mathcal{L}$-autoregressive) fuzzy process in $t \in [a,b]$ if every $\mathcal{F}_{t,h}$ is a affine function for all $0 < |h| < \delta$, that is, if there exist functions q_t and r_t such that

$$[F(t+h)]_\alpha = q_t(h)[F(t)]_\alpha + r_t(h), \forall \alpha \in [0,1].$$

Let $F : [a,b] \to \mathbb{R}_{\mathsf{F}}$ be a $\mathcal{F}$-autoregressive fuzzy process. According to [42], the function F is $\mathcal{F}$-interactive differentiable ($\mathcal{F}$-differentiable for short) at $t_0 \in (a,b)$ if there exists a fuzzy number $F'_{\mathcal{F}}(t_0)$ such that

$$F'_{\mathcal{F}}(t_0) = \lim_{h\to 0} \frac{F(t_0+h) -_{\mathcal{F}_{t,h}} F(t_0)}{h},$$

where the limit above is given in terms of the metric d_∞. Additionally, $F'_{\mathcal{F}}(t_0)$ is called $\mathcal{F}$-interactive derivative of F at t_0. At the endpoints of $[a,b]$, we consider only the one-sided derivative. If F is additionally an $\mathcal{L}$-autoregressive fuzzy process, then we speak of $\mathcal{L}$-differentiability and use the symbol $F'_{\mathcal{L}}(t_0)$ instead of $F'_{\mathcal{F}}(t_0)$.

Theorem 3 *[42] Let $F : [a,b] \to \mathbb{R}_{\mathsf{F}}$ be $\mathcal{F}$-differentiable at $t_0 \in [a,b]$, where the corresponding family of functions $\mathcal{F}_{t,h} : \mathbb{R} \to \mathbb{R}$ is monotone continuously differentiable for each h, and $[F(t)]_\alpha = [f_\alpha^-(t), f_\alpha^+(t)]$ for all $\alpha \in [0,1]$. If f_α^- and f_α^+ are differentiable at t_0, then*

$$[F'_{\mathcal{F}}(t_0)]_\alpha = \begin{cases} \left[(f_\alpha^-)'(t_0), (f_\alpha^+)'(t_0)\right] & \text{if } \quad \mathcal{F}'_{t,h}(z) > 1, \forall (z,h) \in V \\ \left[(f_\alpha^+)'(t_0), (f_\alpha^-)'(t_0)\right] & \text{if } 0 < \mathcal{F}'_{t,h}(z) \le 1, \forall (z,h) \in V \\ \{(f_\alpha^-)'(t_0)\} & \text{if } \quad \mathcal{F}'_{t,h}(z) \le 0, \forall (z,h) \in V \end{cases}.$$

where $V = [F(t)]_\alpha \times (-\delta, 0) \cup (0, \delta)$.

Since $\mathcal{F}_{t,h}$ is monotone, one can easily see that $\mathcal{F}'_{t,h}(z) \ge 1$ for all $(z,h) \in V$ implies that F is expansive at t. On the other hand, if $0 < \mathcal{F}'_{t,h}(z) \le 1$ for all $(z,h) \in V$, then F is contractive at t. The next proposition reveals that $\mathcal{F}$-differentiability implies in continuity.

Proposition 7 *[42] If $F : [a,b] \to \mathbb{R}_{\mathsf{F}}$ is $\mathcal{F}$-differentiable at t_0, then F is continuous at t_0.*

For $\mathcal{L}$-autoregressive fuzzy process, Theorem 3 produces the following immediate corollary.

Corollary 2 *[6] Let $F : [a,b] \to \mathbb{R}_{\mathsf{F}}$ be $\mathcal{L}$-differentiable at t_0 and $F_\alpha(t_0) = [F(t_0)]_\alpha = [f_\alpha^-(t_0), f_\alpha^+(t_0)]$, for all $\alpha \in [0,1]$. If f_α^- and f_α^+ are differentiable at t_0, then*

$$[F'_{\mathcal{L}}(t_0)]_\alpha = \begin{cases} [(f_\alpha^-)'(t_0), (f_\alpha^+)'(t_0)] \text{ if } & q_{t_0}(h) \geq 1, \forall h \in W \\ [(f_\alpha^+)'(t_0), (f_\alpha^-)'(t_0)] \text{ if } 0 < q_{t_0}(h) \leq 1, \forall h \in W \\ [(f_\alpha^-)'(t_0), (f_\alpha^-)'(t_0)] \text{ if } & q_{t_0}(h) < 0, \forall h \in W \end{cases}$$

where $W = (-\delta, 0) \cup (0, \delta)$.

It is worth noting that, in Corollary 2, the function F is expansive when $q(h) \geq 1$ and F is contractive when $0 < q_{t_0}(h) \leq 1$. The case where $q_{t_0}(h) < 0$ is only possible if the function F is real.

Example 2 [6] Let $\lambda \in \mathbb{R}$. If $F : [a, b] \longrightarrow \mathbb{R}_\mathsf{F} \setminus \mathbb{R}$ is $\mathcal{L}$-autoregressive, then $\lambda F(t + h) = q_t(h)\lambda F(t) + \lambda r_t(h)$. Additionally, if $F : [a, b] \longrightarrow \mathbb{R}_\mathsf{F} \setminus \mathbb{R}$ is $\mathcal{L}$-differentiable at $t \in [a, b]$, then $(\lambda F)'_{\mathcal{L}}(t) = \lambda F'_{\mathcal{L}}(t)$.

Example 3 [6] Let $A \in \mathbb{R}_\mathsf{F} \setminus \mathbb{R}$, $g : [a, b] \to \mathbb{R} \setminus \{0\}$ be differentiable on $[a, b]$, and $F : [a, b] \to \mathbb{R}_\mathsf{F}$ be such that $F(t) = g(t)A$. The function can be viewed as $\mathcal{L}$-autoregressive fuzzy process with $F(t + h) = q_t(h)F(t)$ and $q_t(h) = \frac{g(t+h)}{g(t)}$ for all $t \in [a, b]$. In this case, F is $\mathcal{L}$-differentiable at every $t \in [a, b]$ with

$$F'_{\mathcal{L}}(t) = g'(t)A, \ \ \forall t \in [a, b].$$

Using the concept of $\mathcal{F}$-derivative, we can now consider fuzzy initial value problems (FIVPs) of the form

$$\begin{cases} X'_{\mathcal{F}}(t) = F(t, X(t)) \\ X(0) = X_0 \end{cases}, \tag{17}$$

where $F : [a, b] \times \mathbb{R}_\mathsf{F} \to \mathbb{R}_\mathsf{F}$ is a fuzzy function and X_0 is a fuzzy number. A $\mathcal{F}$-interactive fuzzy process $X : [a, b] \to \mathbb{R}_\mathsf{F}$ is a solution of (17) if $X'_{\mathcal{F}}(t)$ exists at every $t \in [a, b]$ and satisfies (17).

Since the notion of $\mathcal{F}$-differentiability is defined in terms of interactive arithmetic operations, we can consider the existence of interactivity in both sides of the fuzzy differential equation (17). In order to illustrate this approach, in what follows, we present an application to a basic market model that describes commodity markets and was developed by Frank Bass during the 1960s [7]. This classic model is given by

$$x'(t) = (1 - x(t))g(x(t)), \tag{18}$$

where $x(t) \in \mathbb{R}$ is the normalized number of consumers subscribing to a service at time t, and $g(x(t))$ is the feedback term.

The choice of a product by a consumer depends on his perception of the product which is usually expressed in an imprecise and subjective way. An important use of (18) is in the forecast before the product launch, when there is still not enough data and, therefore, there is uncertainty in the rate of change. For this reason, we will use the interactive derivative in its study. The system (18) becomes

$$X'_{\mathcal{F}}(t) = (1 -_{\mathcal{F}} X(t)) \cdot_{\mathcal{F}} G(X(t)), \tag{19}$$

where, in this case, $X(t) \in \mathbb{R}_{\mathsf{F}}$ is the normalized number of consumers subscribing to a service at time t, and $G(X(t))$ is the feedback term.

Market feedback implies that the attractiveness of a product depends on the number of users of the product [3]. Thus, in the following example, we will study the model with linear feedback, more specifically, $G(X(t)) = \lambda X(t)$.

Example 4 *(Markets with linear feedback)* Consider the network effects where the attractiveness of the product decreases due to customers, so the market stagnates after a strong initial increase. Rewriting the model (19), we have

$$X'_{\mathcal{F}}(t) = \lambda X(t) \cdot_{\mathcal{F}} (1 -_{\mathcal{F}} X(t)), \;\; X(0) = X_0 \in \mathbb{R}_{\mathsf{F}}, \tag{20}$$

where $0 < \lambda \in \mathbb{R}$ represents the force of the imitative behavior in the market and is called the coefficient of imitation. We are assuming that X is a locally $\mathcal{F}$-autocorrelated fuzzy process, that is,

$$[X(t+h)]_\alpha = \mathcal{F}_{t,h}([X(t)]_\alpha),$$

for all $\alpha \in [0, 1]$ and for each $t \in \mathbb{R}$, where $\mathcal{F}_{t,h} : \mathbb{R} \to \mathbb{R}$ is a family of continuous functions that represents the interactivity of the process at instants t and $t + h$, for sufficiently small $|h|$.

Note that $\lambda X(t)$ and $1 - X(t)$ are $\mathcal{L}$-interactive with $\mathcal{L}(x) = 1 + \frac{x}{\lambda}$. Therefore, we consider the $\mathcal{L}$-interactivity arithmetic in the field $F(t, X(t))$ of FIVP (20). So,

$$[F(t, X(t))]_\alpha = [\lambda X(t) \cdot_{\mathcal{L}} (1 - X(t))]_\alpha =$$

$$\begin{cases} [\lambda x_\alpha^-(t)(1 - x_\alpha^-(t)), \lambda x_\alpha^+(t)(1 - x_\alpha^+(t))] \text{ if } x_\alpha^-(t) < x_\alpha^+(t) < \frac{1}{2} \\ [\min\{\lambda x_\alpha^-(t)(1 - x_\alpha^-(t)), \lambda x_\alpha^+(t)(1 - x_\alpha^+(t))\}, \frac{\lambda}{4}] \text{ if } x_\alpha^-(t) \le \frac{1}{2} \le x_\alpha^+(t) \\ [\lambda x_\alpha^+(t)(1 - x_\alpha^+(t)), \lambda x_\alpha^-(t)(1 - x_\alpha^-(t))] \text{ if } x_\alpha^+(t) > x_\alpha^-(t) > \frac{1}{2} \end{cases},$$

where $\lambda > 0$, $[X(t)]_\alpha = [x_\alpha^-(t), x_\alpha^+(t)]$, and

$$[F(t, X(t))]_\alpha = [f_\alpha^-(t, x_\alpha^-(t), x_\alpha^+(t)), f_\alpha^+(t, x_\alpha^-(t), x_\alpha^+(t))].$$

- **(Contractive market model)** To consider the case where the uncertainty disappears over time we must consider $0 < \mathcal{F}'_{t,h}(z) < 1$. Although we do not know the exact value of $\mathcal{F}'_{t,h}(z)$, we just need to get it between 0 and 1. In this case, by Theorem 3, for all $\alpha \in [0, 1]$, we have

$$\begin{cases} [(x_\alpha^+)'(t), (x_\alpha^-)'(t)] = [f_\alpha^-(t, x_\alpha^-, x_\alpha^+), f_\alpha^+(t, x_\alpha^-, x_\alpha^+)] = [F(t, X)]_\alpha \\ [X(0)]_\alpha = [x_{0\alpha}^-, x_{0\alpha}^+] \end{cases}. \tag{21}$$

A graphical representation of the solution of system (21) with $\lambda = 0.9$ and $X_0 = (0.1; 0.27; 0.42)$ is depicted in Fig. 2a.

- **(Expansive market model)** When we do not have much information about the market it is reasonable to assume that the uncertainty in the solution increases over time which corresponds to impose that $\mathcal{F}'_{t,h}(z) > 1$. In this case, by Theorem 3, we have, for all $\alpha \in [0, 1]$

$$\begin{cases} [(x_\alpha^-)'(t), (x_\alpha^+)'(t)] = [f_\alpha^-(t, x_\alpha^-, x_\alpha^+), f_\alpha^+(t, x_\alpha^-, x_\alpha^+)] = [F(t, X)]_\alpha \\ [X(0)]_\alpha = [x_{0\alpha}^-, x_{0\alpha}^+] \end{cases}. \quad (22)$$

A graphical representation of the solution of system (22) with $\lambda = 0.9$ and $X_0 = (0.1; 0.27; 0.42)$ is depicted in Fig. 2b.
The Bass model is used to predict the speed of growth of the potential market for new products and because these are new products, initial sales data are missing. In this model, we consider fuzzy relationships between dependent and independent variables, since the phenomenon has an uncertain characteristic and fuzzy relationships have an advantage over the classic one whenever little data is available. Thinking about the performance of the forecast, we see that the fuzzy system (20) can reveal the best and worst sales volume results for all cases. What is not expected with the classic model that can be seen in the 1-level of fuzzy solutions. The contractive case (Fig. 2a) indicates that the classic model has a good performance in forecasting the number of consumers over time. On the other hand, the expansive case (Fig. 2b) reveals that it is possible that the number of consumers is already declining, in contrast to the classic model that points to growth in consumption.
Additionally, in this example, we are going to compare the interactive approach with that given by gH-differentiability.
- **(via gH_2 derivative)** Solving the same problem via gH_2-deferenciability, we have

$$\begin{cases} [(x_\alpha^-)'(t), (x_\alpha^+)'(t)] = [x_\alpha^+(t)\lambda(1 - x_\alpha^-(t)), x_\alpha^-(t)\lambda(1 - x_\alpha^+(t))] \\ [X(0)]_\alpha = [x_{0\alpha}^-, x_{0\alpha}^+] \end{cases}. \quad (23)$$

A solution of the system (23) with $\lambda = 0.9$ and $X_0 = (0.1; 0.27; 0.42)$ can be seen in Fig. 3a.
- **(via gH_1 derivative)** Solving the same problem via gH_1-deferenciability, we have

$$\begin{cases} [(x_\alpha^+)'(t), (x_\alpha^-)'(t)] = [x_\alpha^+(t)\lambda(1 - x_\alpha^-(t)), x_\alpha^-(t)\lambda(1 - x_\alpha^+(t))] \\ [X(0)]_\alpha = [x_{0\alpha}^-, x_{0\alpha}^+] \end{cases}. \quad (24)$$

A solution of the system (24) with $\lambda = 0.9$ and $X_0 = (0.1; 0.27; 0.42)$ can be viewed in Fig. 3b.
In Fig. 3a, we can see that the fuzzy solution via gH_2-differentiability is narrower than the solution via $\mathcal{F}$-interactivity. This fact occurs due to the different approaches used in the multiplication operation at $\lambda X(t) \cdot (1 - X(t))$. On the other hand, in Fig. 3b, we can see that the fuzzy solution via gH_1-differentiability is wider than that obtained via $\mathcal{F}$-interactivity. In general, when using gH-differentiability, the standard multiplication operation is used. In our opinion, this is inconsistent, since non-interactivity is given through the Zadeh extension principle, and in

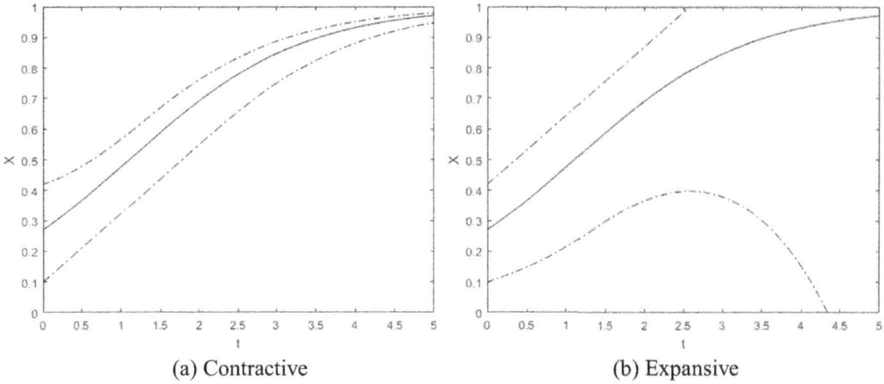

Fig. 2 The continuous line stands the 1-level, whereas the dashed-dotted lines stand for the 0-level of the solutions of systems (21) and (22), respectively. The parameters used were $\lambda = 0.9$, and $X_0 = (0.1; 0.27; 0.42)$

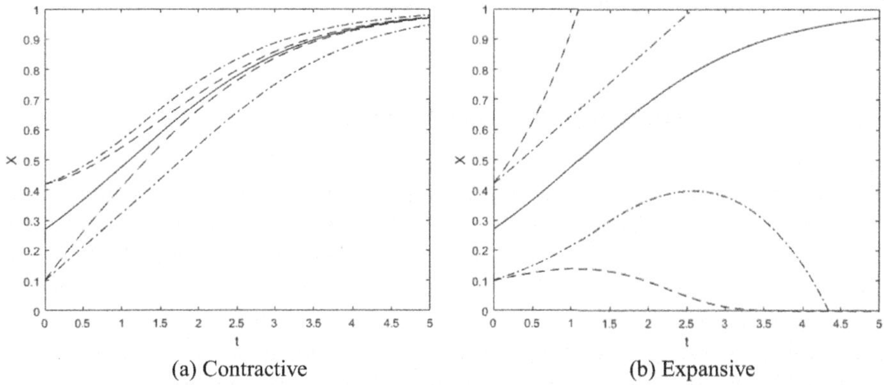

Fig. 3 The continuous line stands the 1-level of solutions of the systems (21), (23), (22), and (24), respectively, whereas the dashed lines stand for the 0-level of the solutions of systems (23) and (24), respectively, and the dashed-dotted lines stand for the 0-level of the solutions of systems (21) and (22), respectively. The parameters used were $\lambda = 0.9$, and $X_0 = (0.1; 0.27; 0.42)$

this case the only fuzzy process differentiable is the real function, i.e., $F(t) \in \mathbb{R}$, $\forall t \in [a, b]$. So, we believe that the solution via $\mathcal{F}$-interactivity is more consistent than the one obtained via gH_2-differentiability.

Other interesting applications of calculus with fuzzy interactivity can be found in [39, 41].

One of the biggest obstacles to establishing a completely satisfactory theory of calculus for fuzzy number-valued functions lies in the fact that the space $\mathbb{R}_{\mathsf{F}}$ equipped with standard arithmetic operations is not a vector space. Based on the idea of linear interactivity between fuzzy numbers, it is possible to define a subset $\mathbb{R}_{\mathsf{F}(A)}$ of $\mathbb{R}_{\mathsf{F}}$ with an algebraic structure and a norm such that $\mathbb{R}_{\mathsf{F}(A)}$ forms a Banach space that is isomorphic to $\mathbb{R}^2$. This approach allows us to employ the well-established concepts of Fréchet derivative and Riemann integral for Banach spaces. Recall that the notion of Fréchet derivative is defined in terms of linear continuous operators and does not consider the following limit:

$$\lim_{h\to 0} \frac{F(t+h)\text{“}-\text{”}F(t)h}{.}$$

Usual approaches to fuzzy calculus consider such a limit, however, the choice of the subtraction "−" at the limit is not unanimous in fuzzy calculus theory.

The next section introduces another approach to fuzzy calculus that considers Banach spaces composed by fuzzy numbers that are interactive with each other, so that the vector addition corresponds to an interactive operation. The operations in this space also correspond to the arithmetic operations for interactive fuzzy numbers but a different way than that presented in this section. The concepts of differential and integral in this space correspond to the classical notions of Fréchet derivative and Riemann integral in Banach spaces.

5 A-Linearly Interactive Fuzzy Processes

We begin this section by considering a simple fuzzy function such as that of the introductory section. Let $g : \mathbb{R} \to \mathbb{R}$ be a differentiable function and let $F(t) = g(t)A$, $A \in \mathbb{R}_{\mathsf{F}} \setminus \mathbb{R}$. The fuzzy function F is not Hukuhara differentiable if $|g(\cdot)|$ is a strictly decreasing function. Thus, Hukuhara differentiability may not be useful in simple situations. In contrast, the function F is gH-differentiable [8] and $\mathcal{F}$-differentiable at t for $g(t) \neq 0$ (see Example 3). These derivatives coincide and are given by $g'(t)A$, which is very intuitive result. On the other hand, the usual notions of fuzzy Aumann and Riemann integrals of the function F coincide since the function F is continuous with respect to the metric d_∞. Despite the fact the $gH/\mathcal{F}$-derivative of F is given in terms of the derivative of g, usually, these fuzzy integrals are not given in terms of the integral of the function g. For example, if $g(t) = t$, then we have that fuzzy Aumann integral of F over $[-1, 1]$ is given by $\frac{1}{2}(A - A) \neq 0$ (see Example 1) and not by $A \int_{-1}^{1} s ds = 0$. This last observation reveals that the same intuitive result for the gH-derivative is not valid for usual notions of fuzzy integrals. This example is one of the motivation to development a calculus theory for certain fuzzy functions, based on the notions of Fréchet derivative and Riemann integral on Banach spaces, in which the corresponding derivative and integral of F is given in terms of the derivative and integral of g, respectively. The concepts and results presented in this section can be found in [21, 43, 44].

This section deals with fuzzy functions of the form $F(t) = q(t)A + r(t)$, where $A \in \mathbb{R}_{\mathsf{F}}$ and $q, r : \mathbb{R} \to \mathbb{R}$. For a fixed fuzzy number A, consider the operator ψ_A : $\mathbb{R}^2 \to \mathbb{R}_{\mathsf{F}}$ that associates each vector $(q, r) \in \mathbb{R}^2$ to the fuzzy number

$$\psi_A(q, r) = qA + r, \ \forall q, r \in \mathbb{R}.$$

We denote the range of the operator ψ_A by the symbol $\mathbb{R}_{\mathsf{F}(A)}$, that is,

$$\mathbb{R}_{\mathsf{F}(A)} = \{\psi_A(q, r) \mid (q, r) \in \mathbb{R}^2\}.$$

Note that $\mathbb{R}$ can be embedded in $\mathbb{R}_{\mathsf{F}(A)}$ since every real number r can be identified with the fuzzy number $\psi_A(0, r) \in \mathbb{R}_{\mathsf{F}(A)}$. Thus, we have $\mathbb{R} \subseteq \mathbb{R}_{\mathsf{F}(A)}$. Moreover, one can prove that if $B \in \mathbb{R}_{\mathsf{F}(A)} \setminus \mathbb{R}$, then $\mathbb{R}_{\mathsf{F}(A)} = \mathbb{R}_{\mathsf{F}(B)}$ [21].

The next proposition establishes that the operator ψ_A does not satisfies the linearity property with respect to the standard addition.

Proposition 8 *[21] Let $(q, r), (u, v) \in \mathbb{R}^2$ and $A \in \mathbb{R}_{\mathsf{F}}$. The operator ψ_A satisfies the following properties:*

1. $\psi_A(q + u, r + v) \subseteq \psi_A(q, r) + \psi_A(u, v)$;
2. $\psi_A(q + u, r + v) = \psi_A(q, r) + \psi_A(u, v)$ *if* $\frac{u}{q} \geq 0$ *for* $q \neq 0$.

Below, we provide the definition of symmetric fuzzy numbers which is a fundamental concept in the characterization of the function ψ_A.

Definition 13 Let $A \in \mathbb{R}_{\mathsf{F}}$ and $x \in \mathbb{R}$. The fuzzy number A is symmetric with respect to x if $A(x - y) = A(x + y), \ \forall y \in \mathbb{R}$. We say that A is non-symmetric if there exists no x such that A is symmetric.

Proposition 9 *[21] Given $A \in \mathbb{R}_{\mathsf{F}}$, the operator ψ_A is injective if and only if A is non-symmetric. In addition, if A is symmetric with respect to x and $A \notin \mathbb{R}$, then the inverse image of $\psi_A(q, r)$ is given by $\psi_A^{-1}(\psi_A(q, r)) = \{(q, r), (-q, 2qx + r)\}$.*

In view of Proposition 9, if the fuzzy number A is non-symmetric, then the operator $\psi_A : \mathbb{R}^2 \to \mathbb{R}_{\mathsf{F}(A)}$ is a bijection that can be used to induce an algebraic structure of vector space over $\mathbb{R}_{\mathsf{F}(A)}$ as follows.

Corollary 3 (Vector Addition and Scalar Product in $\mathbb{R}_{\mathsf{F}(A)}$) *[21] Let $A \in \mathbb{R}_{\mathsf{F}}$ be non-symmetric. The set $\mathbb{R}_{\mathsf{F}(A)}$ is a vector space associated to the scalar field $\mathbb{R}$ with the vector addition and the scalar product defined, for all $B, C \in \mathbb{R}_{\mathsf{F}}$ and $\eta \in \mathbb{R}$, by*

(i) $B +_{\psi_A} C = \psi_A(\psi_A^{-1}(B) + \psi_A^{-1}(C)) = (q_B + q_C)A + (r_B + r_C)$
(ii) $\eta \cdot_{\psi_A} B = \psi_A(\eta\psi_A^{-1}(B)) = (\eta q_B)A + (\eta r_B)$.

where $B = q_B A + r_B$ and $C = q_C A + r_C$. Additionally, $\mathbb{R}_{\mathsf{F}(A)}$ is isomorphic to $\mathbb{R}^2$ via the linear isomorphism ψ_A.

It is worth noting that, for $q, r \in \mathbb{R}$, we have $q \cdot_{\psi_A} A = q \cdot A$ and $qA +_{\psi_A} r = qA + r$, where $+$ and $\cdot$ denote respectively the standard addition and the scalar product of fuzzy numbers. In other words, the scalar product $\cdot_{\psi_A}$ coincides to standard scalar product in $\mathbb{R}_{\mathsf{F}}$. Thus, for notational convenience, we may simply use the same notation of the standard product in the vector space $\mathbb{R}_{\mathsf{F}(A)}$. Moreover, the addition $+_{\psi_A}$ can be viewed as an interactive addition. More precisely, for every $B, C \in \mathbb{R}_{\mathsf{F}(A)}$ there exist $q, u, r, v \in \mathbb{R}$ such that $B = qA + r$ and $C = uA + v$ and, in this case, we can consider that B and C are interactive with respect to the joint possibility distribution J defined levelwise as follows:

$$[J]_\alpha = \{(qx + r, ux + v) \mid x \in [A]_\alpha\}$$

for all $\alpha \in [0, 1]$. Since every $[J]_\alpha$ is a non-empty, connected, and compact set of $\mathbb{R}^2$, from Theorem 2, we have that

$$\begin{aligned}[B +_J C]_\alpha &= \{(q + u)x + (r + v) \mid x \in [A]_\alpha\} \\ &= (q + u)[A]_\alpha + (r + v) \\ &= [\psi_A(q + u, r + v)]_\alpha\end{aligned} \tag{25}$$

for all $\alpha \in [0, 1]$. Hence, the J-interactive sum $B +_J C$ is equal to $\psi_A(q + u, r + v)$. The next corollary states that, for every non-symmetric fuzzy number, we can define a Banach space that is isomorphic to the real Banach space $(\mathbb{R}^2, \| \cdot \|)$.

Corollary 4 *[21] Let $\| \cdot \|$ be a norm of $\mathbb{R}^2$ and $A \in \mathbb{R}_\mathsf{F}$ be non-symmetric. The real vector space $(\mathbb{R}_{\mathsf{F}(A)}, +_{\psi_A}, \cdot)$ forms a Banach space with the norm $\| \cdot \|_{\psi_A}$ given by*

$$\| B \|_{\psi_A} = \| \psi_A^{-1}(B) \|$$

for all $B \in \mathbb{R}_{\mathsf{F}(A)}$.

Given $A \in \mathbb{R}_\mathsf{F}$, a function $F : [a, b] \longrightarrow \mathbb{R}_{\mathsf{F}(A)} \subset \mathbb{R}_\mathsf{F}$ is said to be an A-linearly interactive fuzzy process. In other words, an A-linearly interactive fuzzy process is a fuzzy number-valued function F of the form

$$F(t) = q(t)A + r(t), \quad \forall t \in [a, b]$$

where $q, r : [a, b] \to \mathbb{R}$. From Corollary 4, if A is non-symmetric, then the space $\mathbb{R}_{\mathsf{F}(A)}$ becomes a Banach space and, in this case, we can use the notions of Fréchet derivative and Riemann integral on Banach spaces [26, 53]. For the case where A is symmetric with respect to some $x \in \mathbb{R}$, the function F can be uniformly approximated as closely as desired by another $\tilde{A}$-linearly interactive fuzzy process such that $\tilde{A}$ is non-symmetric. This result is stated in the following proposition.

Proposition 10 *[18] The set of non-symmetric fuzzy numbers with respect to the metric d_∞ is an open set of $\mathbb{R}_\mathsf{F}$ and dense in $\mathbb{R}_\mathsf{F}$. Moreover, any function $F : [a, b] \to \mathbb{R}_{\mathsf{F}(A)}$ can be uniformly approximated as closely as desired by an $\tilde{A}$-linearly interactive fuzzy process, where $\tilde{A}$ is non-symmetric, that is, for every $\epsilon > 0$ there exists a non-symmetric $\tilde{A} \in \mathbb{R}_\mathsf{F}$ and a function $\tilde{F} : [a, b] \to \mathbb{R}_{F(\tilde{A})}$ such that*

$$\sup_{t \in [a,b]} d_\infty(F(t), \tilde{F}(t)) < \epsilon.$$

Proposition 10 indicates that, from a practical point of view, it is sufficient to consider the case where $\mathbb{R}_{\mathsf{F}(A)}$ is a Banach space or, equivalently, A-linearly interactive fuzzy process with A being a non-symmetric fuzzy number. From now on, unless stated otherwise, we assume that $\mathbb{R}_{\mathsf{F}(A)}$ is a Banach space.

If F is an A-linearly interactive fuzzy process, then there exist unique functions $q, r : [a, b] \to \mathbb{R}$ such that $F(t) = q(t)A + r(t)$, $\forall t \in [a, b]$. Thus, we have that $F = \psi \circ p$ where $p(t) = (q(t), r(t))$ for all $t \in [a, b]$. In fact, since $\psi : \mathbb{R}^2 \longrightarrow \mathbb{R}_{\mathsf{F}(A)}$ is isomorphism to $(\mathbb{R}^2, \|\cdot\|)$, we have $F = \psi \circ p \Longleftrightarrow p = \psi^{-1} \circ F$. Using the equivalence above, we can easily show that [21, 43]

- if q and r are bounded functions then F is bounded.
- if q and r are continuous functions at $t \in [a, b]$, then F is continuous at $t \in [a, b]$.
- if q and r are continuous functions in $[a, b]$, then F is continuous in $[a, b]$.

The next proposition establishes a connection between the notion of continuity with respect to the metric d_∞ and the metric d_{ψ_A} induced by $\|\cdot\|_{\psi_A}$.

Proposition 11 *[19] Let $A \in \mathbb{R}_{\mathsf{F}}$ be non-symmetric. If the A-linearly interactive process $F : [a, b] \longrightarrow \mathbb{R}_{\mathsf{F}(A)}$ is continuous w.r.t. the induced metric d_{ψ_A}, then F is continuous w.r.t. to d_∞.*

For each fuzzy function $F : [a, b] \longrightarrow \mathbb{R}_{\mathsf{F}(A)}$, we can define a family of multi-valued functions $F_\alpha : [a, b] \to \mathcal{K}$ given by $F_\alpha(t) = [F(t)]_\alpha$, where $\mathcal{K}$ denotes the family of compact subsets of $\mathbb{R}$. That is,

$$[F(t)]_\alpha = [f_\alpha^-(t), f_\alpha^+(t)] = q(t)[a_\alpha^-, a_\alpha^+] + r(t), \ \ \forall \alpha \in [0, 1],$$

where $[A]_\alpha = [a_\alpha^-, a_\alpha^+]$, $\forall \alpha \in [0, 1]$.

5.1 Differential and Integral Calculus for A-Linearly Interactive Fuzzy Processes

Since $\mathbb{R}_{\mathsf{F}(A)}$ is a Banach space if A is non-symmetric by Corollary 4, we can define the Fréchet derivative for functions $F : [a, b] \to \mathbb{R}_{\mathsf{F}(A)}$.

Definition 14 An A-linearly interactive fuzzy process $F : [a, b] \to \mathbb{R}_{\mathsf{F}(A)}$ is Fréchet differentiable at $t \in [a, b]$ if there exists a continuous linear operator $F'[t] : \mathbb{R} \to \mathbb{R}_{\mathsf{F}(A)}$ such that

$$F(t + h) = F(t) +_{\psi_A} F'[t](h) +_{\psi_A} \omega(h),$$

with $\lim\limits_{h\to 0} \dfrac{\|\omega(h)\|_{\psi_A}}{|h|} = 0$. At the end points, we consider only the one-sided of the limit.

From Definition 14, one can show that if F is Fréchet differentiable at t then $F'[t](h) = hC$, where $C = F'[t](1)$.

The next theorem characterizes the Fréchet derivative of an A-linearly interactive fuzzy process.

Theorem 4 *[21] Let $A \in \mathbb{R}_{\mathsf{F}}$ be non-symmetric and $F : [a, b] \to \mathbb{R}_{\mathsf{F}(A)}$ such that $F(t) = q(t)A + r$, for all $t \in [a, b]$. The function F is Fréchet differentiable at $t \in (a, b)$ if, and only if, $p : [a, b] \to \mathbb{R}^2$ given by $p(t) = (q(t), r(t))$ is differentiable at t. Additionally, $f'[t](h) = q'(t)hA + r'(t)h$ for all $h \in \mathbb{R}$.*

The Fréchet derivative of functions between Banach spaces satisfy some desired properties that are stated in the next two propositions [21, 53].

Proposition 12 *Let $A \in \mathbb{R}_{\mathsf{F}}$ be non-symmetric. If $F : [a, b] \longrightarrow \mathbb{R}_{\mathsf{F}(A)}$ is Fréchet differentiable in $[a, b]$, then F is continuous in $[a, b]$.*

Proposition 13 *Let $A \in \mathbb{R}_{\mathsf{F}}$ be non-symmetric. If $F, G : [a, b] \longrightarrow \mathbb{R}_{\mathsf{F}(A)}$ are Fréchet differentiable in $[a, b]$, then $(F +_{\psi_A} G)'[t] = F'[t] +_{\psi_A} G'[t]$ and $(\lambda \cdot F)'[t] = \lambda \cdot F'[t]$ for all $t \in [a, b]$ and $\lambda \in \mathbb{R}$.*

Since $(\mathbb{R}_{\mathsf{F}(A)}, +_{\psi_A}, \cdot)$ is a Banach space, one can define a derivative of F at a point t by means of a limit such as in calculus for real-valued functions [44]. More precisely, let $A \in \mathbb{R}_{\mathsf{F}}$ be non-symmetric and $F : [a, b] \to \mathbb{R}_{\mathsf{F}(A)}$ such that $F(t) = q(t)A + r(t)$. We say that F is ψ-differentiable or, simply, differentiable at $t \in [a, b]$ if there exists $F'(t) \in \mathbb{R}_{\mathsf{F}(A)}$ such that

$$\lim_{h \to 0} \frac{1}{h} \cdot (F(t + h) -_{\psi_A} F(t)) = F'(t), \tag{26}$$

in sense of d_{ψ_A}. Again, we consider only the one-sided of the limit at the endpoints a and b. We say that F is (ψ-)differentiable in $[a, b]$ if F is (ψ-)differentiable at every $t \in [a, b]$. As we can observe in the next theorem, the notions of Fréchet differentiability and ψ-differentiability have a close relation.

Theorem 5 *[19] A function $F : [a, b] \to \mathbb{R}_{\mathsf{F}(A)}$ is F-differentiable at $t \in [a, b]$ if, and only if, F is ψ-differentiable at t. Moreover,*

$$F'[t](1) = F'(t) = \psi(q'(t), r'(t)) = q'(t)A + r'(t),$$

where $F'[t]$ is the Fréchet derivative of F at t.

Theorem 5 reveals that the ψ-differentiability and Fréchet differentiability are equivalent concepts. Thus, many useful results such as Propositions 12 and 13, that are inherited from the classical theory of Fréchet derivatives on Banach spaces, can also be established with respect to ψ-differentiability using this equivalence. Moreover, as one can note in Theorems 4 and 5, the notions of Fréchet differentiability are given in terms of the derivative of two real-valued functions. The next remark states that we can also define the notion of derivative for A-linearly interactive fuzzy processes with A being symmetric in terms of derivatives of real functions.

Remark 3 *(Derivative of $F : \mathbb{R} \to \mathbb{R}_{\mathcal{F}(A)}$ with symmetric A)* [43] Let A_1, A_2 be arbitrary fuzzy numbers and let $q_i, r_i : \mathbb{R} \to \mathbb{R}$, for $i = 1, 2$, be differentiable functions at $t_0 \in \mathbb{R}$ such that $\psi_{A_1}(q_1(t), r_1(t)) = \psi_{A_2}(q_2(t), r_2(t))$ for all $t \in B(t_0, \delta) := \{u \in \mathbb{R} : |u - t_0| < \delta\}$, $\delta > 0$. We have that [43]

$$\begin{aligned} &\psi_{A_1}(q_1'(t_0), r_1'(t_0)) = \psi_{A_2}(q_2'(t_0), r_2'(t_0)) \\ \text{or} \quad &q_1'(t_0)A_1 + r_1'(t_0) = q_2'(t_0)A_2 + r_2'(t_0). \end{aligned}$$

Moreover, if there exists $t \in B(t_0, \delta)$ such that $f(t) \notin \mathbb{R}$, then $\mathbb{R}_{\mathcal{F}(A_1)} = \mathbb{R}_{\mathcal{F}(A_2)}$. These equalities show that the Fréchet derivative and, consequently, the ψ-derivative of F does not depend on the choice of the fuzzy number A or of the coefficients $q(\cdot)$ and $r(\cdot)$. It is worth noting that there are unique functions q and r such that $F(t) = q(t)A + r(t)$ when A is non-symmetric. However, that is not the case when A is symmetric with respect to some $x \in \mathbb{R}$. For example, let $A = (-1; 0; 1)$ and $q_1(t) = 1$ and $q_2(t) = -1$, we have that $q_1(t)A = A = q_2(t)A$ for all $t \in \mathbb{R}$. Thus, for $F : \mathbb{R} \to \mathbb{R}_{\mathcal{F}(A)}$ with symmetric A, if there exist real functions q and r such that q and r is differentiable at some t_0 and $F(t) = q(t)A + r(t)$ for every t in a neighboor of t_0, then we can define the derivative of F at t_0 as being $F'(t_0) = q'(t_0)A + r'(t_0)$.

Based on the limit (26), we can recursively define derivative of higher order of A-linearly interactive fuzzy processes, with A being a non-symmetric fuzzy number. Thus, from Theorem 5, let $F : [a, b] \to \mathbb{R}_{\mathcal{F}(A)}$ be such that $F(t) = q(t)A + r(t)$ for all $t \in [a, b]$. The function F has derivative of order $n \geq 1$ at $t \in [a, b]$ if, and only if, the functions $q, r : [a, b] \to \mathbb{R}$ have derivatives of order n at t. In this case, we have that the nth order derivative of F at t is given by

$$F^{(n)}(t) = \psi_A(q^{(n)}(t), r^{(n)}(t)) = q^{(n)}(t)A + r^{(n)}(t).$$

From comments above, we have that the function $F(t) = g(t)A$, that is one of the motivations to the development of this theory of calculus on $\mathbb{R}_{\mathcal{F}(A)}$, has derivative given by $F'(t) = g'(t)A$ if g is differentiable at t.

Since $\mathbb{R}_{\mathcal{F}(A)}$ forms a Banach space when A is non-symmetric, one can define integral for A-linearly interactive fuzzy process by means of the definition of Riemann integral on Banach spaces [26].

Definition 15 *(Riemann Integral in $\mathbb{R}_{\mathcal{F}(A)}$)* [43] Let A be a non-symmetric fuzzy number. A function $F : [a, b] \to \mathbb{R}_{\mathcal{F}(A)}$ is said to be Riemann integrable if there exists $S \in \mathbb{R}_{\mathcal{F}(A)}$ such that for every $\epsilon > 0$ there exists $\delta > 0$ such that for all partition $a = x_0 < x_1 < \ldots < x_n = b$ with $x_k - x_{k-1} < \delta$, $k = 1, \ldots, n$, we have

$$\left\| \sum_{k=1}^{n} F(t_k)(x_k - x_{k-1}) \ -_{\psi_A} \ S \right\|_{\psi_A} < \epsilon$$

where $t_k \in [x_{k-1}, x_k]$, $k = 1, \ldots, n$, and the summation is given in terms of the (interactive) addition $+_{\psi_A}$. In this case, we denote S by $\int_a^b F(s)ds$.

The Riemann integral of A-interactive fuzzy processes (with non-symmetric A) can be characterized by the Riemann integrals of two real-valued functions as follows [43].

Theorem 6 *[43] Let $A \in \mathbb{R}_{\mathsf{F}}$ be non-symmetric and $F : [a, b] \longrightarrow \mathbb{R}_{\mathsf{F}(A)}$ be an A-linearly interactive fuzzy process such that $F(t) = q(t)A + r(t)$ for all $t \in [a, b]$. Then F is Riemann integrable on $[a, b]$ if, and only if, $q, r : [a, b] \to \mathbb{R}$ are Riemann integrable on $[a, b]$. In this case,*

$$\int_a^b F(t)dt = \left(\int_a^b q(t)dt\right)A + \left(\int_a^b r(t)dt\right).$$

It is worth noting that the notion of Riemann integral given in Definition 15 is distinct that one present in Definition 8. Both integrals are defined by means of the same limit but the addition operations in the summations are different for each one. More precisely, the definition of Riemann integral for A-linearly interactive fuzzy processes is given in terms of interactive addition $(+_{\psi_A})$ whereas the one presented in Definition 8 is given in terms of standard addition of fuzzy numbers. Since the addition $+_{\psi_A}$ can be viewed as an interactive arithmetic operation (see Equation (25)), we also refers to Riemann integral of Definition 15 as **interactive Riemann integral**. Similarly, we say that the fuzzy integral given in Definition 8 is non-interactive. Thus, these both notions of integrals may not coincide. The next example illustrates this fact.

Example 5 Let $A \in \mathbb{R}_{\mathsf{F}}$ be non-symmetric. Consider the function $F : [-1, 1] \to \mathbb{R}_{\mathsf{F}}$ given by

$$F(t) = tA + e^t$$

The derivative of F at $t \in [-1, 1]$ is

$$F'(t) = A + e^t, \qquad \forall t \in [-1, 1\,]$$

By Theorem 6, the interactive Riemann integral of F over $[-1, 1]$ is

$$\int_{-1}^{1} F(t)dt = \left(\int_{-1}^{1} tdt\right)A + \left(\int_{-1}^{1} e^t dt\right) = 0.A + (e - e^{-1}) = e - e^{-1} \in \mathbb{R}.$$

In contrast, if we consider the notion of fuzzy Riemann integral introduced in Definition 8, then we obtain that the integral of F over $[-1, 1]$ is equal to $\frac{1}{2}(A - A) + (e - e^{-1})$.

Let A be a non-symmetric fuzzy number. Since ψ_A is injective by Proposition 9, for every $F : [a, b] \to \mathbb{R}_{\mathsf{F}(A)}$ there exist unique functions q, r such that $F(t) = q(t)A + r(t)$ for all $t \in [a, b]$. This implies that for every non-symmetric A, the Definition 15 is indeed well-defined. The next proposition ensures that the Riemann integral of F does not depend on the choice of A.

Proposition 14 *[43] Let A_1 and A_2 be non-symmetric fuzzy numbers and let $q_i, r_i : [a,b] \to \mathbb{R}$, $i = 1, 2$, be integrable functions such that*

$$\psi_{A_1}(q_1(s), r_1(s)) = \psi_{A_2}(q_2(s), r_2(s))$$

for all $s \in [a,b]$. The following equality holds true:

$$\psi_{A_1}\left(\int_a^b q_1(s)ds, \int_a^b r_1(s)ds\right) = \psi_{A_2}\left(\int_a^b q_2(s)ds, \int_a^b r_2(s)ds\right)$$
$$\text{or} \quad \int_a^b q_1(s)ds A_1 + \int_a^b r_1(s)ds = \int_a^b q_2(s)ds A_2 + \int_a^b r_2(s)ds.$$

The next theorem can be viewed as a version of the mean value theorem for integrals of the classical theory of calculus.

Theorem 7 *[44] Let $A \in \mathbb{R}_{\mathsf{F}}$ be non-symmetric and $F : [a,b] \longrightarrow \mathbb{R}_{\mathsf{F}(A)}$ be a continuous A-linearly interactive fuzzy process given by $F(t) = \psi_A(q(t), r(t))$ for all $t \in [a,b]$. There exists $C \in \mathbb{R}_{\mathsf{F}(A)}$ such that*

$$\int_a^b F(t)dt = (b-a)C,$$

where $C = q(t_1)A + r(t_2)$ for some $t_1, t_2 \in (a,b)$.

Below, we provide a fuzzy version of the fundamental theorems of the calculus for A-linearly interactive fuzzy processes.

Theorem 8 *[43] Let $A \in \mathbb{R}_{\mathsf{F}}$ be non-symmetric and $F : [a,b] \longrightarrow \mathbb{R}_{\mathsf{F}(A)}$ be an A-linearly interactive fuzzy process given by $F(t) = q(t)A + r(t)$. If $q, r : [a,b] \longrightarrow \mathbb{R}$ are continuous, then F is (interactive) Riemann integrable on $[a,b]$ and the integral function*

$$G(t) = \int_a^t F(s)ds$$

is differentiable with

$$G'(t) = F(t), \forall t \in [a,b].$$

Theorem 9 *[43] Let $A \in \mathbb{R}_{\mathsf{F}}$ be non-symmetric and $F : [a,b] \longrightarrow \mathbb{R}_{\mathsf{F}(A)}$ be a differentiable A-linearly interactive fuzzy process. If F' is (interactive) Riemann integrable, then*

$$\int_a^t F'(s)ds = F(t) -_{\psi_A} F(a),$$

for all $t \in [a,b]$.

The next example emphasizes the difference between the concepts of interactive and non-interactive fuzzy integrals in light of Theorem 8.

Example 6 Let A be a non-symmetric fuzzy number and let $F(t) = \cos(t)A$ for $0 \leq t \leq \pi$.

- **Via interactive Riemann integral (Definition** 15): From Theorem 6, we have

$$G_1(t) = \int_0^t F(s)ds = \sin(t)A, \quad 0 \leq t \leq \pi.$$

In this case, the diameter of G_1 is strictly decreasing in $\left(\frac{\pi}{2}, \pi\right]$. Such a property is not possible when we consider the fuzzy Aumann integral [27]. Moreover, the function G_1 is not Hukuhara differentiable in $\left(\frac{\pi}{2}, \pi\right]$.

- **Via non-interactive Riemann integral (Definition** 8): Note that F is a continuous function and, for all $\alpha \in [0, 1]$, we have

$$[F(t)]_\alpha = \begin{cases} \left[\cos(t)a_\alpha^-, \cos(t)a_\alpha^+\right], & 0 \leq t \leq \frac{\pi}{2} \\ \left[\cos(t)a_\alpha^+, \cos(t)a_\alpha^-\right], & \frac{\pi}{2} < t \leq \pi \end{cases}.$$

Let G_2 be the function defined for all $t \in [0, \pi]$ as follows:

$$G_2(t) = (FA)\int_0^t F(s)ds.$$

From Proposition 5, we have that the fuzzy Aumann integral and the non-interactive fuzzy Riemann integral (given in Definition 8) coincide. Thus, by Eq. (5), for all $t \in \left[0, \frac{\pi}{2}\right]$ and $\alpha \in [0, 1]$ we obtain

$$\begin{aligned} [G_2(t)]_\alpha &= \left[\int_0^t \cos(s)a_\alpha^- ds, \int_0^t \cos(s)a_\alpha^+ ds\right] \\ &= \left[\sin(t)a_\alpha^-, \sin(t)a_\alpha^+\right]. \end{aligned}$$

Now, for every $t \in \left(\frac{\pi}{2}, \pi\right]$ and $\alpha \in [0, 1]$ we obtain $[G_2(t)]_\alpha =$

$$\begin{aligned} &= \left[\int_0^{\frac{\pi}{2}} \cos(s)a_\alpha^- ds + \int_{\frac{\pi}{2}}^t \cos(s)a_\alpha^+ ds, \int_0^{\frac{\pi}{2}} \cos(s)a_\alpha^+ ds + \int_{\frac{\pi}{2}}^t \cos(s)a_\alpha^- ds\right] \\ &= \left[a_\alpha^- + (\sin(t) - 1)a_\alpha^+, a_\alpha^+ + (\sin(t) - 1)a_\alpha^-\right]. \end{aligned}$$

Therefore, we have that

$$G_2(t) = \begin{cases} \sin(t)A, & 0 \leq t \leq \frac{\pi}{2} \\ A + (\sin(t) - 1)A, & \frac{\pi}{2} < t \leq \pi \end{cases}.$$

The diameter of G_2 is an increasing function and is given by

$$diam(G_2(t)) = \begin{cases} \sin(t)diam(A) & , \quad 0 \le t \le \frac{\pi}{2} \\ (2 - \sin(t))diam(A) , & \frac{\pi}{2} < t \le \pi \end{cases}.$$

The function G_2 is Hukuhara differentiable, that is, gH_1-differentiable in $[0, \pi]$ and $(G_2)'_{gH_1}(t) = cos(t)A$ for all $t \in [0, \pi]$.

As we can note in the example above, the theory of calculus for A-linearly interactive fuzzy processes presented in this subsection is indeed different from that based on Hukuhara derivative and Aumann integral [27, 34, 35].

5.2 *Fuzzy Differential Equations for A-Linearly Interactive Processes*

Here, we consider fuzzy initial value problem (FIVP) of the form

$$\begin{cases} X^{'}(t) = F(t, X(t)) \\ X(a) = X_0 \end{cases} \tag{27}$$

where $A \in \mathbb{R}_{\mathsf{F}}$ is non-symmetric, $F : [a, b] \times \mathbb{R}_{\mathsf{F}(A)} \to \mathbb{R}_{\mathsf{F}(A)}$ is a continuous fuzzy process, and $X_0 \in \mathbb{R}_{\mathsf{F}(A)}$. An A-linearly interactive fuzzy process $X : [a, b] \to \mathbb{R}_{\mathsf{F}(A)}$ is a solution to the problem (27) if x is (ψ-)differentiable in $[a, b]$ and satisfies (27). In this case, the solution of FIVP (27) describes a trajectory on $\mathbb{R}_{\mathsf{F}(A)}$.

Lemma 1 *[43] Let $A \in \mathbb{R}_{\mathsf{F}}$ be non-symmetric. The A-linearly interactive fuzzy process $X : [a, b] \longrightarrow \mathbb{R}_{\mathsf{F}(A)}$ is a solution of (27) if, and only if, F is continuous and X satisfies the integral equation*

$$X(t) = X_0 +_{\psi_A} \int_a^t F(s, X(s))ds, \tag{28}$$

for all $t \in [a, b]$.

We denote by $C\left([a, b], \mathbb{R}_{\mathsf{F}(A)}\right)$ the space of all continuous A-linearly interactive fuzzy processes $X : [a, b] \longrightarrow \mathbb{R}_{\mathsf{F}(A)}$ with respect to the metric induced by the norm $\| \cdot \|_{\psi_A}$. We also denote

$$C_\rho = \{X \in C\left([a, b], \mathbb{R}_{\mathsf{F}(A)}\right) : \|x\|_{\psi_A} \le \rho\}.$$

Proposition 15 *[43] Let $\delta > 0$, $F : [a, a + \delta] \times C_\rho \to \mathbb{R}_{\mathsf{F}(A)}$ be continuous and satisfies the Lipschitz condition*

$$\|F(t, X) -_{\psi_A} F(t, Y)\|_{\psi_A} \le L\|X -_{\psi_A} Y\|_{\psi_A},$$

for $t \in [a, a+\delta]$ and $x, y \in C_\rho$. Then the FIVP (27) *has a unique solution $X(\cdot)$ defined in an interval $[a, a+k]$, where $k > 0$ is sufficiently small, and $X(t)$ is given by Eq.* (28) *for all $t \in [a, a+k]$.*

In what follows, we present some methods to solve fuzzy linear differential equations in $\mathbb{R}_{\mathsf{F}(A)}$.

5.2.1 Method of Variation of the Constants for Fuzzy Linear Equations

Consider the FIVP given by

$$\begin{cases} X'(t) = a(t)X(t) +_{\psi_A} G(t) \\ X(t_0) = X_0 \end{cases} \tag{29}$$

where A is a non-symmetric fuzzy number, $X_0 \in \mathbb{R}_{\mathsf{F}(A)}$, a is a scalar continuous function and X, G are continuous trajectories on $\mathbb{R}_{\mathsf{F}(A)}$.

The solution of (29) is given by the variation of the constants formula [43]:

$$X(t) = e^{\int_{t_0}^{t} a(u)du}\left(X_0 +_{\psi_A} \int_{t_0}^{t} e^{-\int_{t_0}^{s} a(u)du} G(s)ds\right). \tag{30}$$

If we replace $X(t) = q(t)A + r(t)$, $G(t) = g_1(t)A + g_2(t)$ and $X_0 = q_0A + r_0$ in Eq. (30), by Theorems 4 and 6, we have

$$\begin{cases} q(t) = e^{(\int_{t_0}^{t} a(s)\,ds)}\left(q_0 + \int_{t_0}^{t} g_1(s)e^{(-\int_{t_0}^{s} a(u)\,du)}ds\right) \\ r(t) = e^{(\int_{t_0}^{t} a(s)\,ds)}\left(r_0 + \int_{t_0}^{t} g_2(s)e^{(-\int_{t_0}^{s} a(u)\,du)}ds\right) \end{cases},$$

which is the variation of the constants formula for two classic differential equations.

Thus, the solution (30) can be written by

$$\begin{aligned} X(t) = &\left(e^{(\int_{t_0}^{t} a(s)\,ds)}\left(q_0 + \int_{t_0}^{t} g_1(s)e^{(-\int_{t_0}^{s} a(u)\,du)}ds\right)\right)A \\ &+ \left(e^{(\int_{t_0}^{t} a(s)\,ds)}\left(r_0 + \int_{t_0}^{t} g_2(s)e^{(-\int_{t_0}^{s} a(u)\,du)}ds\right)\right). \end{aligned}$$

To ilustrate the formula (30) we return to the market model of Frank Bass.

Another way of examining the uncertainty in forecasting the number of consumers of a product to be launched is to consider the Bass model (19) in space $\mathbb{R}_{\mathsf{F}(A)}$, where A is a non-symmetric fuzzy number. Hence,

$$X'(t) = (\tilde{1} -_{\psi_A} X(t))G(X(t)),$$

where $X(t) \in \mathbb{R}_{\mathsf{F}(A)}$ is the normalized number of consumers subscribing to a service at time t, $G(X(t)) \in \mathbb{R}$ is the feedback term, $X'(t)$ refers to the (ψ-)derivative, and $\tilde{1} = 1A + 1$. Next, we present the case where there is no feedback, that is, $G(X(t)) = a \in \mathbb{R}$.

Example 7 *(Markets without feedback)* This model is a market model that is not concerned with competition, but with how the market evolves as a whole. Thus, in this case, all suppliers are seen as a single supplier [3]. An example of this type of model is the mobile and internet markets. The rate at which customers are signing up for subscriptions is given by

$$X'(t) = a(\tilde{1} -_{\psi_A} X(t)), \;\; X(0) = X_0 \in \mathbb{R}_{\mathsf{F}(A)}, \tag{31}$$

where $a \in \mathbb{R}$ is the adaptation rate for the service and $\tilde{1} = 1A + 1$. The adaptation rate is constant and is the same for all potential customers.

Note that, in this model, the desire to buy a product does not depend on how many already own the product, that is, the market feedback does not exist or is so weak that it can be neglected [3]. If $X(t) = q(t)A + r(t)$, then, by Eq. (30), the solution of (31) is given by

$$X(t) = \left(1 - (1 - q_0)e^{-at}\right)A + \left(1 - (1 - r_0)e^{-at}\right). \tag{32}$$

Figure 4 exhibits the solution X of the system (31) given by Eq. (32) with $a = 0.15$, $q_0 = 0.5$, $r_0 = 0.3$ and $A = (0.1; 0.27; 0.42)$. We can see that the evolution of the market (i.e. $\lim_{t\to\infty} X(t)$) is such that $X(t) = A + 1 = \tilde{1}$.

In the next example, instead of considering a constant rate of adaptation, we will consider it exponentially decreasing, that is, $G(X(t)) = a_0e^{-\gamma t}$.

Example 8 *(Time-dependent adaptation rate)* If we expect the adaptation rate to be exponentially decreasing, that is, $a(t) = a_0e^{-\gamma t}$, then

$$X'(t) = a_0e^{-\gamma t}(\tilde{1} -_{\psi_A} X(t)), \;\; X(0) = 0A + 0 = 0, \;\text{and}\; \tilde{1} = 1A + 1. \tag{33}$$

If $X(t) = q(t)A + r(t)$, then, by Eq. (30), the solution of FIVP (33) is given by

$$X(t) = \left(1 - e^{\left[-\frac{a_0}{\gamma}(1-e^{-\gamma t})\right]}\right)A + \left(1 - e^{\left[-\frac{a_0}{\gamma}(1-e^{-\gamma t})\right]}\right). \tag{34}$$

Figure 5 exhibits the solution X given as in Eq. (34) with $a_0 = 0.15$, $\gamma = 0.1$, and $A = (0.1; 0.27; 0.42)$.

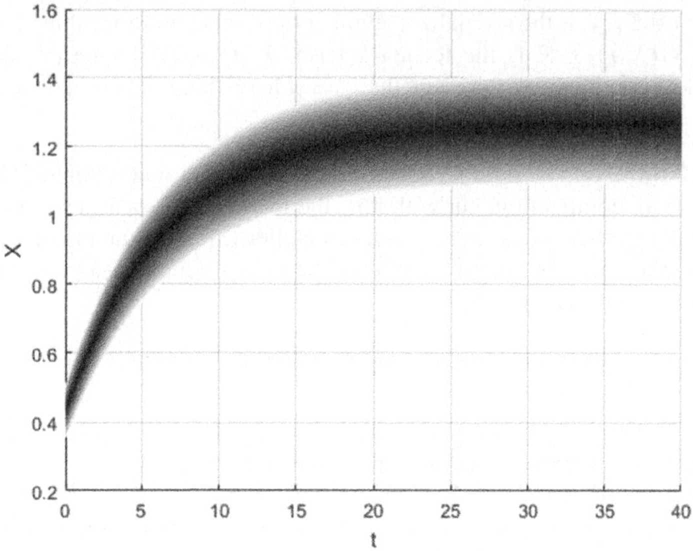

Fig. 4 Graphical representation of the fuzzy process X given as in Eq. (32) with $a = 0.15, q_0 = 0.5$, $r_0 = 0.3$ and $A = (0.1; 0.27; 0.42)$

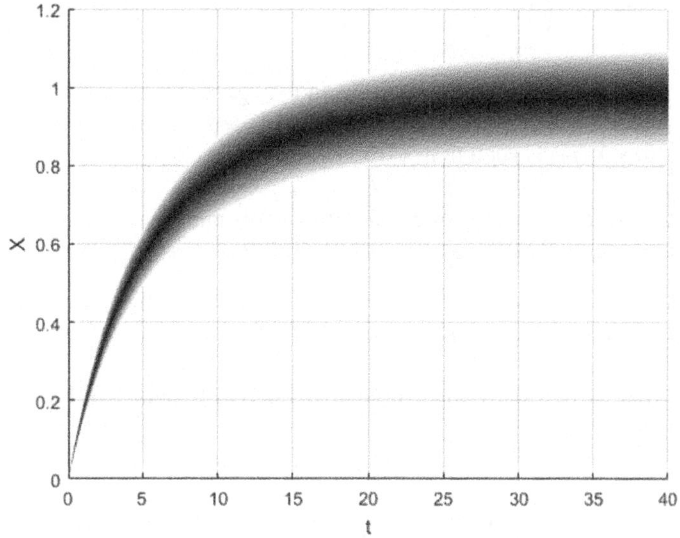

Fig. 5 Graphical representation of the fuzzy process X given as in Eq. (34) with $a_0 = 0.15, \gamma = 0.1$, and $A = (0.1; 0.27; 0.42)$

5.2.2 Second Order Homogeneous Linear Equations with Constants Coefficients

Let $A \in \mathbb{R}_{\mathsf{F}}$ be non-symmetric. Consider the FIVP given by

$$\begin{cases} aX''(t) +_{\psi_A} bX'(t) +_{\psi_A} cX(t) = 0 \\ X(t_0) = X_0 \in \mathbb{R}_{\mathsf{F}(A)} \\ X'(t_0) = X_0' \in \mathbb{R}_{\mathsf{F}(A)}, \end{cases} \tag{35}$$

where $a, b, c \in \mathbb{R}$ and $X(t) \in \mathbb{R}_{\mathsf{F}(A)}$. Let us seek for solutions of the type $X(t) = q(t)A + r(t)$, where $q(t) = r(t) = e^{\gamma t}$. Recall that the second order derivative of $X(t) = q(t)A + r(t)$ is given by $X''(t) = q''(t)A + r''(t)$. Thus, the fuzzy differential equation in FIVP (35), can be rewritten as follows

$$(aq''(t) + bq'(t) + cq(t))A + (ar''(t) + br'(t) + cr(t)) = 0.$$

Grouping the terms, we obtain the following system of differential equations:

$$\begin{cases} aq''(t) + bq'(t) + cq(t) = 0 \\ ar''(t) + br'(t) + cr(t) = 0 \end{cases}$$

Thus, we have to solve two classical systems of homogeneous differential equations for the real functions q and r, which have the following characteristic function:

$$a\gamma^2 + b\gamma + c = 0. \tag{36}$$

As in the classic case, here we can have three situations:

- If $\gamma_1 \in \mathbb{R}$ and $\gamma_2 \in \mathbb{R}$ are the roofs of the characteristic equation (36) and $\gamma_1 \neq \gamma_2$, we have that the solution of (35) is given by [43]

$$X(t) = (c_1 e^{\gamma_1 t} + c_2 e^{\gamma_2 t})A + (d_1 e^{\gamma_1 t} + d_2 e^{\gamma_2 t}). \tag{37}$$

- If $\gamma_1 = \gamma_2 \in \mathbb{R} = \gamma$ is the unique roof of the characteristic equation (36), we have that the solution of (35) is given by

$$X(t) = (c_1 e^{\gamma t} + c_2 t e^{\gamma t})A + (d_1 e^{\gamma t} + d_2 t e^{\gamma t}). \tag{38}$$

- If $\gamma = m \pm ni$ is a complex roof of the characteristic equation (36), we have that the solution of (35) is given by

$$X(t) = e^{mt}(c_1 \cos nt + c_2 \sin nt)A + e^{mt}(d_1 \cos nt + d_2 \sin nt). \tag{39}$$

5.2.3 Second Order Linear Nonhomogeneous Differential Equations with Constants Coefficients

Consider the FIVP given by

$$\begin{cases} a \cdot_{\psi_A} X''(t) +_{\psi_A} b \cdot_{\psi_A} X'(t) +_{\psi_A} c \cdot_{\psi_A} X(t) = G(t) \\ X(t_0) = X_0 \in \mathbb{R}_{\mathcal{F}(A)} \\ X'(t_0) = X'_0 \in \mathbb{R}_{\mathcal{F}(A)}, \end{cases} \tag{40}$$

where $a, b, c \in \mathbb{R}$ and $X(t), G(t) \in \mathbb{R}_{\mathcal{F}(A)}$, for all t. Note that X_0 and X'_0 are interactive [49].

Indeterminate Coefficient Method is a systematic way of determining the general form of the particular solution $X_p(t)$ based on the nonhomogeneous term $G(t) = g_1(t)A + g_2(t)$ in (40).

This method basically consists of using the indeterminate coefficient method in the following differential equations that are obtained considering a solution X of the form $X(t) = q(t)A + r(t)$. Thus, q and r are obtained by solving

$$aq''(t) + bq'(t) + cq(t) = g_1(t) \tag{41}$$

$$ar''(t) + br'(t) + cr(t) = g_2(t) \tag{42}$$

where $q(t) = q_h(t) + q_p(t)$ and $r(t) = r_h(t) + r_p(t)$ with q_h and r_h being the homogeneous solutions and q_p and r_p being the particular solutions of (41) and (42), respectively.

Next, we study the equilibrium in the market model.

Example 9 The development and launch of new products are characterized by the lack of the necessary information of value and physical indicators for accurate pricing. The statistical data on the demand for such a product in the market is very limiting due to its novelty [30]. In this example, we will study the market equilibrium in terms of fuzzy price function $P : \mathbb{R} \to \mathbb{R}_{\mathcal{F}(A)}$. Market equilibrium occurs when the quantity of demand

$$D(t) = a -_{\psi_A} b \cdot P(t) +_{\psi_A} c \cdot P'(t) +_{\psi_A} d \cdot P''(t)$$

is equals to the quantity of supplies

$$S(t) = -f +_{\psi_A} g \cdot P(t),$$

where $0 < a, b, f, g \in \mathbb{R}$, and $c, d \in \mathbb{R}$. Parameter c is positive when buyers expect the price will rise, so they prefer to increase consumption. In the opposite case, $c < 0$, buyers reduce their purchases in anticipation of a lower price in the future. Another factor that guides consumers is the rate d at which this market price change occurs [48]. We will establish the time trajectory of the market price, assuming the equilibrium in each instant t, that is, $D(t) = S(t)$, $\forall t \in \mathbb{R}$.

Simplifying the system, we obtain

$$d \cdot P''(t) +_{\psi_A} c \cdot P'(t) -_{\psi_A} (b+g) \cdot P(t) = -(a+f). \tag{43}$$

To solve the linear differential equation (43) with constants coefficients, we must first find the particular solution based on $G(t) = 0A - (a+f)$. In this case, a particular solution is given by the function $P_p(t) = 0A + \frac{a+f}{b+g}$.

Now, we need to find the homogeneous solution. To this end, we need the roots of the characteristic function $d\gamma^2 + c\gamma - (b+g) = 0$.

1. If $c^2 > -4d(b+g)$, then we have two distinct real roots, say γ_1 and γ_2. Thus, by Eq. (37), the solution of homogeneous equation is
$$P_h(t) = (c_1 e^{\gamma_1 t} + c_2 e^{\gamma_2 t})A + (d_1 e^{\gamma_1 t} + d_2 e^{\gamma_2 t}).$$
2. If $c^2 = -4d(b+g)$, then we have a single real root γ. Thus, by Eq. (38), the solution of homogeneous equation is
$$P_h(t) = (c_1 e^{\gamma t} + c_2 t e^{\gamma t})A + (d_1 e^{\gamma t} + d_2 t e^{\gamma t}).$$
3. If $c^2 < -4d(b+g)$, then we have complex roots, say $\gamma_1 = m + ni$ and $\gamma_2 = m - ni$. Thus, by Eq. (39), the solution of homogeneous equation is
$$P_h(t) = e^{mt}(c_1 e^{nit} + c_2 e^{-nit})A + e^{mt}(d_1 e^{nit} + d_2 e^{-nit}).$$

Therefore, the solution of (43) is given by

$$P(t) = P_h(t) +_{\psi_A} P_p(t).$$

Consider the second order differential equation given by

$$2P''(t) +_{\psi_A} P(t)' +_{\psi_A} 3P(t) = -4$$

with $P'(0) = 1A + 1$, $P(0) = 2A + 2$ and $A = (0.1; 0.27; 0.42)$. The solution of this FIVP is given by

$$\begin{aligned} P(t) = & \frac{e^{-t/4}}{69}\left(22\sqrt{23}\sin\frac{\sqrt{23}t}{4} + 230\cos\frac{\sqrt{23}t}{4} - 92\right)A \\ & + \frac{e^{-t/4}}{69}\left(22\sqrt{23}\sin\frac{\sqrt{23}t}{4} + 230\cos\frac{\sqrt{23}t}{4} - 92\right) \end{aligned} \tag{44}$$

The graphical representation of the solution P can be seen in Fig. 6.

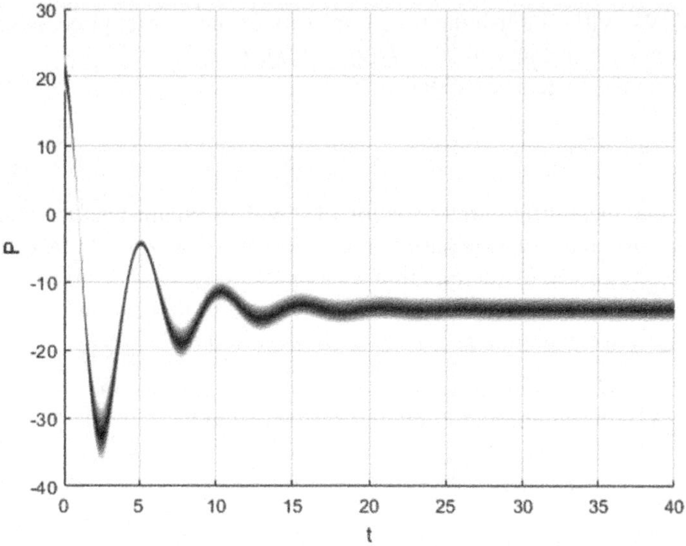

Fig. 6 Graphical representation of fuzzy price function (P) given as in Eq. (44) with $A = (0.1; 0.27; 0.42)$

The price fluctuation is the result of the assimilation of new information and speculation from other markets. These can be broad in the beginning, as is the case in Fig. 6 and then disappear over time, resulting in nothing. Or it can generate a growing perception about the new price level as new information is assimilated, which would be the case where the fluctuation grows over time.

Remark 4 Note that if $d > 0$, then $-4d(b + g)$ is always negative and, consequently, only the first case with two distinct real roots is possible. If $c < 0$, all three cases are possible. If $d < 0$ in the second case, both roots will be negative. When we have $c, d < 0$ in the second case, we have that the root is negative. In third case, this condition produces a negative m as well.

5.3 *The Fuzzy Laplace Transform for A-Linearly Interactive Process*

In this section we present the fuzzy Laplace transform for A-linearly interactive fuzzy processes, which was initially studied by [40]. The Laplace transform is given by means of improper integrals of the form $\int_a^{+\infty} F(t)dt$ which is defined as being the limit of the definite integral $\int_a^b F(t)dt$ when $b \to \infty$.

Theorem 10 *[40] Let A be a non-symmetric fuzzy number and $F : [a, +\infty) \longrightarrow \mathbb{R}_{\mathsf{F}(A)}$ be a function such that $F(t) = q(t)A + r(t)$. The function F is improper fuzzy Riemann integrable on $[a, +\infty)$ if, and only if, the functions $q, r : [a, +\infty) \longrightarrow \mathbb{R}$ are improper Riemann integrable over $[a, +\infty)$. Additionally,*

$$\int_a^{+\infty} F(t)dt = \left(\int_a^{+\infty} q(t)dt\right) A + \int_a^{+\infty} r(t)dt. \tag{45}$$

An important consequence of the Theorem 10 is that it is possible to calculate the fuzzy Laplace transform of the function $F : [a, +\infty) \longrightarrow \mathbb{R}_{\mathsf{F}(A)}$ in terms of the Laplace transform of real valued functions. More precisely, let $A \in \mathbb{R}_{\mathsf{F}}$ be non-symmetric, the fuzzy Laplace transform of a A-linearly interactive fuzzy process $F : [0, +\infty) \longrightarrow \mathbb{R}_{\mathsf{F}(A)}$ is defined as follows [40, 46]:

$$\mathbf{L}\{F(t)\} = \int_0^{+\infty} F(t)e^{-st}dt \tag{46}$$

whenever the limit exists.

Suppose that the fuzzy process F is differentiable of order $n \geq 0$, which is given by $F^{(n)}(t) = q^{(n)}(t)A + r^{(n)}(t)$. The Laplace transform of $F^{(n)}$ exists if, and only if, the Laplace of the real-valued functions $q^{(n)}$ and $r^{(n)}$ exist and, in this case, we have

$$\mathbf{L}\left\{F^{(n)}(t)\right\} = L\left\{q^{(n)}(t)\right\} A + L\left\{r^{(n)}(t)\right\}. \tag{47}$$

where L denotes the usual Laplace transform for real-valued functions. The next proposition states that the Laplace transform for A-linearly interactive fuzzy processes is a linear integral operator.

Proposition 16 *[40] Let $F, G : [0, \infty[\rightarrow \mathbb{R}_{\mathsf{F}(A)}$ be functions such that their fuzzy Laplace transform exist and let c a real number, then:*

$$\mathbf{L}\left\{F(t) +_{\psi_A} cG(t)\right\} = \mathbf{L}\{F(t)\} +_{\psi_A} c\mathbf{L}\{G(t)\}.$$

5.3.1 Fuzzy Initial Value Problem in $\mathbb{R}_{\mathsf{F}(A)}$

Consider the linear nth order FIVP given by

$$\begin{cases} X^{(n)}(t) = \sum_{i=0}^{n-1} k_i X^{(i)}(t) + f(t) \\ \mathbf{X}(t_0) = \left(X(t_0), X'(t_0), \ldots, X^{(n-1)}(t_0)\right) = (A_0, ..., A_{n-1}) \end{cases}, \tag{48}$$

where A is a non-symmetric fuzzy number, $f : [0, +\infty) \longrightarrow \mathbb{R}$, $k_i \in \mathbb{R}$ are constant coefficients, and $\mathbf{X}(t_0)$ is an initial condition vector whose elements $A_i \in \mathbb{R}_{\mathsf{F}(A)}$, for $i = 0, \ldots, n-1$.

Now, consider that $X(t) = q(t)A + r(t)$, that is, $X(t)$ is A-linearly interactive fuzzy process with non-symmetric $A \in \mathbb{R}_{\mathcal{F}}$, and $q, r : [a, b] \longrightarrow \mathbb{R}$ are continuous functions. Applying the fuzzy Laplace transform to both sides of the equation in (48), we have

$$L\left\{q^{(n)}(t)\right\} A + L\left\{r^{(n)}(t)\right\} = \sum_{i=0}^{n-1} k_i \left[L\left\{q^{(i)}(t)\right\} A + L\left\{r^{(i)}(t)\right\}\right] + L\left\{f(t)\right\}.$$

In this way, we obtain two algebraic systems

$$\begin{cases} L\left\{q^{(n)}(t)\right\} = \displaystyle\sum_{i=0}^{n-1} k_i L\left\{q^{(i)}(t)\right\} \\ \mathbf{q}(t_0) = \left(q(t_0), q'(t_0), \ldots, q^{(n-1)}(t_0)\right) \end{cases} \tag{49}$$

and

$$\begin{cases} L\left\{r^{(n)}(t)\right\} = \displaystyle\sum_{i=0}^{n-1} k_i L\left\{r^{(i)}(t)\right\} + L\left\{f(t)\right\} \\ \mathbf{r}(t_0) = \left(r(t_0), r'(t_0), \ldots, r^{(n-1)}(t_0)\right) \end{cases}, \tag{50}$$

where the vectors $\mathbf{q}(t_0)$, $\mathbf{r}(t_0)$ are the initial conditions composed by the coefficients $q_i, r_i \in \mathbb{R}$, for $i = 0, \ldots, n-1$, that are associated with each fuzzy initial condition of the generic nth order FIVP (48).

In addition, by using derivative proprieties of (classical) Laplace transform [31] on systems (49) and (50), we have

$$s^n L\left\{q(t)\right\} - \sum_{i=0}^{n-1} s^i\, q^{(n-1-i)}(t_0) = \sum_{i=0}^{n-1} k_i \left(s^i L\left\{q(t)\right\} - \sum_{j=0}^{i-1} s^j\, q^{(i-1-j)}(t_0)\right) \tag{51}$$

and

$$s^n L\left\{r(t)\right\} - \sum_{i=0}^{n-1} s^i\, r^{(n-1-i)}(t_0) = \sum_{i=0}^{n-1} k_i \left(s^i L\left\{r(t)\right\} - \sum_{j=0}^{i-1} s^j\, r^{(i-1-j)}(t_0)\right) + L\left\{f(t)\right\}. \tag{52}$$

Finally, by solving (51) and (52), using the (classical) inverse Laplace transform [31], we obtain the real functions $q(t)$ and $r(t)$ and, therefore, the fuzzy interactive solution $X(t) \in \mathbb{R}_{\mathcal{F}(A)}$ for all $t \in [a, b]$. It is worth noting that you can solve the FIVP (40) using the fuzzy Laplace transform.

Remark 5 The great advantage of using Laplace transform to solve FIVP is that the formula (47) transform the FIVP into two algebraic systems (49) and (50).

Example 10 Consider the linear homogeneous fuzzy differential equation

$$\begin{cases} X''(t) +_{\psi_A} 4X'(t) +_{\psi_A} 3X(t) = 10e^t \\ X(0) = 3A + 3 \in \mathbb{R}_{\mathsf{F}(A)} \\ X'(0) = 1A + 1 \in \mathbb{R}_{\mathsf{F}(A)}, \end{cases} \tag{53}$$

where $A \in \mathbb{R}_{\mathsf{F}}$ is non-symmetric and $X(t) \in \mathbb{R}_{\mathsf{F}(A)}$, that is, $X(t) = q(t)A + r(t)$.

Applying the fuzzy Laplace transform to (53), we obtain

$$\begin{aligned} L\{q''(t)\} + 4L\{q'(t)\} + 3L\{q(t)\} = 0 \\ q(0) = 3 \quad q'(0) = 1 \end{aligned} \tag{54}$$

and

$$\begin{aligned} L\{r''(t)\} + 4L\{r'(t)\} + 3L\{r(t)\} = L\{10e^t\} \\ r(0) = 3 \quad r'(0) = 1. \end{aligned} \tag{55}$$

By (55), we have

$$\begin{aligned} \left[s^2 L\{r(t)\} - sr(0) - r'(0)\right] + 4[sL\{r(t)\} - r(0)] + 3L\{r(t)\} = \frac{10}{s-1} \Leftrightarrow \\ \left[s^2 + 4s + 3\right] L\{r(t)\} = (4+s)r(0) + r'(0) + \frac{10}{s-1}. \end{aligned}$$

Substituting the initial conditions in (55), we have

$$\begin{aligned} \left[s^2 + 4s + 3\right] L\{r(t)\} = 3s + 13 + \frac{10}{s-1} \Leftrightarrow \\ L\{r(t)\} = \frac{3\,s^2 + 10\,s - 3}{(s-1)(s^2 + 4\,s + 3)} = \frac{5}{2(s+1)} - \frac{3}{4(s+3)} + \frac{5}{4(s-1)}. \end{aligned}$$

Taking the inverse of the transform we obtain

$$\begin{aligned} r(t) = L^{-1}\{L\{r(t)\}\} = L^{-1}\left\{\frac{5}{2(s+1)} - \frac{3}{4(s+3)} + \frac{5}{4(s-1)}\right\} = \\ \frac{1}{4}e^{-3t}(10e^{2t} + 5e^{4t} - 3). \end{aligned}$$

Similarly, for system (54), we obtain

$$q(t) = 5e^{-t} - 2e^{-3t}.$$

Therefore, the fuzzy solution of (53) is given by

$$X(t) = (5e^{-t} - 2e^{-3t})A + \left(\frac{1}{4}e^{-3t}(10e^{2t} + 5e^{4t} - 3)\right).$$

5.4 Linear Fuzzy Partial Differential Equation in $\mathbb{R}_{\mathsf{F}(A)}$

Although this subsection focuses on the one-dimensional case, we can motivate the use of interactive arithmetic operations in the study of fuzzy partial differential equations (FPDEs) by considering the Dirichlet problem for the two-dimensional Laplace equation given as follows. Let Ω be a bounded region of $\mathbb{R}^2$ and $f : \partial\Omega \to \mathbb{R}$, where $\partial\Omega$ denotes the boundary (or frontier) of Ω. The goal is to determine $u : \overline{\Omega} \to \mathbb{R}$ that satisfies the following conditions:

(i) $u(x, y)$ is continuous at every $(x, y) \in \overline{\Omega}$ (continuity)

(ii) $\dfrac{\partial^2 u}{\partial x^2} + \dfrac{\partial^2 u}{\partial y^2} = 0$ (u is harmonic)

(iii) $u = f$ in $\partial\Omega$ (boundary condition)

where $\overline{\Omega}$ denotes the closure of Ω. One of the central issue for obtaining a fuzzy version of this problem is how to define the harmonic condition (ii), since it involves the sum of two fuzzy quantities being equal to zero. Note that this equality is only possible if we consider an interactive addition such as the operation $+_{\psi_A}$ defined in the space $\mathbb{R}_{\mathsf{F}(A)}$ for some non-symmetric fuzzy number A.

Returning to the one-dimensional case, in this subsection, we deal with linear fuzzy partial differential equations of the form

$$a\frac{\partial^2 U}{\partial x^2} +_{\psi_A} b\frac{\partial^2 U}{\partial x \partial t} +_{\psi_A} c\frac{\partial^2 U}{\partial t^2} +_{\psi_A} d\frac{\partial U}{\partial x} +_{\psi_A} e\frac{\partial U}{\partial t} +_{\psi_A} fU +_{\psi_A} G = 0, \quad (56)$$

where the coefficients a, b, c, d, e, f are real functions of (x, t) and $G(x, t) \in \mathbb{R}_{\mathsf{F}(A)}$ is given by $G(x, t) = g_1(x, t)A + g_2(x, t)$. Here, we consider fuzzy partial differential equations (FPDE) in the space $\mathbb{R}_{\mathsf{F}(A)}$, with $A \in \mathbb{R}_{\mathsf{F}}$ non-symmetric, so that the fuzzy solution of (56) is of the form $U(x, t) = q(x, t)A + r(x, t)$. Thus, we obtain the decoupled system for real functions $q(x, t)$ and $r(x, t)$:

$$a\frac{\partial^2 q}{\partial x^2} + b\frac{\partial^2 q}{\partial x \partial t} + c\frac{\partial^2 q}{\partial t^2} + d\frac{\partial q}{\partial x} + e\frac{\partial q}{\partial t} + fq + g_1 = 0 \quad (57)$$

and

$$a\frac{\partial^2 r}{\partial x^2} + b\frac{\partial^2 r}{\partial x \partial t} + c\frac{\partial^2 r}{\partial t^2} + d\frac{\partial r}{\partial x} + e\frac{\partial r}{\partial t} + fr + g_2 = 0, \quad (58)$$

such that the fuzzy solution of (56) is given by the real solutions of (57) and (58). Thus, we can use any classical method to resolve (57) and (58). The fuzzy solution of (56) is given by $U(x, t) = q(x, t)A + r(x, t)$.

In the next example, we use the fuzzy Laplace transform method to obtain U. According to Theorem 10, the fuzzy Laplace transform with respect to t is given through classical Laplace transforms for $q(x, t)$ and $r(x, t)$, that is,

$$\begin{aligned}
\mathbf{L}\{U(x,t)\} &= L\{q(x,t)\}A + L\{r(x,t)\} \\
\mathbf{L}\{U(0,t)\} &= L\{q(0,t)\}A + L\{r(0,t)\} \\
\mathbf{L}\{U(x,t)\} &= L\{q(x,t)\}A + L\{r(x,t)\}.
\end{aligned}$$

Before proceeding, we emphasize that our approach is entirely different from that involving the gH-derivative, where FPDE is resolved through α-levels. The latter needs the Negoita and Ralescu Stacking Theorem, which makes its use difficult. Salahshour and Hagui [37] used gH-derivative to solve the fuzzy heat equation via the transformed Laplace fuzzy. Allahviranloo et al. [1] also proposed numerical methods for solving fuzzy linear partial differential equations under gH derivative. In our approach, the numerical methods are the classic ones applied to obtain two real functions (q and r).

Example 11 *(The Fuzzy heat equation)* Consider the following fuzzy heat equation:

$$\frac{\partial U}{\partial t} -_{\psi_A} c^2 \frac{\partial^2 U}{\partial x^2} = 0 \Leftrightarrow \frac{\partial U}{\partial t} = c^2 \frac{\partial^2 U}{\partial x^2}, \quad c \neq 0 \tag{59}$$

with the conditions

$$\begin{aligned}
U(x,0) &= q(x,0)A + r(x,0) \\
U(0,t) &= q(0,t)A + r(0,t) \\
|U(x,t)| &= |q(x,t)A + r(x,t)| \text{ bounded,}
\end{aligned}$$

where $A \in \mathbb{R}_\mathsf{F}$ is non-symmetric. By Theorem 10, the fuzzy problem (59) is equivalent to the classic cases:

$$\begin{cases} \dfrac{\partial q}{\partial t} = c^2 \dfrac{\partial^2 q}{\partial x^2} \\ q(x,0) \\ q(0,t) \\ |q(x,t)| \text{ bounded,} \end{cases} \tag{60}$$

and

$$\begin{cases} \dfrac{\partial r}{\partial t} = c^2 \dfrac{\partial^2 r}{\partial x^2} \\ r(x,0) \\ r(0,t) \\ |r(x,t)| \text{ bounded.} \end{cases} \tag{61}$$

To solve (60), we can apply the Laplace transform with respect to t

$$sL\{q(x,t)\} - q(x,0) = c^2 \frac{\partial^2 L\{q(x,t)\}}{\partial x^2} \Leftrightarrow \frac{\partial^2 L\{q(x,t)\}}{\partial x^2} - \frac{s}{c^2} L\{q(x,t)\} = -\frac{q(x,0)}{c^2}. \tag{62}$$

For the sake of simplicity, we assume that $q(x,0) = q_0 \in \mathbb{R}$ and $q(0,t) = q_1 \in \mathbb{R}$. Thus, the general solution of (62) is given by

$$L\{q(x,t)\} = k_1(s)e^{\frac{\sqrt{s}}{c}x} + k_2(s)e^{\frac{-\sqrt{s}}{c}x} + \frac{q_0}{s}.$$

Using the condition that $q(x,t)$ is bounded, we have that $L\{q(x,t)\}$ is bounded. Thus, $k_1(s) = 0$. By $q(0,t) = q_1$ we have $\frac{q_1}{s} = L\{q_1\} = L\{q(0,t)\} = k_2(s)e^0 + \frac{q_0}{s}$. So that $k_2(s) = \frac{(q_1 - q_0)}{s}$. Therefore,

$$L\{q(x,t)\} = \frac{(q_1 - q_0)}{s} e^{\frac{-\sqrt{s}}{c}x} + \frac{q_0}{s} \Rightarrow$$

$$q(x,t) = L^{-1}\{L\{q(x,t)\}\} = (q_1 - q_0)L^{-1}\left\{\frac{e^{\frac{-\sqrt{s}}{c}x}}{s}\right\} + q_0 L^{-1}\left\{\frac{1}{s}\right\} =$$

$$= (q_1 - q_0)\left[1 - erf\left(\frac{x}{2c\sqrt{t}}\right)\right] + q_0.$$

Thus,

$$q(x,t) = q_1 - (q_1 - q_0)erf\left(\frac{x}{2c\sqrt{t}}\right).$$

Similarly, taking $r(x,0) = r_0$ and $r(0,t) = r_1$ in Eq. (61), we have

$$r(x,t) = r_1 - (r_0 - r_1)erf\left(\frac{x}{2c\sqrt{t}}\right).$$

Finally, the fuzzy solution of (59) is

$$U(x,t) = \left(q_1 + (q_0 - q_1)erf\left(\frac{x}{2c\sqrt{t}}\right)\right)A + \left(r_1 + (r_0 - r_1)erf\left(\frac{x}{2c\sqrt{t}}\right)\right),$$

where $A \in \mathbb{R}_{\mathsf{F}}$ can be interpreted as a temperature linearly correlated with the initial temperature of the object and erf is the error function [25].

In the example above, we highlight that the expression $\frac{\partial U}{\partial t} -_{\psi_A} c^2 \frac{\partial^2 U}{\partial x^2} = 0$ only makes sense if we consider interactive arithmetic operations.

Another interesting approach for fuzzy heat equation can be found in Bertone et al. [10].

Below, we present an interactive population model in which there is a dispersion term and a growth term. This example using the gH-derivative would make the model very complex because we need translate the problem in α-levels.

Example 12 *(Skellam populational model* [45]*)* The model proposed by Skellam in 1951 suggest that for a population reproducing with rate a and spreading over space may be descripted by

$$\left\{ \frac{\partial U}{\partial t} = d\frac{\partial^2 U}{\partial x^2} +_{\psi_A} aU \right.$$

where $U(x,t)$ is the population at position x, in time t and d is called dispersion rate. The growth term aU, increases the density locally and causes a faster spatial spread in the population than that diffusion process only. Additionally, suppose that at t_0 the population is concentrated entirely at $x = 0$, that is, $U(0,0) = P_0 \in \mathbb{R}_\mathsf{F}$.

We suppose that solution is over $\mathbb{R}_{\mathsf{F}(A)}$, where A is non-symmetric fuzzy number linearly correlated with initial condition P_0. So, $U(x,t) = q(x,t)A + r(x,t)$ and $P_0 = q_0 A + r_0$. Now, by Eqs. (57) and (58), we have

$$\frac{\partial q}{\partial t} = d\frac{\partial^2 q}{\partial x^2} + aq, \quad q(0,0) = q_0 \in \mathbb{R}$$

and

$$\frac{\partial r}{\partial t} = d\frac{\partial^2 r}{\partial x^2} + ar, \quad r(0,0) = r_0 \in \mathbb{R}.$$

Again, each equation of the system above is classic, since q and r are real functions. So, for resolve each equations we can adopt any classical method. According to L. Edelstein-Keshet [17], we have the solution for $q(x,t)$ and $r(x,t)$ and, therefore, the fuzzy solution is given by

$$U(x,t) = \left(\frac{q_0}{2\sqrt{\pi dt}} e^{(at - \frac{x^2}{4dt})}\right) A + \left(\frac{r_0}{2\sqrt{\pi dt}} e^{(at - \frac{x^2}{4dt})}\right).$$

We finish this chapter by commenting that several classes of fuzzy partial differential equations in $\mathbb{R}_{\mathsf{F}(A)}$, such as hyperbolic and elliptic, can be studied using the classic theory through Eqs. (57) and (58).

Acknowledgements This work was partially supported by CNPq under grants no. 306546/2017-5 and 313313/2020-2, and Fapesp under grant no. 2020/09838-0.
In particular, Estevão Esmi was supported by CNPq under grant no. 313313/2020-2 and Fapesp under grant no. 2020/09838-0. Laecio C. de Barros was supported by CNPq under grant no. 306546/2017-5.

References

1. Allahviranloo, T., Afshar Kermani, M.: Numerical methods for fuzzy linear partial differential equations under new definition for derivative. Iran. J. Fuzzy Syst. **7**(3), 33–50 (2010)
2. Allahviranloo, T., Ahmadi, M.B.: Fuzzy Laplace transforms. Soft Comput. **14**(3), 235 (2010)
3. Audestad, J.A.: Some dynamic market models (2015). arXiv:1511.07203
4. Barros, L.C., Bassanezi, R.C., Lodwick, W.A.: First Course in Fuzzy Logic, Fuzzy Dynamical Systems, and Biomathematics. Springer (2016)
5. Barros, L.C., Bassanezi, R.C., Tonelli, P.A.: On the continuity of the Zadeh's extension. Proc. Seventh IFSA World Congress **2**, 3–8 (1997)

6. Barros, L.C., Santo Pedro, F.: Fuzzy differential equations with interactive derivative. Fuzzy Sets Syst. **309**, 64–80 (2017)
7. Bass, F.M.: A new product growth for model consumer durables. Manage. Sci. **15**(5), 215–227 (1969)
8. Bede, B.: Mathematics of Fuzzy Sets and Fuzzy Logic. Springer, Berlin Heidelberg, Berlin, Heidelberg (2013)
9. Bede, B., Stefanini, L., et al.: Generalized differentiability of fuzzy-valued functions. Fuzzy Sets Syst. **230**(1), 119–141 (2013)
10. Bertone, A.M., Jafelice, R.M., Barros, L.C., Bassanezi, R.C.: On fuzzy solutions for partial differential equations. Fuzzy Sets Syst. **219**, 68–80 (2013)
11. Cabral, V.M., Barros, L.C.: Fuzzy differential equation with completely correlated parameters. Fuzzy Sets Syst. **265**, 86–98 (2015)
12. Carlsson, C., Fullér, R., et al.: Additions of completely correlated fuzzy numbers. In: 2004 IEEE International Conference on Fuzzy Systems, vol. 1, pp. 535–539. IEEE (2004)
13. Chalco-Cano, Y., Roman-Flores, H.: On new solutions of fuzzy differential equations. Chaos, Solitons Fractals **38**(1), 112–119 (2008)
14. Diamond, P., Kloeden, P.: Metric topology of fuzzy numbers and fuzzy analysis. In: Fundamentals of Fuzzy Sets, pp. 583–641. Springer (2000)
15. Dubois, D., Prade, H.: Additions of interactive fuzzy numbers. IEEE Trans. Automatic Control **26**(4), 926–936 (1981)
16. Dubois, D., Prade, H.: Fuzzy numbers: an overview. In: Readings in Fuzzy Sets for Intelligent Systems, pp. 112–148. Elsevier (1993)
17. Edelstein-Keshet, L.: Mathematical Models in Biology. SIAM (2005)
18. Esmi, E., Barros, L.C., Santo Pedro, F., Laiate, B.: Banach spaces generated by strongly linearly independent fuzzy numbers. Fuzzy Sets Syst. (2020)
19. Esmi, E., Laiate, B., Barros, L.C., Santo Pedro, F.: Banach spaces generated by strongly linearly independent fuzzy numbers (2020). Submitted for publication
20. Esmi, E., Sánchez, D.E., Wasques, V.F., Barros, L.C.: Solutions of higher order linear fuzzy differential equations with interactive fuzzy values. Fuzzy Sets Syst. (2020)
21. Esmi, E., Santo Pedro, F., Barros, L.C., Lodwick, W.: Fréchet derivative for linearly correlated fuzzy function. Inf. Sci. **435**, 150–160 (2018)
22. Fullér, R.: On generalization of nguyen's theorem: a short survey of recent developments. In: Advances in Soft Computing, Intelligent Robotics and Control, pp. 183–190. Springer (2014)
23. Fullér, R., Majlender, P.: On interactive fuzzy numbers. Fuzzy Sets Syst. **143**(3), 355–369 (2004)
24. Gal, S.G.: Approximation theory in fuzzy setting. In: Handbook of Analytic-Computational Methods in Applied Mathematics, pp. 617–666 (2000)
25. Glaisher, J.: Xxxii. on a class of definite integrals. London, Edinburgh, Dublin Philosoph. Mag. J. Sci. **42**(280), 294–302 (1871)
26. Gordon, R.: Riemann integration in banach spaces. Rocky Mountain J. Math. **21**(3), 923–949 (1991)
27. Kaleva, O.: Fuzzy differential equations. Fuzzy Sets Syst. **24**(3), 301–317 (1987)
28. Klir, G.J.: Fuzzy arithmetic with requisite constraints. Fuzzy Sets Syst. **91**(2), 165–175 (1997)
29. Klir, G.J., Yuan, B.: Fuzzy Sets and Fuzzy Logic: Theory and Applications. Prentice Hall, Upper Saddle River, NY (1995)
30. Mutovkina, N.Y., Borodulin, A.: A neuro-fuzzy pricing model in conditions of market uncertainty. In: International Conference of Artificial Intelligence, Medical Engineering, Education, pp. 461–470. Springer (2019)
31. Nagle, R.K., Saff, E.B., Snider, A.D., West, B.: Fundamentals of Differential Equations and Boundary Value Problems. Addison-Wesley Reading (1996)
32. Negoita, C.V., Ralescu, D.A.: Applications of Fuzzy Sets to Systems Analysis. Wiley, New York (1975)
33. Nguyen, H.T.: A note on the extension principle for fuzzy sets. J. Math. Anal. Appl. **64**(2), 369–380 (1978)

34. Puri, M.L., Ralescu, D.A.: Differentials of fuzzy functions. J. Math. Anal. Appl. **91**(2), 552–558 (1983)
35. Puri, M.L., Ralescu, D.A.: Fuzzy random variables. J. Math. Anal. Appl. **114**(2), 409–422 (1986)
36. Salahshour, S., Allahviranloo, T.: Applications of fuzzy Laplace transforms. Soft Comput. **17**(1), 145–158 (2013)
37. Salahshour, S., Haghi, E.: Solving fuzzy heat equation by fuzzy Laplace transforms. In: International Conference on Information Processing and Management of Uncertainty in Knowledge-Based Systems, pp. 512–521. Springer (2010)
38. Salgado, S.A.B., Barros, L.C., Esmi, E., Eduardo Sánchez, D.: Solution of a fuzzy differential equation with interactivity via Laplace transform. J. Intell. Fuzzy Syst. **37**(2), 2495–2501 (2019)
39. Salgado, S.A.B., Esmi, E., Eduardo Sánchez, D., Barros, L.C.: Solving interactive fuzzy initial value problem via fuzzy Laplace transform. Comput. Appl. Math. (2020). Accepted for publication
40. Salgado, S.A.B., Sánchez, D.E., Barros, L.C., Esmi, E.: Fuzzy Laplace transform and fréchet derivative for fuzzy interactive process (2020). Submitted for publication
41. Sánchez, D.E., Barros, L.C., Esmi, E.: On interactive fuzzy boundary value problems. Fuzzy Sets Syst. **358**, 84–96 (2019)
42. Santo Pedro, F., Barros, L.C., Esmi, E.: Population growth model via interactive fuzzy differential equation. Inf. Sci. **481**, 160–173 (2019)
43. Santo Pedro, F., Esmi, E., Barros, L.C.: Calculus for linearly correlated fuzzy function using fréchet derivative and riemann integral. Inf. Sci. **512**, 219–237 (2020)
44. Shen, Y.: First-order linear fuzzy differential equations on the space of linearly correlated fuzzy numbers. Fuzzy Sets Syst. (2020)
45. Skellam, J.G.: Random dispersal in theoretical populations. Biometrika **38**(1/2), 196–218 (1951)
46. Son, N.T.K., Thao, H.T.P., Dong, N.P., Long, H.V.: Fractional calculus of linear correlated fuzzy-valued functions related to fréchet differentiability. Fuzzy Sets Syst. (2020)
47. Stefanini, L., Bede, B.: Generalized hukuhara differentiability of interval-valued functions and interval differential equations. Nonlinear Anal. Theory, Methods Appl. **71**(3–4), 1311–1328 (2009)
48. Todorova, T.: Problems Book to Accompany Mathematics for Economists. Wiley (2010)
49. Wasques, V.F., Esmi, E., Barros, L.C., Sussner, P.: The generalized fuzzy derivative is interactive. Inf. Sci. **519**, 93–109 (2020)
50. Wasques, V.F., Esmi, E., Barros, L.C., Sussner, P.: Numerical solution for fuzzy initial value problems via interactive arithmetic: application to chemical reactions. Int. J. Comput. Intell. Syst. **13**(1), 1517–1529 (2020)
51. Zadeh, L.A.: Fuzzy sets. Inf. Control **8**(3), 338–353 (1965)
52. Zadeh, L.A.: The concept of a linguistic variable and its application to approximate reasoning - i. Inf. Sci. **8**, 199–249 (1975)
53. Zeidler, E.: Applied Functional Analysis: Main Principles and Their Applications, vol. 109. Springer Science & Business Media (2012)

The Transform Method to Solve Fuzzy Differential Equation via Differential Inclusions

Xiao-Ming Liu, Ling Hong, and Jun Jiang

Abstract This chapter dedicates to offer a numerical method, named transform method, to solve fuzzy differential equation (FDE) via differential inclusions. The method acquires response solutions with their membership distribution functions. To do that, the FDE in form of differential inclusions is transformed into the governing equation of membership degree and the membership distribution of the fuzzy solution is composed by the membership degrees solved from the governing equation. In the procedure, no comparison or data storage is required, which makes the method possess high computational efficiency and low memory cost. Since the governing equation of the membership degree is derived from the master equation of fuzzy dynamics, the method is validated by theoretically proving the equivalence of the solution of the FDE via differential inclusions and that of the fuzzy master equation. Furthermore, the transform method is verified by comparing the numerical solution with the analytical solution in two examples. At last, with the help of the method, the response solutions are obtained for the Mathieu system and the rotor/stator contact system with fuzzy uncertainties.

X.-M. Liu (✉)
State Key Laboratory of Compressor Technology, Hefei General Machinery Research Institute, Hefei, Anhui 230031, People's Republic of China
e-mail: liuxiaoming_4109@163.com

L. Hong · J. Jiang
State Key Laboratory for Strength and Vibration of Mechanical Structures, Xi'an Jiaotong University, No.28 Xianning West Road, Xi'an, Shaanxi 710049, People's Republic of China
e-mail: hongling@mail.xjtu.edu.cn

J. Jiang
e-mail: jun.jiang@mail.xjtu.edu.cn

T. Allahviranloo and S. Salahshour (eds.), *Advances in Fuzzy Integral and Differential Equations*, Studies in Fuzziness and Soft Computing 412,
https://doi.org/10.1007/978-3-030-73711-5_2

1 Introduction

It is well-known that ordinary differential equation (ODE) is a powerful tool to describe the evolution of deterministic dynamical system. When fuzzy uncertainty is considered, the prevailing way is modeling the system as a fuzzy differential equation (FDE), $\dot{\boldsymbol{X}}(t) = \boldsymbol{F}(t, \boldsymbol{X})$. The FDE can be rewritten into the form via differential inclusions, $\dot{\boldsymbol{x}}(t) \in \boldsymbol{F}(t, \boldsymbol{x})$, where $\dot{\boldsymbol{x}}(t)$ is an ordinary derivative and $\boldsymbol{F}(t, \boldsymbol{x})$ is a family of differential inclusions [1–14]. The solution is a set of functions whose derivatives satisfy the inclusions for all $\boldsymbol{x}$ and t. It preserves the main properties of the corresponding deterministic system such as stability, periodicity and bifurcation [5–7, 9, 13].

Since it is very difficult to get the analytical solution of FDE except for a few linear systems, numerical methods become the main tools to study fuzzy dynamical systems. For instance, based on the arithmetic operations of interactive fuzzy numbers via sup-J extension principle, Euler's and Runge-Kutta methods were respectively adapted to find numerical solutions of the bidimensional FDEs in [15, 16]. Most numerical methods of FDE construct the solution with attainable sets by solving a set of ODEs [4, 10, 17–23]. It means that a huge number of ODEs are required to be numerically integrated and their results in each time step have to be stored for comparison to outline the boundaries of the attainable sets. Thus, solving FDE numerically is a complex, compute-intensive and memory-intensive task. There are developments of numerical methods to solve FDE committed to either reducing the complexity or the computation intensity or the memory cost [5–7, 10–12]. For example, by assuming that $\boldsymbol{F}(t, \boldsymbol{x})$ satisfies the Lipshitz condition, [14, 24] calculated only the smallest closed subset using the smallest closed envelope of $\boldsymbol{F}(t, \boldsymbol{x})$ instead of the entire $\boldsymbol{F}(t, \boldsymbol{x})$, which reduced the number of the ODEs integrated and also the data stored to some extent but increased the complexity since the maximized Hamiltonian must be achieved to find the qualified ODEs in every time step.

Except for modeling the system as a FDE, Friedman and Sandler [25–27] studied fuzzy dynamical system based on the min-max fuzzy logic rules by introducing the definitions of fuzzy state and fuzzy velocity. They derived the master equation of fuzzy dynamics and then the governing equation of the fuzzy state's membership degree as the analog of the Fokker-Planck-Kolmogorov (FPK) equation of statistical dynamics. The membership degree can be solved from the governing equation and a pack of such membership degrees outlined the membership distribution of the fuzzy state.

This chapter focus on solving FDE via differential inclusions by associating fuzzy master equation with fuzzy differential inclusions. After regarding the attainable set of $\boldsymbol{F}(t, \boldsymbol{x})$ as the fuzzy velocity [25–27], the FDE in form of differential inclusions can be rewritten as the form of the governing equation of the membership degree. Given an initial point in the state space and the partial derivatives of the initial membership degree of the state with respect to the state variables at that point, the corresponding response can be obtained with its location in the state space and its membership degree at the required moment. Theoretically we prove that the solution obtained from the fuzzy master equation is equivalent to that from the FDE via differential inclusions under some proper constraints. Consequently, a multitude of membership degrees solved from the governing equation can be used to compose the membership distribution of the FDE's fuzzy solution via differential inclusions. Then, the transform method solving FDE via differential inclusions is described. The numerical method includes the procedure to transform FDE from the form of differential inclusions into the governing equation of the membership degree and the approach to numerically solve the governing equation which is a partial differential equation. To examine the accuracy of the method, two examples are computed and the results are compared with their analytical solutions. Then more complex fuzzy dynamical systems, the fuzzy Mathieu system and the fuzzy rotor/stator contact system, are studied respectively with the help of the proposed method.

2 The Method

2.1 The Transform Method

Given an initial value problem

$$\begin{cases} \dot{\boldsymbol{x}} & = \boldsymbol{f}(t, \boldsymbol{x}, \boldsymbol{w}) \\ \boldsymbol{x}(0) & = \boldsymbol{x}_0 \end{cases}, \tag{1}$$

where $t \in [0, T]$, $\boldsymbol{x}_0 \in \Upsilon \subset R^n$, $\boldsymbol{w} \in \Omega \subset R^m$ and the vector-valued function $\boldsymbol{f}$ is continuous on $[0, T] \times R^n \times \Omega$. Denote by $(\mathcal{F}_C)^p(S)$ the space of convex and compact fuzzy sets on $S \subset R^p$. Consider the parameter $\boldsymbol{w}$ and the initial state $\boldsymbol{x}_0$ are replaced by fuzzy sets $\boldsymbol{X}_0 \in (\mathcal{F}_C)^n(\Upsilon)$ and $\boldsymbol{W} \in (\mathcal{F}_C)^m(\Omega)$ respectively because of fuzzy uncertainties, applying the Zadeh's extension principle [28, 29] on $\boldsymbol{f}$ can get a fuzzy initial value problem

$$\begin{cases} \dot{\boldsymbol{X}} & = \widehat{\boldsymbol{f}}(t, \boldsymbol{X}, \boldsymbol{W}) \\ \boldsymbol{X}(0) & = \boldsymbol{X}_0 \end{cases}, \tag{2}$$

where $\widehat{\boldsymbol{f}} : [0, T] \times (\mathcal{F}_C)^n \times (\mathcal{F}_C)^m(\Omega) \to (\mathcal{F}_C)^n$. Because of the continuity of $\boldsymbol{f}$, $\widehat{\boldsymbol{f}}$ is continuous on the Hausdorff measure space and $[\widehat{\boldsymbol{f}}(t, \boldsymbol{X}, \boldsymbol{W})]^\alpha = \boldsymbol{f}(t, [\boldsymbol{X}]^\alpha, [\boldsymbol{W}]^\alpha) = \{\boldsymbol{f}(t, x, w) | x \in [\boldsymbol{X}]^\alpha, w \in [\boldsymbol{W}]^\alpha\}$ [30, 31]. then (2) can be represented by the fuzzy differential equation via differential inclusions:

$$\begin{cases} \dot{\boldsymbol{x}} & \in \boldsymbol{f}(t, \boldsymbol{x}, [\boldsymbol{W}]^\alpha) \\ \boldsymbol{x}(0) & \in [\boldsymbol{X}_0]^\alpha \end{cases} \tag{3}$$

where $\alpha \in [0, 1]$. If there exists an absolutely continuous function $\boldsymbol{x}(\cdot)$ satisfying above inclusions a.e. in $[0, T]$, then $\boldsymbol{x}(\cdot)$ is a solution of (3). On the other hand, $\boldsymbol{x}(\cdot, \boldsymbol{x}_0, \boldsymbol{c}(\cdot))$ is the solution of the initial value problem

$$\begin{cases} \dot{\boldsymbol{x}} & = \boldsymbol{f}(t, \boldsymbol{x}, \boldsymbol{c}(t)) \\ \boldsymbol{x}(0) & = \boldsymbol{x}_0 \end{cases}, \tag{4}$$

where $\boldsymbol{c}(\cdot) : [0, T] \to \Omega$ is a measurable function satisfying $\boldsymbol{c}(t) \in [\boldsymbol{W}]^\alpha$ a.e. in $[0, T]$. Denote by $\mathcal{M}([0, T], [\boldsymbol{W}]^\alpha)$ the set of measurable functions $\{\boldsymbol{c}(\cdot)|\boldsymbol{c}(t) \in [\boldsymbol{W}]^\alpha, t \in [0, T]\}$ and $\Sigma_\alpha(\cdot, [\boldsymbol{X}_0]^\alpha, [\boldsymbol{W}]^\alpha)$ the set of all solutions of (3), then we have the conclusion [14, 24, 32]

$$\Sigma_\alpha(\cdot, [\boldsymbol{X}_0]^\alpha, [\boldsymbol{W}]^\alpha) = \{\boldsymbol{x}(\cdot, \boldsymbol{x}_0, \boldsymbol{c}(\cdot))|\boldsymbol{x}_0 \in [\boldsymbol{X}_0]^\alpha, \boldsymbol{c}(\cdot) \in \mathcal{M}([0, T], [\boldsymbol{W}]^\alpha)\}. \tag{5}$$

Let Υ be an open subset of R^n and H a mapping from $[0, T] \times \Upsilon$ into the compact and convex subset of R^n. If there exist $b, T, M > 0$ such that H maps $[0, T] \times (\boldsymbol{x}_0 + (b + MT)B^n)$ into the ball of radius M, where B^n is the unit ball of R^n and $(\boldsymbol{x}_0 + (b + MT)B^n) \subset \Upsilon$, then the *boundedness assumption* holds. Since $\widehat{\boldsymbol{f}}$ is continuous, from the theorem in the literature [5], when the boundedness assumption holds for all $\boldsymbol{x}_0 \in [\boldsymbol{X}_0]^0$ and

$$\begin{cases} \dot{\boldsymbol{x}} & \in \boldsymbol{f}(t, \boldsymbol{x}, [\boldsymbol{W}]^0) \\ \boldsymbol{x}(0) & = \boldsymbol{x}_0 \in [\boldsymbol{X}_0]^0 \end{cases}, \tag{6}$$

the fuzzy differential equation (2) has a unique compact fuzzy solution $\Sigma(\cdot, \boldsymbol{X}_0, \boldsymbol{W})$ via differential inclusions and for any $\alpha \in [0, 1]$

$$[\Sigma(\cdot, \boldsymbol{X}_0, \boldsymbol{W})]^\alpha = \Sigma_\alpha(\cdot, [\boldsymbol{X}_0]^\alpha, [\boldsymbol{W}]^\alpha). \tag{7}$$

Define the attainable sets of $\Sigma(\cdot, \boldsymbol{X}_0, \boldsymbol{W})$ as

$$\mathcal{A}_\alpha(t, [\boldsymbol{X}_0]^\alpha, [\boldsymbol{W}]^\alpha) = \{\boldsymbol{x}(t, \boldsymbol{x}_0, \boldsymbol{c}(t))|\boldsymbol{x}(\cdot, \boldsymbol{x}_0, \boldsymbol{c}(\cdot)) \in [\Sigma(\cdot, \boldsymbol{X}_0, \boldsymbol{W})]^\alpha\}, \tag{8}$$

where $t \in [0, T]$. Accordingly, the family of attainable sets $\{\mathcal{A}_\alpha(t, [\boldsymbol{X}_0]^\alpha, [\boldsymbol{W}]^\alpha)|\alpha \in [0, 1]\}$ defines a nonempty compact fuzzy set $\mathcal{A}(t, \boldsymbol{X}_0, \boldsymbol{W})$. Furthermore, if $\widehat{\boldsymbol{f}}$ is concave, $\mathcal{A}_\alpha(t, [\boldsymbol{X}_0]^\alpha, [\boldsymbol{W}]^\alpha)$ are convex and $\mathcal{A}(t, \boldsymbol{X}_0, \boldsymbol{W})$ is nonempty compact and convex [14].

To numerically solve the fuzzy differential equation via differential inclusions, we can search the family of the attainable sets $\{\mathcal{A}_\alpha(t, [\boldsymbol{X}_0]^\alpha, [\boldsymbol{W}]^\alpha)|\alpha \in [0, 1]\}$ at any required moment $t \in [0, T]$. That is, we have to consider all initial conditions $\boldsymbol{x}_0 \in [\boldsymbol{X}_0]^\alpha$ and all measurable functions $\boldsymbol{c}(\cdot) \in \mathcal{M}([0, T], [\boldsymbol{W}]^\alpha)$, then collect the set of all solutions $\boldsymbol{x}(t, \boldsymbol{x}_0, \boldsymbol{c}(t))$ by numerically integrating (4), which can be a very difficult and burdensome task. Herein we will identify the fuzzy set $\mathcal{A}(t, \boldsymbol{X}_0, \boldsymbol{W})$ by its membership function $\mu_{\mathcal{A}(t,X_0,W)}: R^n \to [0,1]$. To do that, the transforming of (3) is required.

From the point view of dynamics, (3) is re-interpreted as below:

(i) $\boldsymbol{x}$ is the state of the system.
(ii) $\dot{\boldsymbol{x}}$ is the velocity of the change of the state, which is recorded as $\boldsymbol{u}$.
(iii) At time t and state $\boldsymbol{x}$, the family of the sets $\{\boldsymbol{f}(t,\boldsymbol{x},[\boldsymbol{W}]^\alpha)|\alpha \in [0,1]\}$ defines a fuzzy set of the velocity of the state's change, written as $\boldsymbol{U}(t,\boldsymbol{x}) \in (\mathcal{F}_C)^n$. Following the extension principle [28, 29], the membership function of $\boldsymbol{U}(t,\boldsymbol{x})$ is

$$\mu_{U(t,x)}(\boldsymbol{u}) = \begin{cases} \sup\limits_{w \in f^{-1}(u)} \{\mu_W(\boldsymbol{w})\}, & \text{if } \boldsymbol{f}^{-1}(\boldsymbol{u}) \neq \emptyset \\ 0, & \text{if } \boldsymbol{f}^{-1}(\boldsymbol{u}) = \emptyset \end{cases}, \tag{9}$$

where $\boldsymbol{f}^{-1}(\boldsymbol{u}) = \{\boldsymbol{w}|\boldsymbol{w} \in \Omega, \boldsymbol{f}(\boldsymbol{w}) = \boldsymbol{u}\}$. $\mu_{U(t,x)}(\boldsymbol{u})$ represents the possibility that $\boldsymbol{u}$ is the velocity of the state's change at $(t,\boldsymbol{x})$.
(iv) The solution of the system is a fuzzy process of the state, written as $\mathcal{X}(\cdot)$. Its value at time t, $\mathcal{X}(t)$, is a fuzzy set, which represents the response of the system at t and can be identified by its membership function $\mu_{\mathcal{X}(t)}: R^n \to [0,1]$. The set of the membership degrees at different states $\{(\boldsymbol{x},\mu_{\mathcal{X}(t)}(\boldsymbol{x}))|\boldsymbol{x} \in [\boldsymbol{X}]^0\}$ is referred to as the membership distribution of the response at time t.

Then the evolution of the fuzzy system can be described by the fuzzy master equation (10) under some conditions following the conclusions in [25] which are recited below.

Theorem 1 *([25])* $[\mathcal{X}(\cdot)]^0$ *is an n-dimensional manifold describing the set of all possible states of the system. At any given time* $t \in [0,T]$, *the family of* $\{[\mathcal{X}(\cdot)]^\alpha|\alpha \in [0,1]\}$ *defines a fuzzy set* $\mathcal{X}(t) \in (\mathcal{F}_C)^n(\Upsilon(t))$ *where* $\Upsilon(t) \subset R^n$. $\mu_{\mathcal{X}(t)}: R^n \to [0,1]$ *is the membership function of* $\mathcal{X}(t)$. *At any given state* $\boldsymbol{x} \in \Upsilon(t)$ *and time* $t \in [0,T]$, *the velocity of the change of the state is a fuzzy set* $\boldsymbol{U}(t,\boldsymbol{x}) \in (\mathcal{F}_C)^n$. *Let* $\mu_{U(t,x)}: R^n \to [0,1]$ *be the membership function of* $\boldsymbol{U}(t,\boldsymbol{x})$. *If there exists a constant* $B(\epsilon)$ *for all* $t \in [0,T]$, $\boldsymbol{x} \in \Upsilon(t)$ *and* $\epsilon > 0$ *such that* $\mu_{U(t,x)}(\boldsymbol{u}) \leq \epsilon$ *whenever* $||\boldsymbol{u}|| \geq B(\epsilon)$, *among which* $\boldsymbol{u} \in R^n$, *then the evolution of the membership distribution of the state can be determined by the fuzzy master equation*

$$\mu_{\mathcal{X}(t+\delta)}(\boldsymbol{x}) = \sup_{\boldsymbol{u}}\{min\{\mu_{U(t,x-u\delta)}(\boldsymbol{u}), \mu_{\mathcal{X}(t)}(\boldsymbol{x}-\boldsymbol{u}\delta)\}\}, \tag{10}$$

where δ *is a very short time. Further, if* $\sup\limits_{u \in R^n}\{\mu_{U(t,x)}(\boldsymbol{u})\} = 1$, $\mu_{\mathcal{X}(t)}(\boldsymbol{x})$ *is continuous in* t.

Notice that, if $\boldsymbol{U}(t,\boldsymbol{x})$ is identified by (9), $\sup\limits_{u \in R^n}\{\mu_{U(t,x)}(\boldsymbol{u})\} = 1$ is true. Moreover, if $\boldsymbol{f}$ is continuous in $(t,\boldsymbol{x},\boldsymbol{w})$, we can immediately reach the constant $B(\epsilon)$ for all $t \in [0,T]$, $\boldsymbol{x} \in \Upsilon(t)$ and $\epsilon > 0$.

With regard to the relationship between $\mathcal{X}(\cdot)$ and $\mathcal{A}(\cdot,\boldsymbol{X}_0,\boldsymbol{W})$, we give a theorem followed by its proof.

Theorem 2 *Let* $W \in (\mathcal{F}_C)^m(\Omega)$ *and* $X_0 \in (\mathcal{F}_C)^n(\Upsilon)$, *where* $\Omega \supset [W]^0$ *and* $\Upsilon \supset [X_0]^0$ *are open subsets of* R^m *and* R^n *respectively.* $\widehat{f} : [0, T] \times (\mathcal{F}_C)^n \times (\mathcal{F}_C)^m(\Omega) \to (\mathcal{F}_C)^n$ *is obtained by applying the Zadeh's extension principle to* $f : [0, T] \times R^n \times \Omega \to R^n$. *If*

(i) the function f *is continuous in* $(t, x, w) \in [0, T] \times R^n \times \Omega$,
(ii) the boundedness assumption holds for all $x_0 \in [X_0]^0$ *and*

$$\begin{cases} \dot{x} & \in f(t, x, [W]^0) \\ x(0) & = x_0 \in [X_0]^0 \end{cases}, \tag{11}$$

then the solution of the fuzzy differential equation via differential inclusions (3) can be identified by the fuzzy master equation (10), which means $\mathcal{A}(t, X_0, W) = \mathcal{X}(t)$ *for all* $t \in [0, T]$.

Proof Obviously, both $\mathcal{A}(0, X_0, W)$ and $\mathcal{X}(0)$ equal X_0 at time $t = 0$. Assume at time $t \in [0, T)$ there is $\mathcal{A}(t, X_0, W) = \mathcal{X}(t)$, i.e.,

$$\mathcal{A}_\alpha(t, [X_0]^\alpha, [W]^\alpha) = [\mathcal{X}(t)]^\alpha = \begin{cases} \{x \in R^n | \mu_{\mathcal{X}(t)}(x) \geq \alpha\}, & \alpha \in (0, 1] \\ cl\{x \in R^n | \mu_{\mathcal{X}(t)}(x) > \alpha\}, & \alpha = 0 \end{cases}. \tag{12}$$

After a very short time δ such that $t + \delta \in (0, T]$, if the equivalence $\mathcal{A}(t + \delta, X_0, W) = \mathcal{X}(t + \delta)$ can be deduced from the presume $\mathcal{A}(t, X_0, W) = \mathcal{X}(t)$, the conclusion of the theorem is true. Next, we argue $\mathcal{A}(t + \delta, X_0, W) = \mathcal{X}(t + \delta)$ by demonstrating $\mathcal{A}(t + \delta, X_0, W)$ and $\mathcal{X}(t + \delta)$ have the same α-levels for all $\alpha \in [0, 1]$.

For $y \in \mathcal{A}_\alpha(t + \delta, [X_0]^\alpha, [W]^\alpha)$, there is a solution $x(\cdot, x_0, c(\cdot)) \in \Sigma_\alpha(\cdot, [X_0]^\alpha, [W]^\alpha)$ such that $x_0 \in [X_0]^\alpha$, $c(\cdot) \in \mathcal{M}([0, T], [W]^\alpha)$ and $x(t + \delta, x_0, c(t + \delta)) = y$. Let $u = f(t, x(t, x_0, c(t)), c(t))$. While $\delta \to 0$, we have

$$\begin{aligned} x(t, x_0, c(t)) &= x(t + \delta, x_0, c(t + \delta)) - u\delta \\ &= y - u\delta. \end{aligned}$$

Since

$$\begin{aligned} [\mathcal{X}(t)]^\alpha &= \mathcal{A}_\alpha(t, [X_0]^\alpha, [W]^\alpha) \\ &= \{x(t, x_0, c(t)) | x(\cdot, x_0, c(\cdot)) \in \Sigma_\alpha(\cdot, [X_0]^\alpha, [W]^\alpha)\}, \end{aligned}$$

$y - u\delta \in [\mathcal{X}(t)]^\alpha$, i.e., $\mu_{\mathcal{X}(t)}(y - u\delta) \geq \alpha$. In addition, because $\mu_W(c(t)) \geq \alpha$, $\mu_{U(t, y - u\delta)}(u) \geq \alpha$. Then, from the fuzzy master equation (10), there exists $\mu_{\mathcal{X}(t+\delta)}(y) \geq \alpha$, that is, $y \in [\mathcal{X}(t + \delta)]^\alpha$. Thus, $\mathcal{A}_\alpha(t + \delta, [X_0]^\alpha, [W]^\alpha) \subset [\mathcal{X}(t + \delta)]^\alpha$.

To obtain the conclusion of $[\mathcal{X}(t+\delta)]^{\alpha} \subset \mathcal{A}_{\alpha}(t+\delta, [\boldsymbol{X}_0]^{\alpha}, [\boldsymbol{W}]^{\alpha})$, consider the case of $\alpha > 0$ first. For $\forall \boldsymbol{y} \in [\mathcal{X}(t+\delta)]^{\alpha}$, i.e., $\mu_{\mathcal{X}(t+\delta)}(\boldsymbol{y}) \geq \alpha$, and $\forall \eta_1 \in (0, \alpha)$, from the fuzzy master equation (10), there exists at least one velocity $\boldsymbol{u}$ to ensure

$$\mu_{U(t, \boldsymbol{y}-\boldsymbol{u}\delta)}(\boldsymbol{u}) \geq \alpha - \eta_1, \tag{13}$$

$$\mu_{\mathcal{X}(t)}(\boldsymbol{y} - \boldsymbol{u}\delta) \geq \alpha - \eta_1. \tag{14}$$

Thus, $\boldsymbol{y} - \boldsymbol{u}\delta \in [\mathcal{X}(t)]^{\alpha-\eta_1} = \mathcal{A}_{\alpha-\eta_1}(t, [\boldsymbol{X}_0]^{\alpha-\eta_1}, [\boldsymbol{W}]^{\alpha-\eta_1})$. Besides, from (13) and (9), there must exists $\boldsymbol{w} \in \boldsymbol{f}^{-1}(\boldsymbol{u})$ at $(t, \boldsymbol{y} - \boldsymbol{u}\delta)$ such that $\mu_{\boldsymbol{W}}(\boldsymbol{w}) \geq \alpha - \eta_1 - \eta_2$, i.e., $\boldsymbol{w} \in [\boldsymbol{W}]^{\alpha-\eta_1-\eta_2}$, for $\forall \eta_2 \in (0, \alpha - \eta_1)$. For simplicity, let $\tilde{\alpha} = \alpha - \eta_1 - \eta_2$. Since $\mathcal{A}_{\tilde{\alpha}}(t, [\boldsymbol{X}_0]^{\tilde{\alpha}}, [\boldsymbol{W}]^{\tilde{\alpha}}) = \{\boldsymbol{x}(t, \boldsymbol{x}_0, \boldsymbol{c}(t)) | \boldsymbol{x}(\cdot, \boldsymbol{x}_0, \boldsymbol{c}(\cdot)) \in \Sigma_{\tilde{\alpha}}(\cdot, [\boldsymbol{X}_0]^{\tilde{\alpha}}, [\boldsymbol{W}]^{\tilde{\alpha}})\}$ and $\boldsymbol{x}(\cdot, \boldsymbol{x}_0, \boldsymbol{c}(\cdot))$ is the solution of (4) while $\boldsymbol{x}_0 \in [\boldsymbol{X}_0]^{\tilde{\alpha}}$ and $\boldsymbol{c}(\cdot) \in \mathcal{M}([0, T], [\boldsymbol{W}]^{\tilde{\alpha}})$, $\boldsymbol{y} - \boldsymbol{u}\delta$ is the value of some solution $\boldsymbol{x}(\cdot, \boldsymbol{x}_0, \boldsymbol{c}(\cdot))$ at time t, i.e., $\boldsymbol{y} - \boldsymbol{u}\delta = \boldsymbol{x}(t, \boldsymbol{x}_0, \boldsymbol{c}(t))$. By choosing a continuous function $\boldsymbol{c}'(\cdot) \in \mathcal{M}([0, T], [\boldsymbol{W}]^{\tilde{\alpha}})$ which possesses $\boldsymbol{c}'(t) = \boldsymbol{w}$, we can construct a new function

$$\boldsymbol{c}''(\tau) = \begin{cases} \boldsymbol{c}(\tau) & \text{if } \tau \in [0, t) \\ \boldsymbol{c}'(\tau) & \text{if } \tau \in [t, T] \end{cases}, \tag{15}$$

such that $\boldsymbol{c}''(\cdot) \in \mathcal{M}([0, T], [\boldsymbol{W}]^{\tilde{\alpha}})$. After the substitution of $\boldsymbol{c}''(\cdot)$ into (4), we get a new solution $\boldsymbol{x}(\cdot, \boldsymbol{x}_0, \boldsymbol{c}''(\cdot)) \in \Sigma_{\tilde{\alpha}}(\cdot, [\boldsymbol{X}_0]^{\tilde{\alpha}}, [\boldsymbol{W}]^{\tilde{\alpha}})$. While $\delta \to 0$, we have

$$\begin{aligned} \boldsymbol{x}(t, \boldsymbol{x}_0, \boldsymbol{c}''(t)) &= \boldsymbol{x}(t-\delta, \boldsymbol{x}_0, \boldsymbol{c}''(t-\delta)) + \boldsymbol{f}\left(t-\delta, \boldsymbol{x}(t-\delta, \boldsymbol{x}_0, \boldsymbol{c}''(t-\delta)), \boldsymbol{c}''(t-\delta)\right)\delta \\ &= \boldsymbol{x}(t-\delta, \boldsymbol{x}_0, \boldsymbol{c}(t-\delta)) + \boldsymbol{f}\left(t-\delta, \boldsymbol{x}(t-\delta, \boldsymbol{x}_0, \boldsymbol{c}(t-\delta)), \boldsymbol{c}(t-\delta)\right)\delta \\ &= \boldsymbol{x}(t, \boldsymbol{x}_0, \boldsymbol{c}(t)) \\ &= \boldsymbol{y} - \boldsymbol{u}\delta, \end{aligned} \tag{16}$$

and

$$\begin{aligned} \boldsymbol{x}(t+\delta, \boldsymbol{x}_0, \boldsymbol{c}''(t+\delta)) &= \boldsymbol{x}(t, \boldsymbol{x}_0, \boldsymbol{c}''(t)) + \boldsymbol{f}\left(t, \boldsymbol{x}(t, \boldsymbol{x}_0, \boldsymbol{c}''(t)), \boldsymbol{c}''(t)\right)\delta \\ &= \boldsymbol{y} - \boldsymbol{u}\delta + \boldsymbol{f}(t, \boldsymbol{y} - \boldsymbol{u}\delta, \boldsymbol{w})\delta \\ &= (\boldsymbol{y} - \boldsymbol{u}\delta) + \boldsymbol{u}\delta \\ &= \boldsymbol{y}. \end{aligned} \tag{17}$$

Because $\boldsymbol{x}(t+\delta, \boldsymbol{x}_0, \boldsymbol{c}''(t+\delta)) \in \mathcal{A}_{\tilde{\alpha}}(t+\delta, [\boldsymbol{X}_0]^{\tilde{\alpha}}, [\boldsymbol{W}]^{\tilde{\alpha}})$, there is $\boldsymbol{y} \in \mathcal{A}_{\tilde{\alpha}}(t+\delta, [\boldsymbol{X}_0]^{\tilde{\alpha}}, [\boldsymbol{W}]^{\tilde{\alpha}})$. According to Proposition 4 of [33], since $\mathcal{A}_{\tilde{\alpha}}(t+\delta, [\boldsymbol{X}_0]^{\tilde{\alpha}}, [\boldsymbol{W}]^{\tilde{\alpha}})$ is a compact fuzzy set, $\boldsymbol{y} \in \bigcap_{\tilde{\alpha} \in (0,\alpha)} \mathcal{A}_{\tilde{\alpha}}(t+\delta, [\boldsymbol{X}_0]^{\tilde{\alpha}}, [\boldsymbol{W}]^{\tilde{\alpha}}) = \mathcal{A}_{\alpha}(t+\delta, +[\boldsymbol{X}_0]^{\alpha}, [\boldsymbol{W}]^{\alpha})$, which indicates $[\mathcal{X}(t+\delta)]^{\alpha} \subset \mathcal{A}_{\alpha}(t+\delta, [\boldsymbol{X}_0]^{\alpha}, [\boldsymbol{W}]^{\alpha})$.

For the remaining case of $\alpha = 0$, we have

$$\begin{aligned}[\mathcal{X}(t+\delta)]^0 &= cl\left(\bigcap_{\alpha\in[0,1]}[\mathcal{X}(t+\delta)]^\alpha\right)\\ &\subset cl\left(\bigcap_{\alpha\in[0,1]}\mathcal{A}_\alpha(t+\delta,[\boldsymbol{X}_0]^\alpha,[\boldsymbol{W}]^\alpha)\right)\\ &= \mathcal{A}_0(t+\delta,[\boldsymbol{X}_0]^0,[\boldsymbol{W}]^0).\end{aligned} \tag{18}$$

Therefore, $[\mathcal{X}(t+\delta)]^\alpha \subset \mathcal{A}_\alpha(t+\delta,[\boldsymbol{X}_0]^\alpha,[\boldsymbol{W}]^\alpha)$ for $\forall\alpha\in[0,1]$.

With $\mathcal{A}_\alpha(t+\delta,[\boldsymbol{X}_0]^\alpha,[\boldsymbol{W}]^\alpha)\subset[\mathcal{X}(t+\delta)]^\alpha$ in mind, we have $\mathcal{A}_\alpha(t+\delta,[\boldsymbol{X}_0]^\alpha,[\boldsymbol{W}]^\alpha)=[\mathcal{X}(t+\delta)]^\alpha$ for all $\alpha\in[0,1]$.

Following from the recursive relations, we obtain $\mathcal{A}(t,\boldsymbol{X}_0,\boldsymbol{W})=\mathcal{X}(t)$ for all $t\in[0,T]$.

According to the main theorem in [25], the membership function $\mu_{\mathcal{X}(t)}$ can be obtained by solving a partial differential equation.

Theorem 3 *([25]) $[\mathcal{X}(\cdot)]^0$ is an n-dimensional manifold describing the set of all possible states of the system. At any given time $t\in[0,T]$, the familiy of $\{[\mathcal{X}(\cdot)]^\alpha|\alpha\in[0,1]\}$ defines a fuzzy set $\mathcal{X}(t)\in(\mathcal{F}_C)^n(\Upsilon(t))$ where $\Upsilon(t)\subset R^n$. $\mu_{\mathcal{X}(t)}:R^n\to[0,1]$ is the membership function of $\mathcal{X}(t)$. At any given state $\boldsymbol{x}\in\Upsilon(t)$ and time $t\in[0,T]$, the velocity of the change of the state is a fuzzy set $\boldsymbol{U}(t,\boldsymbol{x})\in(\mathcal{F}_C)^n$. Let $\mu_{U(t,x)}:R^n\to[0,1]$ be the membership function of $\boldsymbol{U}(t,\boldsymbol{x})$. If*

(i) there exists a constant $B(\epsilon)$ for all $t\in[0,T]$, $\boldsymbol{x}\in\Upsilon(t)$ and $\epsilon>0$ such that $\mu_{U(t,x)}(\boldsymbol{u})\le\epsilon$ whenever $||\boldsymbol{u}||\ge B(\epsilon)$, among which $\boldsymbol{u}\in R^n$,
(ii) $\sup_{\boldsymbol{u}\in R^n}\{\mu_{U(t,x)}(\boldsymbol{u})\}=1$,
(iii) $\mu_{U(t,x)}(\boldsymbol{u})$ is piece-wise smooth in t and $\boldsymbol{x}$,
(iv) the initial condition $\mu_{\mathcal{X}(0)}(\boldsymbol{x})$ is piece-wise smooth in $\boldsymbol{x}$,

then the membership degree $\mu_{\mathcal{X}(t)}(\boldsymbol{x})$ is the solution of the equation

$$\frac{\partial\mu_{\mathcal{X}(t)}(\boldsymbol{x})}{\partial t}+\boldsymbol{v}^T(t,\boldsymbol{x})\nabla_{\boldsymbol{x}}\mu_{\mathcal{X}(t)}(\boldsymbol{x})=0. \tag{19}$$

In (19), $\boldsymbol{v}(t,\boldsymbol{x})$ is the velocity solved from the equations

$$\nabla_{\boldsymbol{x}}\mu_{\mathcal{X}(t)}(\boldsymbol{x})=\beta(t,\boldsymbol{x})\nabla_{\boldsymbol{u}}\mu_{U(t,x)}(\boldsymbol{v}(t,\boldsymbol{x})), \tag{20}$$

$$\mu_{\mathcal{X}(t)}(\boldsymbol{x})=\mu_{U(t,x)}(\boldsymbol{v}(t,\boldsymbol{x})) \tag{21}$$

with $\beta(t,\boldsymbol{x})$ being the Lagrange mulitplier. Also, $\boldsymbol{v}(t,\boldsymbol{x})$ is the solution of the equations

$$D_x\mu_{U(t,x)}(\boldsymbol{v}(t,\boldsymbol{x}))=\beta(t,x)\nabla_{\boldsymbol{u}}\mu_{U(t,x)}(\boldsymbol{v}(t,\boldsymbol{x})), \tag{22}$$

$$D_t\mu_{U(t,x)}(\boldsymbol{v}(t,\boldsymbol{x}))=-\beta(t,x)\boldsymbol{v}^T(t,\boldsymbol{x})\nabla_{\boldsymbol{u}}\mu_{U(t,x)}(\boldsymbol{v}(t,\boldsymbol{x})) \tag{23}$$

with the initial conditions as substituting $t = 0$ into (20) and (21). Besides, the velocity $\boldsymbol{v}(t, \boldsymbol{x})$ determines a flow $\boldsymbol{\Gamma}_t(\boldsymbol{x}_0) \in R^n$ on the state space, which is the solution of the following equations

$$\frac{\partial \boldsymbol{\Gamma}_t(\boldsymbol{x}_0)}{\partial t} = \boldsymbol{v}(t, \boldsymbol{\Gamma}_t(\boldsymbol{x}_0)) \tag{24}$$

$$\boldsymbol{\Gamma}_0(\boldsymbol{x}_0) = \boldsymbol{x}_0. \tag{25}$$

Along the flow, the membership degree of the state, $\mu_{\mathcal{X}(t)}(\boldsymbol{\Gamma}_t(\boldsymbol{x}_0))$, remains unchanged, i.e.,

$$\mu_{\mathcal{X}(t)}(\boldsymbol{\Gamma}_t(\boldsymbol{x}_0)) = \mu_{\mathcal{X}(0)}(\boldsymbol{x}_0). \tag{26}$$

Notice that the conditions (i)–(iii) are guaranteed if $\boldsymbol{U}(t, \boldsymbol{x}) = \boldsymbol{f}(t, \boldsymbol{x}, \boldsymbol{W})$ and $\boldsymbol{f}$ is continuous in $(t, \boldsymbol{x}, \boldsymbol{w})$.

If the membership function $\mu_{U(t,x)}$ can be expressed as the form

$$\mu_{U(t,x)}(\boldsymbol{v}) = g\left(\Delta \boldsymbol{u}^T(\boldsymbol{v}, t, \boldsymbol{x}) D(t, \boldsymbol{x}) \Delta \boldsymbol{u}(\boldsymbol{v}, t, \boldsymbol{x})\right), \tag{27}$$

where $\Delta \boldsymbol{u}(\boldsymbol{v}, t, \boldsymbol{x}) = \boldsymbol{v}(t, \boldsymbol{x}) - \boldsymbol{u}^1(t, \boldsymbol{x})$ is the deviation of $\boldsymbol{v}(t, \boldsymbol{x})$ from the most possible velocity $\boldsymbol{u}^1(t, \boldsymbol{x})$ and $D(t, \boldsymbol{x})$ is an $n \times n$ matrix indicating the indeterminacy of the system, Proposition 5.1 of [25] offered the result below.

Theorem 4 *([25]) Under the assumptions of Theorem 3, If $\mu_{U(t,x)} : R^n \to [0, 1]$ is given by (27) in which $D(t, \boldsymbol{x})$ is a positive matrix, the membership degree $\mu_{\mathcal{X}(t)}(\boldsymbol{x})$ can be obtained by first solving the scalar function $\Phi(t, \boldsymbol{x})$ from*

$$\begin{cases} \frac{\partial \Phi(t,\boldsymbol{x})}{\partial t} + (\boldsymbol{u}^1(t, \boldsymbol{x}))^T \nabla_{\boldsymbol{x}} \Phi(t, \boldsymbol{x}) + \Phi(t, \boldsymbol{x}) \sqrt{(\nabla_{\boldsymbol{x}} \Phi(t, \boldsymbol{x}))^T D^{-1}(t, \boldsymbol{x}) \nabla_{\boldsymbol{x}} \Phi(t, \boldsymbol{x})} = 0 \\ \Phi^2(0, \boldsymbol{x}_0) = g^{-1}(\mu_{\mathcal{X}(0)}(\boldsymbol{x}_0)) \end{cases}, \tag{28}$$

and then substituting $\Phi(t, \boldsymbol{x})$ into the equation

$$\mu_{\mathcal{X}(t)}(\boldsymbol{x}) = g(\Phi^2(t, \boldsymbol{x})). \tag{29}$$

To be general, we consider the case that not all components of $\boldsymbol{f}(t, \boldsymbol{x}, [W]^\alpha)$ are subject to fuzzy uncertainties, then, the velocity $\boldsymbol{v}(t, \boldsymbol{x}) = [v_1, \ldots, v_n]^T$ must coincide with $\boldsymbol{u}^1(t, \boldsymbol{x}) = [u_1^1, \ldots, u_n^1]^T$ in those components not influenced. To be simple, assume that only the first k ($k \leq n$) components of $\boldsymbol{f}(t, \boldsymbol{x}, [W]^\alpha)$ are fuzzy numbers (if not, we can rearrange the order of the components in $\boldsymbol{x}$, $\boldsymbol{f}(t, \boldsymbol{x}, [W]^\alpha)$ and $\boldsymbol{v}(t, \boldsymbol{x})$ to achieve the same conclusion), then we have $v_{k+1} = u_{k+1}^1, \ldots, v_n = u_n^1$ and $\Delta u_{k+1} = 0, \ldots, \Delta u_n = 0$ and

$$D(t, \boldsymbol{x}) = \begin{bmatrix} D^*(t, \boldsymbol{x}) & O \\ O & O \end{bmatrix}, \tag{30}$$

where $D^*(t, \boldsymbol{x})$ is a $k \times k$ matrix and Os are null matrices. Let $D^*(t, \boldsymbol{x})$ is a positive-definite matrix, we can get a similar conclusion to that of Theorem 4. Though their derivation processes are very similar, we demonstrate it for the completeness of the content.

Let $\Phi(t, \boldsymbol{x}) = \left(\Delta \boldsymbol{u}^T(\boldsymbol{v}(t, \boldsymbol{x}), t, \boldsymbol{x}) D(t, \boldsymbol{x}) \Delta \boldsymbol{u}(\boldsymbol{v}(t, \boldsymbol{x}), t, \boldsymbol{x})\right)^{1/2}$, then

$$\mu_{U(t,x)}(\boldsymbol{v}(t, \boldsymbol{x})) = g\left(\Phi^2(t, \boldsymbol{x})\right). \tag{31}$$

By substituting above equation into (22) and (23), we have

$$2g'\left(\Phi^2(t, \boldsymbol{x})\right)\Phi(t, \boldsymbol{x})\nabla_x \Phi(t, \boldsymbol{x}) = \beta(t, \boldsymbol{x}) 2g'\left(\Phi^2(t, \boldsymbol{x})\right) D(t, \boldsymbol{x}) \Delta \boldsymbol{u}\left(\boldsymbol{v}(t, \boldsymbol{x}), t, \boldsymbol{x}\right), \tag{32}$$

which is

$$\Phi(t, \boldsymbol{x})\nabla_x \Phi(t, \boldsymbol{x}) = \beta(t, \boldsymbol{x}) D(t, \boldsymbol{x}) \Delta \boldsymbol{u}\left(\boldsymbol{v}(t, \boldsymbol{x}), t, \boldsymbol{x}\right), \tag{33}$$

and

$$\begin{aligned} &2g'\left(\Phi^2(t, \boldsymbol{x})\right)\Phi(t, \boldsymbol{x})\frac{\partial \Phi(t, \boldsymbol{x})}{\partial t} \\ &= -\beta(t, \boldsymbol{x})\left(\Delta \boldsymbol{u}(\boldsymbol{v}(t, \boldsymbol{x}), t, \boldsymbol{x}) + \boldsymbol{u}^1(t, \boldsymbol{x})\right)^T \\ &\quad 2g'\left(\Phi^2(t, \boldsymbol{x})\right) D(t, \boldsymbol{x}) \Delta \boldsymbol{u}\left(\boldsymbol{v}(t, \boldsymbol{x}), t, \boldsymbol{x}\right), \end{aligned} \tag{34}$$

that is

$$\begin{aligned} &\Phi(t, \boldsymbol{x})\frac{\partial \Phi(t, \boldsymbol{x})}{\partial t} \\ &= -\beta(t, \boldsymbol{x})\Delta \boldsymbol{u}^T(\boldsymbol{v}(t, \boldsymbol{x}), t, \boldsymbol{x}) D(t, \boldsymbol{x}) \Delta \boldsymbol{u}\left(\boldsymbol{v}(t, \boldsymbol{x}), t, \boldsymbol{x}\right) \\ &\quad - \beta(t, \boldsymbol{x})(\boldsymbol{u}^1(t, \boldsymbol{x}))^T D(t, \boldsymbol{x}) \Delta \boldsymbol{u}\left(\boldsymbol{v}(t, \boldsymbol{x}), t, \boldsymbol{x}\right). \end{aligned} \tag{35}$$

Following the definition of $\Phi(t, \boldsymbol{x})$ and (33), the first term on the right-hand side of (35) can be transformed as below

$$\begin{aligned} &\beta(t, \boldsymbol{x})\Delta \boldsymbol{u}^T(\boldsymbol{v}(t, \boldsymbol{x}), t, \boldsymbol{x}) D(t, \boldsymbol{x}) \Delta \boldsymbol{u}(\boldsymbol{v}(t, \boldsymbol{x}), t, \boldsymbol{x}) \\ &= \beta(t, \boldsymbol{x})\left[\sqrt{\Delta \boldsymbol{u}^T(\boldsymbol{v}(t, \boldsymbol{x}), t, \boldsymbol{x}) D(t, \boldsymbol{x}) \Delta \boldsymbol{u}(\boldsymbol{v}(t, \boldsymbol{x}), t, \boldsymbol{x})}\right]^2 \\ &= \beta(t, \boldsymbol{x})\Phi(t, \boldsymbol{x})\sqrt{\Delta \boldsymbol{u}^T(\boldsymbol{v}(t, \boldsymbol{x}), t, \boldsymbol{x}) D(t, \boldsymbol{x}) \tilde{D}(t, \boldsymbol{x}) D(t, \boldsymbol{x}) \Delta \boldsymbol{u}(\boldsymbol{v}(t, \boldsymbol{x}), t, \boldsymbol{x})} \\ &= \Phi(t, \boldsymbol{x})\sqrt{(\beta(t, \boldsymbol{x}) D(t, \boldsymbol{x}) \Delta \boldsymbol{u}(\boldsymbol{v}(t, \boldsymbol{x}), t, \boldsymbol{x}))^T \tilde{D}(t, \boldsymbol{x})(\beta(t, \boldsymbol{x}) D(t, \boldsymbol{x}) \Delta \boldsymbol{u}(\boldsymbol{v}(t, \boldsymbol{x}), t, \boldsymbol{x}))} \\ &= \Phi^2(t, \boldsymbol{x})\sqrt{(\nabla_x \Phi(t, \boldsymbol{x}))^T \tilde{D}(t, \boldsymbol{x}) \nabla_x \Phi(t, \boldsymbol{x})}, \end{aligned} \tag{36}$$

where

$$\tilde{D}(t, \boldsymbol{x}) = \begin{bmatrix} D^{*-1}(t, \boldsymbol{x}) & O \\ O & O \end{bmatrix}, \tag{37}$$

and the second term on the right-hand side of (35) equals $\Phi(t, \boldsymbol{x})(\boldsymbol{u}^1(t, \boldsymbol{x}))^T \nabla_x \Phi(t, \boldsymbol{x})$. Thus, (35) becomes

$$\frac{\partial \Phi(t, \boldsymbol{x})}{\partial t} + (\boldsymbol{u}^1(t, \boldsymbol{x}))^T \nabla_x \Phi(t, \boldsymbol{x}) + \Phi(t, \boldsymbol{x})\sqrt{(\nabla_x \Phi(t, \boldsymbol{x}))^T \tilde{D}(t, \boldsymbol{x}) \nabla_x \Phi(t, \boldsymbol{x})} = 0. \tag{38}$$

Adding the initial condition of $\Phi(0, \boldsymbol{x}_0)$ given by

$$\Phi^2(0, \boldsymbol{x}_0) = g^{-1}(\mu_{\mathcal{X}(0)}(\boldsymbol{x}_0)), \tag{39}$$

$\Phi(t, \boldsymbol{x}_0)$ can be solved from (38). Following (21), the membership degree of the state can be obtained by

$$\mu_{\mathcal{X}(t)}(\boldsymbol{x}) = g(\Phi^2(t, \boldsymbol{x})). \tag{40}$$

Therefore, we have the extension of Theorem 4.

Theorem 5 *Under the assumptions of Theorem 3, if* $\mu_{U(t,x)} : R^n \to [0, 1]$ *is given by (27) in which* $D(t, \boldsymbol{x})$ *is the matrix as (30) and* $D^*(t, \boldsymbol{x})$ *is a positive-definite* $k \times k$ *matrix, the membership degree* $\mu_{\mathcal{X}(t)}(\boldsymbol{x})$ *can be obtained by first solving the scalar function* $\Phi(t, \boldsymbol{x})$ *from*

$$\begin{cases} \frac{\partial \Phi(t,x)}{\partial t} + (\boldsymbol{u}^1(t, \boldsymbol{x}))^T \nabla_x \Phi(t, \boldsymbol{x}) + \Phi(t, \boldsymbol{x})\sqrt{(\nabla_x \Phi(t, \boldsymbol{x}))^T \tilde{D}(t, \boldsymbol{x}) \nabla_x \Phi(t, \boldsymbol{x})} = 0 \\ \Phi^2(0, \boldsymbol{x}_0) = g^{-1}(\mu_{\mathcal{X}(0)}(\boldsymbol{x}_0)) \end{cases},$$

where

$$\tilde{D}(t, \boldsymbol{x}) = \begin{bmatrix} D^{*-1}(t, \boldsymbol{x}) & O \\ O & O \end{bmatrix},$$

and then substituting $\Phi(t, \boldsymbol{x})$ *into the equation*

$$\mu_{\mathcal{X}(t)}(\boldsymbol{x}) = g(\Phi^2(t, \boldsymbol{x})).$$

Notice that the fuzzy set $\boldsymbol{U}(t, \boldsymbol{x})$ may degrade to a crisp value at some specific t and $\boldsymbol{x}$. Then, $\Delta \boldsymbol{u}(v(t, \boldsymbol{x}), t, \boldsymbol{x}) = \boldsymbol{0}$ and $D = O$. In that case, let $\tilde{D} = O$ and substitute D and $\tilde{D}$ into the above derivation, the same conclusion can be obtained.

2.1.1 Numerical Solution

For most nonlinear dynamical systems, analytically solving the partial differential equation (38) is a difficult task. Therefore, we will give the numerically solution as the membership distribution of the fuzzy response, $\{(\boldsymbol{x}, \mu_{\mathcal{X}(t)}(\boldsymbol{x}))|\mu_{\mathcal{X}(t)}(\boldsymbol{x}) > 0\}$, by constructing the characteristic equations [34] to convert (38) to a group of ordinary differential equations.

Denoting the left-hand side of (38) by $F(t, \boldsymbol{x}, \Phi, \zeta, \boldsymbol{\psi})$ where $\boldsymbol{x} = [x_1, \ldots, x_n]^T$, $\zeta = \partial\Phi/\partial t$ and $\boldsymbol{\psi} = [\psi_1, \ldots, \psi_n]^T = \nabla_{\boldsymbol{x}}\Phi = [\partial\Phi/\partial x_1, \ldots, \partial\Phi/\partial x_n]^T$, the characteristic equations are

$$\begin{cases} \frac{d\boldsymbol{x}}{dt} = \nabla_{\boldsymbol{\psi}} F \\ \frac{d\zeta}{dt} = -\frac{\partial F}{\partial t} - \frac{\partial F}{\partial \Phi}\zeta \\ \frac{d\boldsymbol{\psi}}{dt} = -\nabla_{\boldsymbol{x}} F - \frac{\partial F}{\partial \Phi}\boldsymbol{\psi} \\ \frac{d\Phi}{dt} = \boldsymbol{\psi}^T \nabla_{\boldsymbol{\psi}} F \end{cases}, \tag{41}$$

that is

$$\begin{cases} \frac{d\boldsymbol{x}}{dt} = \boldsymbol{u}^1 + \frac{\Phi}{q}\tilde{D}\boldsymbol{\psi} \\ \frac{d\zeta}{dt} = -\boldsymbol{\psi}^T \frac{\partial \boldsymbol{u}^1}{\partial t} - \frac{\Phi}{2q}\boldsymbol{\psi}^T \frac{\partial \tilde{D}}{\partial t}\boldsymbol{\psi} - q\zeta \\ \frac{d\boldsymbol{\psi}}{dt} = -\boldsymbol{\psi}^T \frac{\partial \boldsymbol{u}^1}{\partial \boldsymbol{x}} - \frac{\Phi}{2q}\boldsymbol{\psi}^T \frac{\partial \tilde{D}}{\partial \boldsymbol{x}}\boldsymbol{\psi} - q\boldsymbol{\psi} \\ \frac{d\Phi}{dt} = 0 \end{cases}, \tag{42}$$

where $q = \sqrt{\boldsymbol{\psi}^T \tilde{D}\boldsymbol{\psi}}$ and $\boldsymbol{u}^1 = [\boldsymbol{U}]^1 = \boldsymbol{f}(t, \boldsymbol{x}, [\boldsymbol{W}]^1)$.

To solve (42), we also need the initial conditions of $\boldsymbol{x}$, Φ, $\boldsymbol{\psi}$ and ζ. Firstly, choose the initial state $\boldsymbol{x}_0$ from $[\boldsymbol{X}_0]^0$. Then, $\Phi(0, \boldsymbol{x}_0)$ and $\boldsymbol{\psi}(0, \boldsymbol{x}_0)$ are obtained from (39) and its partial derivatives. At last, $\zeta(0, \boldsymbol{x}_0)$ is calculated using (38) with the substitutions of $\boldsymbol{x}_0$, $\Phi(0, \boldsymbol{x}_0)$ and $\boldsymbol{\psi}(0, \boldsymbol{x}_0)$.

The gains of solving (42) are the values of $\boldsymbol{x}$, Φ, ζ and $\boldsymbol{\psi}$ at time t. Particularly, $\boldsymbol{x}$s at different time t form a flow of state determined by the initial state $\boldsymbol{x}_0$, which means $\boldsymbol{x} = \boldsymbol{\Gamma}_t(\boldsymbol{x}_0)$. According to (26), there is $\mu_{\mathcal{X}(t)}(\boldsymbol{x}) = \mu_{\mathcal{X}(t)}(\boldsymbol{\Gamma}_t(\boldsymbol{x}_0)) = \mu_{\boldsymbol{X}_0}(\boldsymbol{x}_0)$. Then, given an initial state $\boldsymbol{x}_0$, we have $\boldsymbol{x}$ and its membership degree $\mu_{\mathcal{X}(t)}(\boldsymbol{x})$ as the outcomes. By sampling $\{\boldsymbol{x}_{0_i}, i = 1, \ldots, N | \boldsymbol{x}_{0_i} \in [\boldsymbol{X}_0]^0\}$, an approximation of $\mathcal{X}(t)$, denoted by $\bar{\mathcal{X}}(t)$, can be identified by the membership distribution $\{(\boldsymbol{x}_i, \mu_{\mathcal{X}(t)}(\boldsymbol{x}_i)), i = 1, \ldots, N | \boldsymbol{x}_i = \boldsymbol{\Gamma}_t(\boldsymbol{x}_{0_i}), \boldsymbol{x}_{0_i} \in [\boldsymbol{X}_0]^0\}$. Thus, for all $\alpha \in [0, 1]$,

$$\begin{aligned} \mathcal{A}_\alpha\{t, [\boldsymbol{X}_0]^\alpha, [\boldsymbol{W}]^\alpha\} &= [\mathcal{X}(t)]^\alpha \\ &\approx [\bar{\mathcal{X}}(t)]^\alpha \\ &= \{\boldsymbol{x}_i, i = 1, \ldots, N | \boldsymbol{x}_i = \boldsymbol{\Gamma}_t(\boldsymbol{x}_{0_i}), \boldsymbol{x}_{0_i} \in [\boldsymbol{X}_0]^\alpha\}. \end{aligned} \tag{43}$$

Apparently, the denser the initial values of $\boldsymbol{x}_0$ are sampled from $[\boldsymbol{X}_0]^0$, the finer the membership distribution of the fuzzy response is characterized.

Next, we show the computational efficiency of the numerical method. Take the Euler method proposed by Hüllermeier in [3, 4] as a standard, recorded below as Euler-Hüllermeier method to distinguish from the Euler method of ordinary differential equation. For the fuzzy initial value problem (3), the attainable sets $\mathcal{A}_\alpha(t + \Delta t, [\boldsymbol{X}_0]^\alpha, [\boldsymbol{W}]^\alpha)$ are approximated by the iterative relationship

$$\mathcal{A}_\alpha(t+\Delta t, [\boldsymbol{X}_0]^\alpha, [\boldsymbol{W}]^\alpha) \approx \mathcal{B}_\alpha(t+\Delta t) = \bigcup_{x\in\mathcal{B}_\alpha(t)} x + \Delta t \boldsymbol{f}(t, \boldsymbol{x}, [\boldsymbol{W}]^\alpha), \alpha \in [0,1], \tag{44}$$

where $\mathcal{B}_\alpha(0) = \mathcal{A}_\alpha(0, [\boldsymbol{X}_0]^\alpha, [\boldsymbol{W}]^\alpha) = [\boldsymbol{X}_0]^\alpha$. In the calculation, $\mathcal{B}_\alpha(t)$ and $\boldsymbol{f}(t, \boldsymbol{x}, [\boldsymbol{W}]^\alpha)$ are further approximated by representative finite data structures $\{\boldsymbol{x}_i, i = 1, \ldots, N|\boldsymbol{x}_i \in \mathcal{B}_\alpha(t)\}$ and $\{\boldsymbol{f}(t, \boldsymbol{x}, \boldsymbol{w}_i), i = 1, \ldots, N|\boldsymbol{w}_i \in [\boldsymbol{W}]^\alpha\}$ respectively. Because the union operation is complex, $\mathcal{B}_\alpha(t+\Delta t)$ is required to be approximated by a new representative discretization $\{\boldsymbol{x}'_i, i = 1, \ldots, N|\boldsymbol{x}'_i \in \mathcal{B}_\alpha(t+\Delta t)\}$ to implement the calculation of the next Δt. Ignoring the time cost of the discretization and assuming $\boldsymbol{x} \in R^n$, the time complexity of one calculation of (44) is $O(nN^2)$. To be comparable, the ordinary differential equation (42) is considered to be integrated by Euler method. Since only the discretization of $[\boldsymbol{X}_0]^\alpha$ is needed and notice that (42) contains $2n+2$ equations, the time complexity of one integration of (42) is $O((2n+2)N)$. Taking the case that $n = 1$ and $N = 1500$ as an example, the time complexities of the transform method is $O(6 \times 10^3)$ which is much less than that of the Euler-Hüllermeier method, $O(2.25 \times 10^6)$. Generally, the more dimensional the problem is, the more samples needed for the discretization. If $n = 2$ and $N = 750000$, then the times complexities become $O(4.5 \times 10^6)$ and $O(1.125 \times 10^{12})$ for the two methods respectively. Apparently, the transform numerical method is more efficient than the Euler-Hüllermeier method. In fact, the step of discretization is a non-trivial task and such step can be omitted by the new method to some extent resulting from the continuity of (42), therefore, the practical efficiency of the transform method can be even higher.

To verify the feasibility of the transform method to solve the fuzzy differential equation via differential inclusions, we give two examples to show the agreements between the results by the proposed method and the analytical solutions.

In the following sections, we simply use $\mathcal{X}(t)$ to represent the numerical solution, unless different implication is specified. 1500 initial states $\boldsymbol{x}_0 \in [\boldsymbol{X}_0]^0$ will be sampled to calculate the membership distribution of the fuzzy set $\mathcal{X}(t)$ and compare the boundaries of $[\mathcal{X}(t)]^\alpha$ with $\mathcal{A}_\alpha\{t, [\boldsymbol{X}_0]^\alpha, [\boldsymbol{W}]^\alpha\}$ to show the correctness of the proposed method.

Example 1 Consider the fuzzy differential equation via differential inclusions

$$\begin{cases} \dot{x}(t) \in -[W]^\alpha x \\ x(0) = x_0 \in [X_0]^\alpha \end{cases}, \tag{45}$$

where W is a triangular fuzzy number as below

$$\mu_W(w) = \begin{cases} 1 - -10|w-2|, & 1.9 < w < 2.1 \\ 0, & \text{otherwise} \end{cases}. \tag{46}$$

Then, we have $u^1 = -[W]^1 x = -2x$ and $v = -wx$. Thus, $\Delta u(v, t, x) = (2-w)x$. Combining with (9), there is

$$\mu_{U(t,x)}(v) = \begin{cases} \sup\limits_{w=-v/x} \{\mu_W(w)\}, & x \neq 0 \\ \sup\limits_{w \in R}\{\mu_W(w)\}, & x = 0 \end{cases}$$
$$= \begin{cases} 1 - -10\sqrt{(\Delta u(v,t,x)/x)^2}, & x \neq 0 \\ 1, & x = 0 \end{cases}. \quad (47)$$

Let

$$D(t,x) = \begin{cases} 1/x^2, & x \neq 0 \\ 0, & x = 0 \end{cases}$$

and $\Phi(t,x) = \sqrt{\Delta u(v,t,x)D(t,x)\Delta u(v,t,x)}$, then $\tilde{D}(t,x) = x^2$ and (47) can be rewritten as

$$\mu_{U(t,x)}(v) = g(\Phi^2(t,x))$$
$$= \begin{cases} 1 - -10\Phi(t,x), & 0 \leq \Phi(t,x) < 0.1 \\ 0, & \text{otherwise} \end{cases}.$$

From (39), we have $\Phi^2(0,x_0) = g^{-1}(\mu_{\mathcal{X}(0)}(x_0))$. Thus, for $x_0 \in [\mathcal{X}(0)]^0$, there is

$$\Phi(0,x_0) = \frac{1 - \mu_{\mathcal{X}(0)}(x_0)}{10}.$$

By substituting u^1 and D into the characteristic equations (42), we obtain

$$\begin{cases} \frac{dx}{dt} = -2x + \Phi sign(\psi)|x| \\ \frac{d\zeta}{dt} = -\frac{\zeta}{|\psi x|} \\ \frac{d\psi}{dt} = 2\psi - \Phi sign(x)|\psi| - \frac{\zeta}{|\psi x|} \\ \frac{d\Phi}{dt} = 0 \end{cases}. \quad (48)$$

Since $\mathcal{X}(0) = X_0$, the initial conditions of (48) are

$$\begin{cases} x_0 \in [X_0]^0 \\ \Phi(0,x_0) = \frac{1-\mu_{X_0}(x_0)}{10} \\ \psi(0,x_0) = -0.1\frac{\partial \mu_{X_0}(x_0)}{\partial x} \\ \zeta(0,x_0) = -u^1(0,x_0)\psi(0,x_0) - \Phi(0,x_0)|\psi(0,x_0)x_0| \end{cases}.$$

1500 points are sampled from $[X_0]^0$ to be x_0, then the membership distribution of the numerical solution $\mathcal{X}(t)$ is composed by 1500 pairs of $(x, \mu_{\mathcal{X}(t)}(x))$ which are obtained by numerically integrating (48) with the fourth order Runge-Kutta method.

Record the α-levels of W and X_0 as $[W]^\alpha = [w_\alpha^-, w_\alpha^+]$ and $[X_0]^\alpha = [x_{0\alpha}^-, x_{0\alpha}^+]$ respectively. The attainable set of (45)'s analytical solution is

$$\mathcal{A}_\alpha(t, [\boldsymbol{X}_0]^\alpha, [\boldsymbol{W}]^\alpha) = [x_{0\alpha}^- e^{-w_\alpha^+ t}, x_{0\alpha}^+ e^{-w_\alpha^- t}]. \tag{49}$$

When X_0 is a trapezoidal fuzzy number

$$\mu_{X_0}(x_0) = \begin{cases} x_0 - 4.5, & 4.5 < x_0 < 5 \\ 1, & 5 \le x_0 \le 5.5 \\ 6.5 - x_0, & 5.5 < x_0 < 6.5 \\ 0, & \text{otherwise} \end{cases}, \tag{50}$$

the numerical solution $\mathcal{X}(t)$ is displayed in Fig. 1a by its membership distribution $\{(x, \mu_{\mathcal{X}(t)}(x))|\mu_{\mathcal{X}(t)}(x) > 0\}$ during the period $t \in [0, 2.5]$, where $x = \Gamma_t(x_0)$.

For clarity, Fig. 1b, $[\mathcal{X}(t)]^0$ and $[\mathcal{X}(t)]^1$ are selected to illustrate the comparisons with their counterparts of the analytical solution, $\mathcal{A}_0(t, [\boldsymbol{X}_0]^0, [\boldsymbol{W}]^0)$ and $\mathcal{A}_1(t, [\boldsymbol{X}_0]^1, [\boldsymbol{W}]^1)$, which shows almost no difference. To be more intuitive, the errors of the numerical solutions by comparing with the analytical solutions are listed in the Table 1.

Example 2 The second example is as following

$$\begin{cases} \dot{x}(t) \in -x + [W]^\alpha \cos t \\ x(0) = x_0 \in [X_0]^\alpha \end{cases}, \tag{51}$$

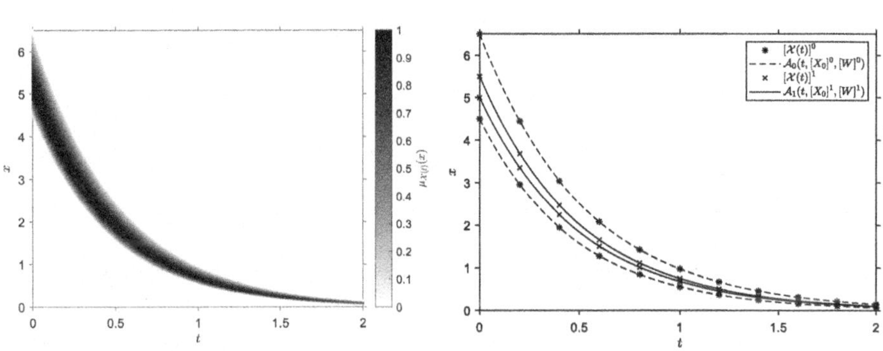

(a) (b)

Fig. 1 The numerical solution and the analytical solution of the fuzzy differential equation via differential inclusions (45) when X_0 is a trapezoidal fuzzy number as (50) (a) the membership distribution $\{(x, \mu_{\mathcal{X}(t)}(x))|\mu_{\mathcal{X}(t)}(x) > 0\}$ during the period $t \in [0, 2.5]$, among which different gray scales from white to black represent different values of $\mu_{\mathcal{X}(t)}(x)$ from 0 to 1, (b) comparisons between the boundaries of the 0/1-level of the numerical solution $\mathcal{X}(t)$ and those of the attainable set $\mathcal{A}_0(t, [\boldsymbol{X}_0]^0, [\boldsymbol{W}]^0)/\mathcal{A}_1(t, [\boldsymbol{X}_0]^1, [\boldsymbol{W}]^1)$ of the analytical solution

Table 1 The errors of $[\mathcal{X}(t)]^0$ and $[\mathcal{X}(t)]^1$ (the numerical solutions of the fuzzy differential equation via differential inclusions (45)) by comparing them with $\mathcal{A}_0(t, [X_0]^0, [W]^0)$ and $\mathcal{A}_1(t, [X_0]^1, [W]^1)$ (the analytical solutions) respectively. Let $[\mathcal{X}(t)]^\alpha = [x_\alpha^-, x_\alpha^+]$ and $\mathcal{A}_\alpha(t, [X_0]^\alpha, [W]^\alpha) = [a_\alpha^-, a_\alpha^+]$. The errors are measured by $e_\alpha^- = |x_\alpha^- - a_\alpha^-|$ and $e_\alpha^+ = |x_\alpha^+ - a_\alpha^+|$

t	e_0^-	e_0^+	e_1^-	e_1^+
0.0	0.0	0.0	0.0	0.0
0.4	3.3×10^{-15}	2.7×10^{-15}	2.7×10^{-15}	8.4×10^{-15}
0.8	2.3×10^{-15}	3.1×10^{-15}	2.0×10^{-15}	5.8×10^{-15}
1.2	2.5×10^{-15}	3.4×10^{-15}	4.2×10^{-15}	1.9×10^{-15}
1.6	4.1×10^{-15}	6.1×10^{-15}	5.2×10^{-15}	4.5×10^{-15}
2.0	3.0×10^{-15}	5.0×10^{-15}	3.7×10^{-15}	3.6×10^{-15}

where W is a triangular fuzzy number

$$\mu_W(w) = \begin{cases} 1 - |w|, & -1 < w < 1 \\ 0, & \text{otherwise} \end{cases}. \tag{52}$$

Then, $u^1 = -x + [W]^1 \cos t = -x$ and $v = -x + w \cos t$. Thus, $\Delta u(v, t, x) = w \cos t$ and

$$\mu_{U(t,x)}(v) = \begin{cases} \sup\limits_{w=(x+v)/\cos x} \{\mu_W(w)\}, & \cos t \neq 0 \\ \sup\limits_{w\in R}\{\mu_W(w)\}, & \cos t = 0 \end{cases}$$
$$= \begin{cases} 1 - \sqrt{(\Delta u(v, t, x)/\cos t)^2}, & \cos t \neq 0 \\ 1, & \cos t = 0 \end{cases}.$$

Let

$$D(t, x) = \begin{cases} \frac{1}{\cos^2 t}, & \cos t \neq 0 \\ 0 & \cos t = 0 \end{cases},$$

and $\Phi(t, x) = \sqrt{\Delta u(v, t, x) D(t, x) \Delta u(v, t, x)}$, then we have $\tilde{D} = \cos^2 t$ and

$$\mu_{U(t,x)}(v) = g(\Phi^2(t, x))$$
$$= \begin{cases} 1 - \Phi(t, x), & 0 \leq \Phi(t, x) < 1 \\ 0, & \text{otherwise} \end{cases}.$$

From (39), there is $\Phi^2(0, x_0) = g^{-1}(\mu_{\mathcal{X}(0)}(x_0))$. Furthermore, for $x_0 \in [\mathcal{X}(0)]^0$, there is

$$\Phi(0, x_0) = 1 - \mu_{\mathcal{X}(0)}(x_0).$$

Substitute u^1 and $\tilde{D}$ into (42), we obtain

$$\begin{cases} \frac{dx}{dt} = -x + \Phi\text{sign}(\psi)|\cos t| \\ \frac{d\zeta}{dt} = \phi|\psi|\text{sign}(\cos t)\sin t - |\psi\cos t|\zeta \\ \frac{d\psi}{dt} = \psi - |\psi\cos t|\psi \\ \frac{d\Phi}{dt} = 0 \end{cases}. \tag{53}$$

Since $\mathcal{X}(0) = X_0$, the initial conditions of (53) are

$$\begin{cases} x_0 \in [X_0]^0 \\ \Phi(0, x_0) = 1 - \mu_{X_0}(x_0) \\ \psi(0, x_0) = -\frac{\partial\mu_{X_0}(x_0)}{\partial x} \\ \zeta(0, x_0) = -u^1(0, x_0)\psi(0, x_0) - \Phi(0, x_0)|\psi(0, x_0)| \end{cases}.$$

When X_0 is a triangular fuzzy number as

$$\mu_{X_0}(x_0) = \begin{cases} 1 - |x_0|, & -1 < x_0 < 1 \\ 0, & \text{otherwise} \end{cases}, \tag{54}$$

taking 1500 points from $[X_0]^0$ as x_0, 1500 pairs of $(x, \mu_{\mathcal{X}(t)}(x))$ are obtained by numerically integrating (53). Then the numerical solution $\mathcal{X}(t)$ is outlined by the membership distribution $\{(x, \mu_{\mathcal{X}(t)}(x))|\mu_{\mathcal{X}(t)}(x) > 0\}$. The result is shown in Fig. 2a as the membership distributions at different time.

Following the literature [14], the attainable set of (51)'s analytical solution is

$$\begin{aligned} &\mathcal{A}_\alpha(t, [\boldsymbol{X}_0]^\alpha, [\boldsymbol{W}]^\alpha) \\ &= \left[(\alpha - 1)e^{-t}(1 + \int_0^t e^\tau|\cos\tau|d\tau), (1 - \alpha)e^{-t}(1 + \int_0^t e^\tau|\cos\tau|d\tau)\right]. \end{aligned} \tag{55}$$

To check the correctness of the numerical solution, $[\mathcal{X}(t)]^0$ and $[\mathcal{X}(t)]^{0.5}$ are compared with $\mathcal{A}_0(t, [\boldsymbol{X}_0]^0, [\boldsymbol{W}]^0)$ and $\mathcal{A}_{0.5}(t, [\boldsymbol{X}_0]^{0.5}, [\boldsymbol{W}]^{0.5})$ separately in Fig. 2b. And the errors of the numerical solutions by comparing with the analytical solutions are listed in the Table 2, which shows a great agreement.

In the next section, we use the proposed numerical method to study the transient response and the evolution of two nonlinear dynamic systems affected by fuzzy uncertainties.

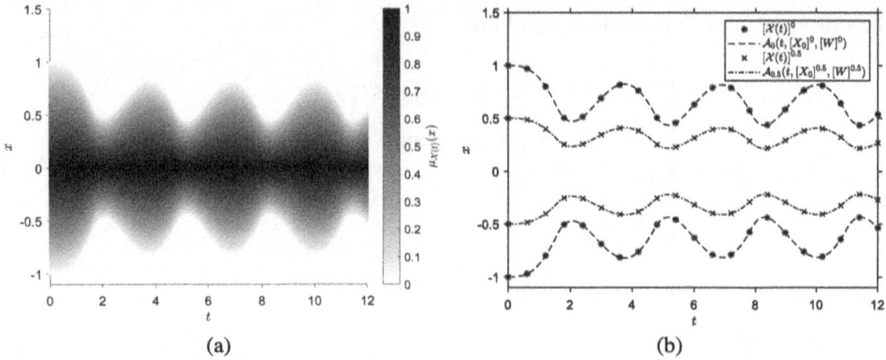

Fig. 2 The numerical solution and the analytical solution of the fuzzy differential equation via differential inclusions (51) when X_0 is a triangular fuzzy number as (54). **a** the membership distribution $\{(x, \mu_{\mathcal{X}(t)}(x)) | \mu_{\mathcal{X}(t)}(x) > 0\}$, among which different gray scales from white to black represent different values of $\mu_{\mathcal{X}(t)}(x)$ from 0 to 1, **b** comparisons between the boundaries of the 0/0.5-level of the numerical solution $\mathcal{X}(t)$ and those of the attainable set $\mathcal{A}_0(t, [\boldsymbol{X}_0]^0, [\boldsymbol{W}]^0)/\mathcal{A}_{0.5}(t, [\boldsymbol{X}_0]^{0.5}, [\boldsymbol{W}]^{0.5})$ of the analytical solution

Table 2 The errors of $[\mathcal{X}(t)]^0$ and $[\mathcal{X}(t)]^{0.5}$ (the numerical solution of the fuzzy differential equation via differential inclusions (51)) by comparing them with $\mathcal{A}_0(t, [X_0]^0, [W]^0)$ and $\mathcal{A}_{0.5}(t, [X_0]^{0.5}, [W]^{0.5})$ (the analytical solutions) respectively. Let $[\mathcal{X}(t)]^\alpha = [x_\alpha^-, x_\alpha^+]$ and $\mathcal{A}_\alpha(t, [X_0]^\alpha, [W]^\alpha) = [a_\alpha^-, a_\alpha^+]$. The errors are measured by $e_\alpha^- = |x_\alpha^- - a_\alpha^-|$ and $e_\alpha^+ = |x_\alpha^+ - a_\alpha^+|$

t	e_0^-	e_0^+	$e_{0.5}^-$	$e_{0.5}^+$
0.0	0.0	0.0	0.0	0.0
2.4	1.2×10^{-4}	1.2×10^{-4}	6.1×10^{-5}	6.1×10^{-5}
4.8	6.3×10^{-5}	6.3×10^{-5}	3.1×10^{-5}	3.1×10^{-5}
7.2	2.0×10^{-6}	2.0×10^{-6}	1.0×10^{-6}	1.0×10^{-6}
9.6	9.0×10^{-6}	9.0×10^{-6}	4.5×10^{-6}	4.5×10^{-6}
12.0	4.8×10^{-5}	4.8×10^{-5}	2.4×10^{-5}	2.4×10^{-5}

3 Mathieu System with Fuzzy Uncertainties

The control equation of the Mathieu system with fuzzy uncertainties is given by a family of differential inclusions

$$\begin{cases} \dot{x}_1(t) &= x_2 \\ \dot{x}_2(t) &\in -25x_1^3 - 0.173x_2 - 2.62x_1 + [W]^\alpha(0.456x_1 + 0.92)(1 - \cos 2t) \\ \boldsymbol{x}(0) &= \boldsymbol{x}_0 \in [\boldsymbol{X}_0]^\alpha \end{cases} . \tag{56}$$

where W is a triangular fuzzy number

$$\mu_W(w) = \begin{cases} 1 - -20|w - 3.9|, & 3.85 < w < 3.95 \\ 0, & \text{otherwise} \end{cases},$$

and $\boldsymbol{X}_0 \in (\mathcal{F}_C)^2(\Upsilon)$ is a fuzzy set with a cone membership distribution centered on the point (a, b) as below

$$\mu_{X_0}(\boldsymbol{x}_0) = \begin{cases} 1 - \frac{\sqrt{(x_{1_0}-a)^2+(x_{2_0}-b)^2}}{0.02}, & \sqrt{(x_{1_0}-a)^2+(x_{2_0}-b)^2} < 0.02 \\ 0, & \text{otherwise} \end{cases}. \tag{57}$$

Then, we have

$$\begin{aligned} \boldsymbol{u}^1 &= \begin{bmatrix} u_1^1 \\ u_2^1 \end{bmatrix} \\ &= \begin{bmatrix} x_2 \\ -25x_1^3 - 0.173x_2 - 2.62x_1 + [W]^1(0.456x_1 + 0.92)(1 - \cos 2t) \end{bmatrix}, \end{aligned}$$

$$\begin{aligned} \boldsymbol{v} &= \begin{bmatrix} v_1 \\ v_2 \end{bmatrix} \\ &= \begin{bmatrix} x_2 \\ -25x_1^3 - 0.173x_2 - 2.62x_1 + w(0.456x_1 + 0.92)(1 - \cos 2t) \end{bmatrix}, \end{aligned}$$

and

$$\Delta\boldsymbol{u}(\boldsymbol{v}, t, \boldsymbol{x}) = \begin{bmatrix} 0 \\ (w - [W]^1)(1 - \cos 2t)(0.456x_1 + 0.92) \end{bmatrix}.$$

Thus,

$$\begin{aligned} \mu_{\boldsymbol{U}(t,\boldsymbol{x})}(\boldsymbol{v}) &= \begin{cases} \sup\limits_{w=\frac{v_2+25x_1^3+0.173x_2+2.62x_1}{(1-\cos 2t)(0.456x_1+0.92)}} \{\mu_W(w)\}, & (1 - \cos 2t)(0.456x_1 + 0.92) \neq 0 \\ \sup\limits_{w\in R} \{\mu_W(w)\}, & \text{otherwise} \end{cases} \\ &= \begin{cases} 1 - -20\sqrt{\frac{||\Delta\boldsymbol{u}(\boldsymbol{v},t,\boldsymbol{x})||^2}{(1-\cos 2t)^2(0.456x_1+0.92)^2}}, & (1 - \cos 2t)(0.456x_1 + 0.92) \neq 0 \\ 1, & \text{otherwise} \end{cases}. \end{aligned}$$

Let

$$D^*(t, \boldsymbol{x}) = \begin{cases} \frac{1}{(1-\cos 2t)^2(0.456x_1+0.92)^2}, & (1 - \cos 2t)(0.456x_1 + 0.92) \neq 0 \\ 0, & \text{otherwise} \end{cases},$$

$$D(t, \boldsymbol{x}) = \begin{bmatrix} 0 & 0 \\ 0 & D^*(t, \boldsymbol{x}) \end{bmatrix},$$

and $\Phi(t, \boldsymbol{x}) = \sqrt{\Delta \boldsymbol{u}^T(\boldsymbol{v}, t, \boldsymbol{x}) D(t, \boldsymbol{x}) \Delta \boldsymbol{u}(\boldsymbol{v}, t, \boldsymbol{x})}$, then we have

$$\tilde{D} = \begin{bmatrix} 0 & 0 \\ 0 & (0.456x_1 + 0.92)^2(1 - \cos 2t)^2 \end{bmatrix},$$

and

$$\begin{aligned} \mu_{U(t,\boldsymbol{x})}(\boldsymbol{v}) &= g(\Phi^2(t, \boldsymbol{x})) \\ &= \begin{cases} 1 - -20\Phi(t, \boldsymbol{x}), & 0 \leq \Phi(t, \boldsymbol{x}) < 0.05 \\ 0, & \text{otherwise} \end{cases}. \end{aligned}$$

Then, for $\boldsymbol{x}_0 \in [\mathcal{X}(0)]^0$, there exists

$$\Phi(0, \boldsymbol{x}_0) = \frac{1 - \mu_{\mathcal{X}(0)}(\boldsymbol{x}_0)}{20}.$$

By substituting $\boldsymbol{u}^1$ and $\tilde{D}$ into the characteristic equations (42), the original system can be transformed into

$$\begin{cases} \frac{dx_1}{dt} = & u_1^1 \\ \frac{dx_2}{dt} = & u_2^1 + 2sign(\psi_2) \sin^2 t |0.456x_1 + 0.92| \\ \frac{d\zeta}{dt} = & -4\psi_2[W]^1 \sin t \cos t (0.456x_1 + 0.92) \\ & \quad -4\Phi \sin t \cos t |\psi_2(0.456x_1 + 0.92)| - q\zeta \\ \frac{d\psi_1}{dt} = & \psi_2 \left(75x_1^2 + 2.62 - 0.912[W]^1 \sin^2 t\right) \\ & \quad -0.912\Phi \sin^2 t |\psi_2| sign(0.456x_1 + 0.92) - q\psi_1 \\ \frac{d\psi_2}{dt} = & -\psi_1 + 0.173\psi_2 - q\psi_2 \\ \frac{dz}{dt} = & 0 \end{cases}$$

where $q = 2\sin^2 t |\psi_2(0.456x_1 + 0.92)|$. Since $\mathcal{X}(0) = \boldsymbol{X}_0$, the initial conditions of the above differential equations are

$$\begin{cases} \boldsymbol{x}_0 \in & [\boldsymbol{X}_0]^0 \\ \Phi(0, \boldsymbol{x}_0) = & \frac{1}{20}(1 - \mu_{X_0}(\boldsymbol{x}_0)) \\ \boldsymbol{\psi}(0, \boldsymbol{x}_0) = & -\frac{1}{20}[\frac{\partial \mu_{X_0}(\boldsymbol{x}_0)}{\partial x_1}, \frac{\partial \mu_{X_0}(\boldsymbol{x}_0)}{\partial x_2}]^T \\ \zeta(0, \boldsymbol{x}_0) = & -u_1^1(0, \boldsymbol{x}_0)\psi_1(0, \boldsymbol{x}_0) - u_2^1(0, x_0)\psi_2(0, \boldsymbol{x}_0) \end{cases}.$$

To better understand the behavior of the fuzzy responses, the global phase portrait of the corresponding deterministic Mathieu system is first drawn in Fig. 3. The so-called corresponding deterministic system is obtained by reducing the

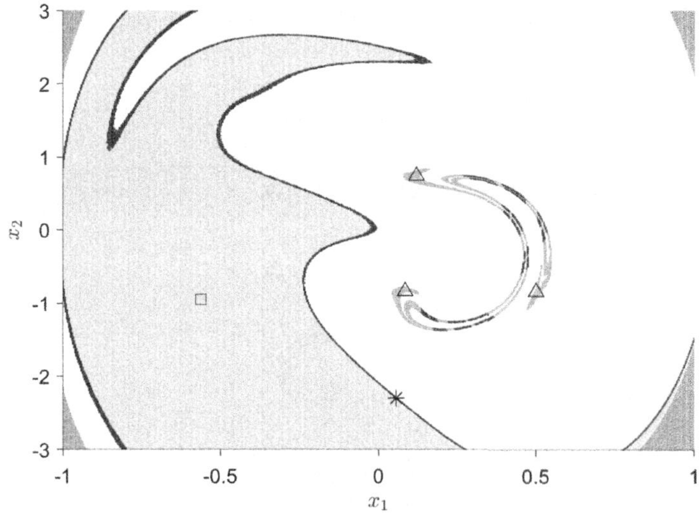

Fig. 3 The global phase portrait of the deterministic Mathieu system by eliminating the uncertainties in the fuzzy system (56). There exists two periodic attractors in the region $[-1, 1] \times [-3, 3]$. '△' represents the period-3 attractor and '□' the period-1 attractor. The white area and light gray area are the basins of period-3 and period-1 attractors respectively. In addition, '$*$' is the saddle standing on the boundary of two basins which is drawn by a black line. The black line segments compose a chaotic saddle in the basin of period-3 attractor. And the gray curve is the unstable manifold of the chaotic saddle linking with the period-3 attractor

uncertainties of (56) to zero, which means $[\boldsymbol{X}_0]^\alpha = [\boldsymbol{X}_0]^1 = \{(a, b)\}$ and $[W]^\alpha = [W]^1 = \{3.9\}$ ($\alpha \in [0, 1]$). If we check the system once every period π, two periodic attractors are found in the domain $[-1, 1] \times [-3, 3]$: a period-1 attractor around $(-0.5615, -0.9493)$ and a period-3 attractor consisting of three points near $(0.5011, -0.8391)$, $(0.1201, 0.7443)$ and $(0.0844, -0.8360)$. In addition, there are a chaotic saddle in the basin of the period-3 attactor and a period-1 saddle on the boundary between the basins of the two attractors.

Let $(a, b) = (0.5011, -0.8391)$, that is, the initial state of the system is very near the period-3 deterministic attractor. To delineate the fuzzy system's transient response, 750000 points are sampled randomly from $[\boldsymbol{X}_0]^0$ to be $\boldsymbol{x}_0$. Then the membership distributions of the solution at the moments of 0, 3π, 6π, 9π, 63π, 64π and 65π are shown sequentially in Fig. 4a–g. In the deterministic dynamical system the adjacency of the initial state to the attractor will make the response stable on the invariant set very quickly. On the contrary, Fig. 4a–d indicate the fuzzy response gradually diffuses along the stable manifold of the deterministic attractor until covers the chaotic saddle and its unstable manifold. Moreover, only the part of the fuzzy response which has the membership degree as 1 is stable on the deterministic period-3 attractor, see Fig. 4e–g, other parts more likely do a chaotic motion containing the deterministic period-3 attractor, the chaotic saddle and the saddle's unstable manifold to the attractor as its backbones.

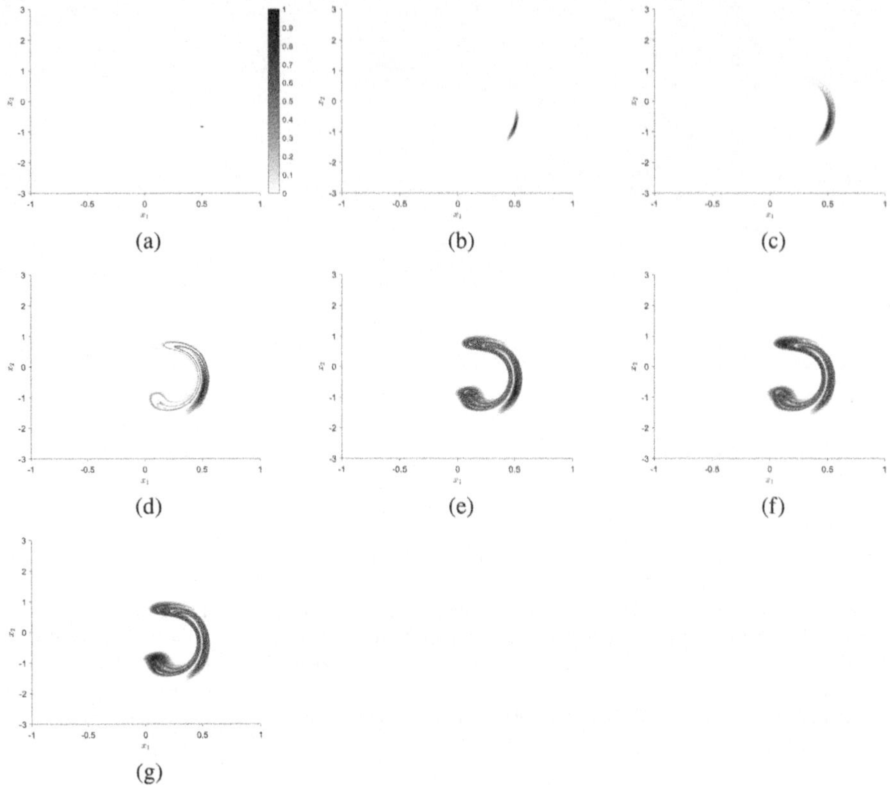

Fig. 4 The membership distributions of $\mathcal{X}(t)$, the fuzzy responses of the fuzzy Methieum system (56) at different moments, where different gray scales from white to black represent different values of $\mu_{\mathcal{X}(t)}(\boldsymbol{x})$ from 0 to 1. **a** $t = 0$, **b** $t = 3\pi$, **c** $t = 6\pi$, **d** $t = 9\pi$, **e** $t = 63\pi$, **f** $t = 64\pi$, **g** $t = 65\pi$

4 Rotor/stator Rubbing System with Fuzzy Uncertainties

Consider the rotor/stator contact system consisting of a Jeffcott rotor and a rigidly fixed non-rotating stator as illustrated in Fig. 5. The rotor is a disk of mass m and radius r_{disk} locating in the middle of a weightless shaft. The distance between the disk's center of mass M and geometrical center O is e. The stiffness of the shaft is k_s. When the rotor is rest, there is a gap of r_0 between the rotor and the stator. When the rotor rotates and contacts with the stator, the contact surface is considered as a spring with the stiffness k_b and the friction is assumed to be a sliding one with the friction coefficient μ. According to the state of contact, the rotor could has four types of motion: no-rub motion, synchronous full annular rub [35–37], partial rub [38, 39] and dry friction backward whirl [40, 41]. The first type has no contact; the second has a full annular contact and the shaft whirls with constant amplitude and the same frequency and direction as the rotation of the disk; the third has intermittent

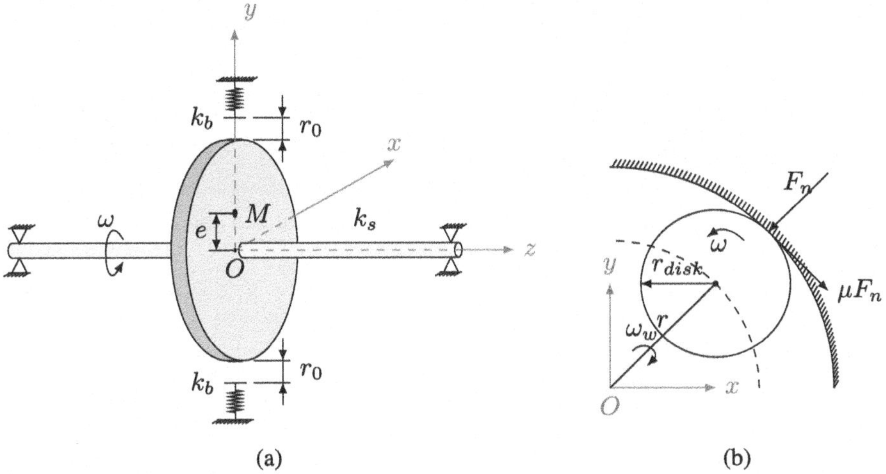

Fig. 5 **a** The diagrammatic sketch of the rotorstator contact system. **b** The forces acted on the rotor while the rub occurs

rubs; the last also has a full annular contact but the shaft whirls backward with slightly fluctuating amplitude and a super-synchronous frequency. Denote by ω and ω_w respectively the angular velocities of the disk and the whirling shaft. Following [42], the rotor/stator contact system can be described by the differential equations

$$\begin{cases} m\ddot{x} + c\dot{x} + k_s x + \Theta k_b(1 - \frac{r_0}{r})\left[x - \mu \text{sign}(v_{rel})y\right] = me\omega^2 \cos \omega\tau \\ m\ddot{y} + c\dot{y} + k_s y + \Theta k_b(1 - \frac{r_0}{r})\left[\mu \text{sign}(v_{rel})x + y\right] = me\omega^2 \sin \omega\tau \end{cases}, \quad (58)$$

and all four motions of the rotor can be obtained from (58). In the equations, Θ is a switch function indicating whether the rub occurs or not, which is defined as

$$\Theta = \begin{cases} 1\,, & r \geq r_0 \\ 0\,, & r < r_0 \end{cases},$$

where $r = \sqrt{x^2 + y^2}$. $v_{rel} = \omega r_{disk} + \omega_w r$, where $\omega_w = (-\dot{x}y + \dot{y}x)/r^2$, is the relative velocity of the rotor to the stator at the contact point, which determines the direction of the friction through the sign function

$$\text{sign}(v_{rel}) = \begin{cases} 1, & v_{rel} \geq 0 \\ -1, & v_{rel} < 0 \end{cases}.$$

Let

$$
\begin{array}{llll}
x_1 = x/e, & x_3 = y/e, & R_0 = r_0/e, & R_{disk} = r_{disk}/e, \\
2\xi = c/\sqrt{k_b m}, & \beta = k_s/k_b, & V_{rel} = v_{rel}/e, & \Omega = \omega/\sqrt{k_b/m}, \\
\omega_b = \omega_w/\sqrt{k_b/m}, & R = \sqrt{x_1^2 + x_3^2}, & t = \tau\sqrt{k_b/m} &
\end{array},
$$

then the governing equations (58) can be nondimensionalized and further converted into the first order differential equations

$$
\begin{cases}
\dot{x_1}(t) = x_2 \\
\dot{x_2}(t) = -\xi x_2 - \beta x_1 - \Theta(1 - \frac{R_0}{R})\left[x_1 - \mu \mathrm{sign}(V_{rel}) x_3\right] + \Omega^2 \cos \Omega t \\
\dot{x_3}(t) = x_4 \\
\dot{x_4}(t) = -\xi x_4 - \beta x_3 - \Theta(1 - \frac{R_0}{R})\left[\mu \mathrm{sign}(V_{rel}) x_1 + x_3\right] + \Omega^2 \sin \Omega t,
\end{cases} \tag{59}
$$

where

$$
V_{rel} = R_{disk}\Omega + R\left(-\frac{x_2 x_3}{R^2} + \frac{x_4 x_1}{R^2}\right),
$$

$$
\mathrm{sign}(V_{rel}) = \begin{cases} 1, & V_{rel} \geq 0 \\ -1, & V_{rel} < 0 \end{cases},
$$

and

$$
\Theta = \begin{cases} 1, & R \geq R_0 \\ 0, & R < R_0 \end{cases}.
$$

As shown in Fig. 6, when $\xi = 0.05$, $\beta = 0.04$, $R_0 = 1.05$, $R_{disk} = 20R_0$, $\mu = 0.20$ and $\Omega = 0.3$, the system has two long-term responses behaving as partial rub and dry friction backward whirl respectively [42]. Among the two solutions, dry friction backward whirl is a dangerous running state will damage the system.

In (59), the amplitude of the external excitation is considered as the constant Ω^2. However, it inevitably fluctuates under the influence of the uncertain drift of the actuator's working conditions. To reflect the reality of the rotor/stator contact system, we model the fluctuation of the amplitude as a fuzzy set $W \in \mathcal{F}_C$ with a triangular membership function

$$
\mu_W(w) = \begin{cases} 1 - |w|/0.06, & -0.06 < w < 0.06 \\ 0, & \text{otherwise} \end{cases}.
$$

To be generic, the initial condition is also considered as a fuzzy set, then (59) can be rewritten as the form of differential inclusions

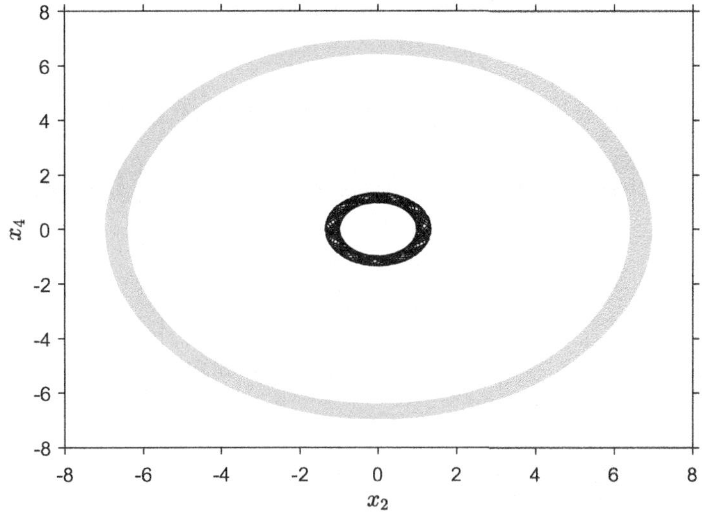

Fig. 6 The rotor/stator contact system (59) has two long-term responses when $\xi = 0.05$, $\beta = 0.04$, $R0 = 1.05$, $R_{disk} = 20R_0$, $\mu = 0.20$ and $\Omega = 0.3$. The black circle represents the partial rub and the gray circle represents the dry friction backward whirl

$$\begin{cases} \dot{x}_1(t) = x_2 \\ \dot{x}_2(t) \in -\left\{\xi x_2 + \beta x_1 + \Theta(1 - \frac{R_0}{R})\left[x_1 - \mu \text{sign}(V_{rel}) x_3\right]\right\} + (\Omega^2 + [W]^\alpha)\cos\Omega t \\ \dot{x}_3(t) = x_4 \\ \dot{x}_4(t) \in -\left\{\xi x_4 + \beta x_3 + \Theta(1 - \frac{R_0}{R})\left[\mu \text{sign}(V_{rel}) x_1 + x_3\right]\right\} + (\Omega^2 + [W]^\alpha)\sin\Omega t \\ \boldsymbol{x}(0) = \boldsymbol{x}_0 \in [\boldsymbol{X_0}]^\alpha \end{cases} . \tag{60}$$

Let $\boldsymbol{X}_0 \in (\mathcal{F}_C)^4$ has a cone membership distribution which is centered at $[\boldsymbol{X}_0]^1 = (0.9344, 0.3769, -0.7149, 0.1583)$, one point of the deterministic solution of the partial rub. Specifically, the membership function of $\boldsymbol{X}_0$ is

$$\mu_{X_0}(\boldsymbol{x_0}) = \begin{cases} 1 - \frac{d}{0.2}, & d < 0.2 \\ 0, & \text{otherwise} \end{cases},$$

where $d = \sqrt{(x_{1_0} - 0.9344)^2 + (x_{2_0} - 0.3769)^2 + (x_{3_0} + 0.7149)^2 + (x_{4_0} - 0.1583)^2}$.

According to (60),

$$\boldsymbol{u}^1 = \begin{bmatrix} u_1^1 \\ u_2^1 \\ u_3^1 \\ u_4^1 \end{bmatrix} = \begin{bmatrix} x_2 \\ -\left\{\xi x_2 + \beta x_1 + \Theta(1 - \frac{R_0}{R})\left[x_1 - \mu \text{sign}(V_{rel}) x_3\right]\right\} + \Omega^2 \cos\Omega t \\ x_4 \\ -\left\{\xi x_4 + \beta x_3 + \Theta(1 - \frac{R_0}{R})\left[\mu \text{sign}(V_{rel}) x_1 + x_3\right]\right\} + \Omega^2 \sin\Omega t \end{bmatrix},$$

and

$$v = \begin{bmatrix} v_1 \\ v_2 \\ v_3 \\ v_4 \end{bmatrix} = \begin{bmatrix} x_2 \\ -\left\{\xi x_2 + \beta x_1 + \Theta(1 - \frac{R_0}{R})\left[x_1 - \mu \text{sign}(V_{rel}) x_3\right]\right\} + (\Omega^2 + w)\cos\Omega t \\ x_4 \\ -\left\{\xi x_4 + \beta x_3 + \Theta(1 - \frac{R_0}{R})\left[\mu \text{sign}(V_{rel}) x_1 + x_3\right]\right\} + (\Omega^2 + w)\sin\Omega t \end{bmatrix}.$$

Then

$$\Delta \boldsymbol{u}(v, t, \boldsymbol{x}) = \begin{bmatrix} \Delta u_1 \\ \Delta u_2 \\ \Delta u_3 \\ \Delta u_4 \end{bmatrix} = \begin{bmatrix} 0 \\ w\cos\Omega t \\ 0 \\ w\sin\Omega t \end{bmatrix}.$$

Thus,

$$\begin{aligned} \mu_{U(t,\boldsymbol{x})}(\boldsymbol{v}) &= \sup_{w^2 = \Delta u_2^2 + \Delta u_4^2} \{\mu_W(w)\} \\ &= \begin{cases} 1 - \frac{\sqrt{\Delta u_2^2 + \Delta u_4^2}}{0.06}, & 0 \le \sqrt{\Delta u_2^2 + \Delta u_4^2} < 0.06 \\ 0, & \text{otherwise} \end{cases}. \end{aligned}$$

Let

$$D = \begin{bmatrix} 0 & 0 & 0 & 0 \\ 0 & 1 & 0 & 0 \\ 0 & 0 & 0 & 0 \\ 0 & 0 & 0 & 1 \end{bmatrix}$$

and $\Phi(t, \boldsymbol{x}) = \left(\Delta \boldsymbol{u}^T(v, t, \boldsymbol{x}) D \Delta \boldsymbol{u}(v, t, \boldsymbol{x})\right)^{1/2}$, then we have $\tilde{D} = D$ and

$$\begin{aligned} \mu_{U(t,\boldsymbol{x})}(\boldsymbol{v}) &= g(\Phi^2(t, \boldsymbol{x})) \\ &= \begin{cases} 1 - \frac{\Phi(t,\boldsymbol{x})}{0.06}, & 0 \le \Phi(t, \boldsymbol{x}) < 0.06 \\ 0, & \text{otherwise} \end{cases}. \end{aligned}$$

For $\boldsymbol{x}_0 \in [\mathcal{X}(0)]^0$, there is $\Phi(0, \boldsymbol{x}_0) = 0.06(1 - \mu_{\mathcal{X}(0)}(\boldsymbol{x}_0))$. By the substitution of $\boldsymbol{u}^1$ and $\tilde{D}$ into the characteristic equations (42), the original system is converted into

$$\begin{cases} \frac{dx_1}{dt} = u_1 \\ \frac{dx_2}{dt} = u_2 + \frac{\Phi\psi_2}{q} \\ \frac{dx_3}{dt} = u_3 \\ \frac{dx_4}{dt} = u_4 + \frac{\Phi\psi_4}{q} \\ \frac{d\zeta}{dt} = \Omega^3\left(\psi_2 \sin\Omega t - \psi_4 \cos\Omega t\right) - q\zeta \\ \frac{d\psi_1}{dt} = \psi_2\left[\beta + \Theta\left(A_1 G + B\right)\right] + \psi_4 \Theta\left(A_1 H + BC\right) - q\psi_1 \\ \frac{d\psi_2}{dt} = -\psi_1 + 2\xi\psi_2 - q\psi_2 \\ \frac{dp_3}{dt} = \psi_2 \Theta\left(A_2 G - BC\right) + \psi_4\left[\beta + \Theta\left(A_2 H + B\right)\right] - q\psi_3 \\ \frac{d\psi_4}{dt} = -\psi_3 + 2\xi\psi_4 - q\psi_4 \\ \frac{d\Phi}{dt} = 0 \end{cases},$$

where

$$\begin{cases} q = \sqrt{\psi_2^2 + \psi_4^2} \\ A_1 = \frac{R_0 x_1}{R^3} \\ A_2 = \frac{R_0 x_3}{R^3} \\ B = 1 - \frac{R_0}{R} \\ C = \mu \mathrm{sign}(V_{rel}) \\ G = x_1 - C \\ H = C + x_3 \end{cases}.$$

Since $\mathcal{X}(0) = \boldsymbol{X}_0$, the initial conditions of the characteristic equations are

$$\begin{cases} \boldsymbol{x}_0 \in [\boldsymbol{X}_0]^0 \\ \Phi(0, \boldsymbol{x}_0) = 0.06\left(1 - \mu_{\boldsymbol{X}_0}(\boldsymbol{x}_0)\right) \\ \boldsymbol{\psi}(0, \boldsymbol{x}_0) = 0.3(\boldsymbol{x}_0 - [0.9344, 0.3769, -0.7149, 0.1583]^T)/d \\ \zeta(0, \boldsymbol{x}_0) = -(\boldsymbol{u}^1(0, \boldsymbol{x}_0))^T \boldsymbol{\psi}(0, \boldsymbol{x}_0) - \Phi(0, \boldsymbol{x}_0)\sqrt{\psi_2^2(0, \boldsymbol{x}_0) + \psi_4^2(0, \boldsymbol{x}_0)} \end{cases}.$$

Let $T = 2\pi/\Omega$, the evolution of the membership distribution of the fuzzy response is studied in the period from $t = 0$ to $t = 60000T$ by randomly sampling 780000 initial states from $[\boldsymbol{X}_0]^0$. Particularly, the membership distributions at the moments $0, 1T, 4T, 7T, 44T, 1200T, 20000T, 40000T$ are shown respectively in Fig. 7a–h as the projections on the plane $x_1 - x_3$. It can be seen from the figures, though the initial state locates near the partial rub solution, only a part of the fuzzy response, which has the membership degree of 1, behaves the same as the deterministic solution. Different from the phenomenon of moving chaotically around the basin interior saddle in the fuzzy Mathieu system, the rest part of the fuzzy response turns to the other long-term deterministic responses of dry friction backward whirl. It means that the uncertainties on the amplitude of the external excitation may cause the transfer to the dangerous behavior even when the system already being in a safe running state.

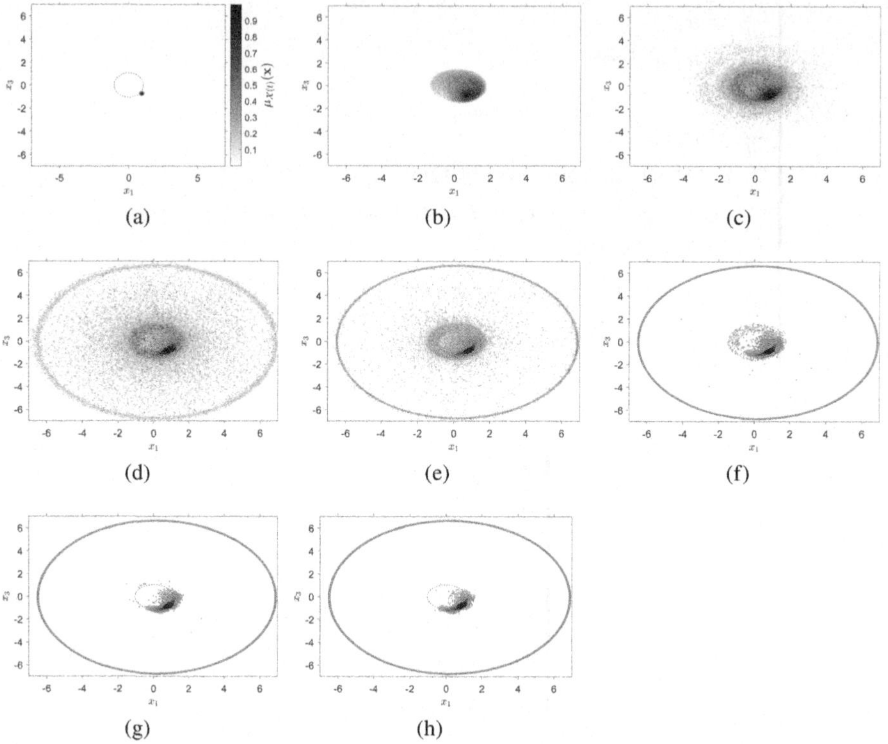

Fig. 7 The membership distributions of $\mathcal{X}(t)$, the fuzzy responses of the rotor/stator rubbing system with fuzzy uncertainties described by (60) at different moments, **a** $t = 0T$, **b** $t = 1T$, **c** $t = 4T$, **d** $t = 7T$, **e** $t = 44T$, **f** $t = 1200T$, **g** $t = 20000T$, **h** $t = 40000T$. The pictures are the projections of the 4-D membership distributions on the x_1-x_3 (replacements) plane, where different gray scales from white to black represent different values of $\mu_{\mathcal{X}(t)}(\boldsymbol{x})$ from 0 to 1. The dashed circle in the pictures represents the inner ring of stator

5 Conclusion

The fuzzy differential equation (FDE) is solved by transforming it to the governing equation of the membership degree of the fuzzy state. The solution obtained in this way is proved to be equal to the solution via differential inclusions. Different from the attainable sets of the solution acquired by most numerical methods of FDE, the solution is outlined by its membership distribution directly. The calculation involves no comparison or data storage, thus is efficient in computing and low cost in storage. From the comparisons of the numerical and the analytical solutions in the two examples, the new method is very accurate to solve FDE via differential inclusions.

With the transform method, the responses of the Mathieu system and the rotor/stator contact system with fuzzy uncertainties are studied. Particularly, for the fuzzy Mathieu system, given the initial fuzzy state around the period-3 deterministic attractor, the fuzzy response covers multiple deterministic dynamical structures including the period-3 attractor, the chaotic saddle in the basin interior and the unstable manifold from the saddle to the attractor; for the fuzzy rotor/stator contact system, given the initial fuzzy state around a point of the deterministic solution of partial rub, the fuzzy response splits into two parts: one stays around the partial rub, the other jumps to the dry friction backward whirl. The studies demonstrate that, even the initial states are around the attractors of the corresponding deterministic systems, different long term responses from deterministic versions can be induced by the fuzzy uncertainties. For some dangerous responses caused by fuzzy uncertainties, such analysis can provide a reference to predict them.

Acknowledgements This study was funded by the National Natural Science Foundation of China (Grant No. 11672218 and 11972274 and 11772243 and 11332008) and the National Key Research and Development Program of China (Grant No. 2019YFB1504601) and the Youth Project of Provincial Natural Science Foundation of Anhui, China (Grant No. 1908085QE234).

References

1. Guo, M., Xue, X., Li, R.: Impulsive functional differential inclusions and fuzzy population models. Fuzzy Sets Syst. **138**(3), 601–615 (2003)
2. Baidosov, V.A.: Fuzzy differential inclusions. J. Appl. Math. Mech. **54**(1), 8–13 (1990)
3. Hüllermeier, E.: An approach to modelling and simulation of uncertain dynamical systems. Int. J. Uncertainty, Fuzziness Knowl. Based Syst. **05**(02), 117–137 (1997)
4. Hüllermeier, E.: Numerical methods for fuzzy initial value problems. Int. J. Uncertainty, Fuzziness Knowl. Based Syst. **07**(05), 439–461 (1999)
5. Diamond, P.: Time-dependent differential inclusions, cocycle attractors and fuzzy differential equations. IEEE Trans. Fuzzy Syst. **7**(6), 734–740 (1999)
6. Diamond, P.: Stability and periodicity in fuzzy differential equations. IEEE Trans. Fuzzy Syst. **8**(5), 583–590 (2000)
7. Diamond, P., Watson, P.: Regularity of solution sets for differential inclusions quasi-concave in a parameter. Appl. Math. Lett. **13**(1), 31–35 (2000)
8. Rzeżuchowski, T., Wąsowski, J.: Differential equations with fuzzy parameters via differential inclusions. J. Math. Anal. Appl. **255**(1), 177–194 (2001)
9. Lakshmikantham, V., Mohapatra, R.N.: Theory of Fuzzy Differential Equations and Inclusions. Taylor & Francis (2003)
10. Abbasbandy, S., Viranloo, T.A., López-Pouso, o., Nieto, J.J.: Numerical methods for fuzzy differential inclusions. Comput. Math. Appl. **48**(10), 1633–1641 (2004)
11. Majumdar, K.K., Majumder, D.D.: Fuzzy differential inclusions in atmospheric and medical cybernetics. IEEE Trans. Syst. Man Cybern. Part B **34**(2), (2004)
12. Abbasbandy, S., Nieto, J.J., Alavi, M.: Tuning of reachable set in one dimensional fuzzy differential inclusions. Chaos, Solitons & Fractals **26**(5), 1337–1341 (2005)
13. Agarwal, R.P., O'Regan, D., Lakshmikantham, V.: Maximal solutions and existence theory for fuzzy differential and integral equations. J. Appl. Anal. **11**(2), 171–186 (2005)
14. Chalco-Cano, Y., Román-Flores, H.: Some remarks on fuzzy differential equations via differential inclusions. Fuzzy Sets Syst. **230**, 3 – 20 (2013). Differential equations over fuzzy spaces—theory, applications, and algorithms

15. Wasques, V.F., Esmi, E., Barros, L.C., Sussner, P.: Numerical solutions for bidimensional initial value problem with interactive fuzzy numbers. In: Fuzzy Information Processing, vol. 831, pp. 84–95. Springer, Cham (2018)
16. Wasques, V.F., Esmi, E., Barros, L.C., Bede, B.: Comparison between numerical solutions of fuzzy initial-value problems via interactive and standard arithmetics. In: Fuzzy Techniques: Theory and Applications. Advances in Intelligent Systems and Computing, vol. 1000, pp. 704–715. Springer, Cham (2019)
17. Allahviranloo, T., Ahmady, N., Ahmady, E.: Numerical solution of fuzzy differential equations by predictor-corrector method. Inform. Sci. **177**(7), 1633–1647 (2007)
18. Bede, B.: Note on "numerical solutions of fuzzy differential equations by predictor–corrector method". Inform. Sci. **178**(7), 1917–1922 (2008)
19. Palligkinis, S.C., Papageorgiou, G., Famelis, I.T.: Runge-kutta methods for fuzzy differential equations. Appl. Math. Comput. **209**(1), 97–105 (2009)
20. Khastan, A., Ivaz, K.: Numerical solution of fuzzy differential equations by nyström method. Chaos, Solitons and Fractals **41**(2), 859–868 (2009)
21. Mondal, S.P., Roy, S., Das, B.: Numerical solution of first-order linear differential equations in fuzzy environment by runge-kutta-fehlberg method and its application. Int. J. Differ. Equ. **2016**, 1–14 (2016)
22. Abu Arqub, O., AL-Smadi, M., Momani, S., Hayat, T.: Numerical solutions of fuzzy differential equations using reproducing kernel hilbert space method. Soft Comput. **20**(8), 3283–3302 (2016)
23. Ahmadian, A., Salahshour, S., Chan, C.S., Baleanu, D.: Numerical solutions of fuzzy differential equations by an efficient runge-kutta method with generalized differentiability. Fuzzy Sets Syst. **331**, 47–67 (2018)
24. Raczynski, S.: Continuous simulation, differential inclusions, uncertainty, and traveling in time. SIMULATION: Trans. Soc. Model. Simul. **80**(2) (2004)
25. Friedman, Y., Sandler, U.: Evolution of systems under fuzzy dynamic laws. Fuzzy Sets Syst. **84**(1), 61–74 (1996)
26. Friedman, Y., Sandler, U.: Fuzzy dynamics as an alternative to statistical mechanics. Fuzzy Sets Syst. **106**(1), 61–74 (1999)
27. Sandler, U., Tsitolovsky, L.: Fuzzy dynamics of brain activity. Fuzzy Sets Syst. **121**(2), 237–245 (2001)
28. Zadeh, L.A.: The concept of a linguistic variable and its application to approximate reasoning–i. Inform. Sci. **8**(3), 199–249 (1975)
29. Nguyen, H.T.: A note on the extension principle for fuzzy sets. J. Math. Anal. Appl. **64**(2), 369–380 (1978)
30. Román-Flores, H., Rojas-Medar, M.: Embedding of level-continuous fuzzy sets on banach spaces. Inform. Sci. **144**(1), 227–247 (2002)
31. Barros, L.C., Bassanezi, R.C., Tonelli, P.A.: On the continuity of the zadeh's extension. Proc. Seventh IFSA World Congress **2**, 3–8 (1997)
32. Aubin, J.P., Cellina, A.: Differential Inclusions: Set-Valued Maps and Viability Theory, 1st edn. Springer, Berlin Heidelberg (1984)
33. Cecconello, M.S., Bassanezi, R.C., Brandão, A.V., Leite, J.: Periodic orbits for fuzzy flows. Fuzzy Sets Syst. **230**, 21–38 (2013)
34. John, F.: Partial Differential Equations. Springer (1991)
35. Black, H.F.: Interaction of a whirling rotor with a vibrating stator across a clearance annulus. J. Mech. Eng. Sci. **10**(1), 1–12 (1968)
36. Yu, J.J., Goldman, P., Bently, D.E., Muzynska, A.: Rotor/seal experimental and analytical study on full annular rub. J. Eng. Gas Turbines Power **124**(2), 340–350 (2002)
37. Jiang, J., Ulbrich, H.: Stability analysis of sliding whirl in a nonlinear jeffcott rotor with cross-coupling stiffness coefficients. Nonlinear Dyn. **24**(3), 269–283 (2001)
38. Jiang, J., Ulbrich, H.: Rub-induced parametric excitation in rotors. J. Mech. Design **101**(4), 640–644 (1979)

39. Ehrich, F.E.: High order subharmonic response of high speed rotors in bearing clearance. J. Vib. Acoust. Stress Reliab. Design **110**(1), 9–16 (1988)
40. Jiang, J., Ulbrich, H.: The physical reason and the analytical condition for the onset of dry whip in rotor-to-stator contact systems. J. Vib. Acoust. Stress Reliab. Design **127**(6), 594–603 (2005)
41. Jiang, J.: The analytical solution and the existence condition of dry friction backward whirl in rotor-to-stator contact systems. J. Vib. Acoust. Stress Reliab. Design **129**(2), 260–264 (2007)
42. Jiang, J.: Determination of the global responses characteristics of a piecewise smooth dynamical system with contact. Nonlinear Dyn. **57**(3), 351–361 (2009)

Complete Controllability of Fuzzy Fractional Evolutions Equation Under Fréchet Derivative in Linear Correlated Fuzzy Spaces

Nguyen Thi Kim Son, Hoang Thi Phuong Thao, Tran Van Bang, and Hoang Viet Long

Abstract In this paper, we study the complete controllability for a class of fuzzy fractional evolution equations in terms of Fréchet–Caputo derivatives in the linear correlated fuzzy space. Sufficient conditions for the complete controllability for the fuzzy fractional evolution equations are considered in both cases of the unique or not unique control variable. Technics of the fuzzy semigroups and the noncompactness measure are used in the linear correlated fuzzy spaces. Some illustrated examples are given.

Keywords Fréchet–Caputo derivatives · Linear correlated fuzzy-valued function · Complete controllability · Laplace transform

MSC 2010 47H10 · 47H04 · 03E72 · 46S40

1 Introduction

Fuzzy differential equations have been arguably one of the most active research directions so far, contributing to the creation of general fuzzy analysis as well as

This research is funded by Vietnam National Foundation for Science and Technology Development (NAFOSTED) under Grant No. 101.02-2018.311.

N. T. K. Son
Faculty of Natural Sciences and Technology, Hanoi Metropolitan University, Hanoi, Vietnam

H. T. P. Thao
VNU University of Languages and International Studies, Viet Nam National University, Hanoi, Vietnam

T. V. Bang
Department of Mathematics, Hanoi Pedagogical University 2, Hanoi, Vietnam

H. V. Long (✉)
Facutly of Information Technology, People's Police University of Technology and Logistics, Bac Ninh, Vietnam

T. Allahviranloo and S. Salahshour (eds.), *Advances in Fuzzy Integral and Differential Equations*, Studies in Fuzziness and Soft Computing 412,
https://doi.org/10.1007/978-3-030-73711-5_3

promoting applied research in the simulation of dynamical systems in nature by fuzzy differential equations. However, when studying fuzzy differential equations, we face significant challenges, due to the non-linear nature of the space of fuzzy numbers and the complexity in defining the differences in this space. There are several important approaches to the study of fuzzy differential equations that have been well analyzed in many previous works. For example, the first approach uses the Hukuhara difference [17]. This was further elaborated by using generalized Hukuhara difference [5]. This approach has created significance in the theoretical study of fuzzy differential equations over the past ten years. However, as mentioned above, the complexity of constructing the derivative concept and the difficulty in verifying the existence of this derivative has significantly reduced the applied studies of the fuzzy differential equation in engineering problems. Recently there has been a new approach based on granular difference through a horizontal representation of the membership function [14]. This approach has the advantage that helps avoid having to apply the conditions of monotony on the different function in constructing granular derivatives. Some studies extending to advanced fuzzy differential equations can be found in [15, 22–24]. An alternative approach does not need to use the Hukuhara difference but is based on the interactive difference presented in [4]. The interactive derivative was obtained by means of fuzzification of the classical derivative operator for standard functions. Recently, [8] introduced the concept of Fréchet derivative for completely correlated fuzzy processes. That is the class of functions $f : \mathbb{R} \rightarrow \mathbb{R}_{\mathcal{F}(A)}$, where $\mathbb{R}_{\mathcal{F}(A)} = \{\Psi_A(q, r)/(q, r) \in \mathbb{R}^2\}$. Here Ψ_A is a linear isomorphism mapping between two structures $\mathbb{R}_{\mathcal{F}(A)}$ and $\mathbb{R}^2$. This approach has many advantages. Firstly, in the case $\mathbb{R}_{\mathcal{F}(A)}$ is Banach space (when A is a nonsymmetric fuzzy number), advanced techniques of analysis on Banach spaces can be applied to research the fuzzy differential equation. Even if $\mathbb{R}_{\mathcal{F}(A)}$ is not Banach space (when A is symmetric fuzzy number), we can still calculate the Fréchet derivative smoothly via mapping $\Psi_A : \mathbb{R}^2 \rightarrow \mathbb{R}_{\mathcal{F}(A)}$.

Controllability is one of the basic properties of controlled systems. This property is well-known from the mathematical theory of systems (Kalman in 1960) as the concept of the reachability of terminal states. This means that it is possible to control a dynamic system from an arbitrary initial state to an arbitrary final state using a set of admissible controls. The concept of controllability plays an important role in the analysis and design of control systems. The basic concepts of control theory in the finite dimension space have been discussed in [9, 11]. Anurag Shukla et al. studied the approximate and complete controllability of semi-linear delay control systems by using fixed point theory in [25]. Klamka and Locha [11] has given some remarks on stochastic controllability.

The semigroup theory of linear operators on Banach space was introduced in 1920 and thrived in 1948 with the Hille—Yosida theorem, and reached its fullness in 1957 with the birth of the book "Semigroups and Functional Analysis" by E. Hille and R. S. Philips. In the years of the 70, 80s of the twentieth century, thanks to research by many universities and many research centers of physics, the semigroup theory on the Banach spaces has reached a perfect state. It has become an important tool in the study of differential equations, functional equations in mathematic, physics, quantum, mechanics, etc. The semigroup theory of the fuzzy valued linear operators was first published by Gal [10] and improved by Son [20]. It plays an important role in the study of the asymptotic behavior of evolution equations with uncertainties.

Recently, the linear correlated fuzzy spaces have been introduced and gradually being finalized (see [8], [16]). A highlight point in these spaces can be mentioned is the flexible application of results on Bannach spaces onto these situations. In this paper, we consider the complete controllability of the following fuzzy fractional evolution equation in the linear correlated fuzzy spaces $n\mathbb{R}_{\mathcal{F}(A)}$ (Definition 2.4)

$$\begin{cases} {}_F^C\mathcal{D}^p \mu(t) = B\mu(t) +_A f(t, \mu(t)) +_A Eu(t) & t \in J = [0, b], \\ \mu(0) = \mu_0. \end{cases} \tag{1}$$

where B is the generator of a C_0 − semigroup $\{T(t)\}_{t \geq 0}$ on $n\mathbb{R}_{\mathcal{F}(A)}$, the input control $u \in L^1\left(J, n\mathbb{R}_{\mathcal{F}(A)}\right)$; the linear operator E and the function $f(.,.)$ satisfy some given conditions

Main results are devided in two specific cases. When the external force f satisfies the Lipschitz condition under the second variable, by using the Banach fixed point theorem, we prove that the problem (4) is completely controllable with a pseudo-number additional set of input data. The second case is when the external force f is loosened, the general Lipschitz inequality is satisfied with the variational coefficient and a growth condition on noncompact measure. By using Karanoselski's fixed point theorem, evaluated according to Holder's inequality, the auxiliary lemmas of the noncompact measure, we can prove the results of the complete controlability of the problem. This problem is considered in $\mathbb{R}_{\mathcal{F}(A)}$ with A is non-symmetric fuzzy number. This is a new spatial structure that needs to build corresponding analytical results for this space such as: the distance d_A and it's invariant property (Proposition 2.2), the translational property, the Laplace transform on $\mathbb{R}_{\mathcal{F}(A)}$ space (Theorem 2.3, Theorem 2.4) and some special properties of this transformation. Based on suitable fixed theorems, we obtain the results for the complete controlability (Theorem 3.1, Theorem 3.2). Some illustrated examples are given.

2 Preliminaries

2.1 The Space of Linear Correlated Fuzzy Numbers

Denote by $\mathbb{R}_{\mathcal{F}}$ the space of all fuzzy numbers on the real line $\mathbb{R}$. According to [6], the characteristic properties of a fuzzy number u are presented via its $\alpha-$ cuts or level sets, which are defined by

$$[u]^\alpha = \begin{cases} \{t \in \mathbb{R} : u(t) \geq \alpha\} & \text{if } \alpha \in (0, 1], \\ \overline{\{t \in \mathbb{R} : u(t) > 0\}} & \text{if } \alpha = 0. \end{cases}$$

In addition, it is well-known that the level sets of u can be rewritten in the parametric form $[u]^\alpha = [u_\alpha^-, u_\alpha^+]$ and the diameter of $[u]^\alpha$ is given by $\text{len}[u]^\alpha = u_\alpha^+ - u_\alpha^-$ for each $\alpha \in [0, 1]$.

The space $(\mathbb{R}_\mathcal{F}, d_\infty)$ endowed with the supremum metric

$$d_\infty(u, v) = \sup_{0\le\alpha\le1} d_H\left([u]^\alpha, [v]^\alpha\right) \quad \text{for all } u, v \in \mathbb{R}_\mathcal{F},$$

is a complete metric space (see [6]).

Definition 2.1 ([8]) For each $A \in \mathbb{R}_\mathcal{F}$, define $\psi_A : \mathbb{R} \times \mathbb{R} \to \mathbb{R}_\mathcal{F}$ by

$$(q, r) \mapsto \psi_A(q, r),$$

where the level sets of $\psi_A(q, r)$ are

$$[\psi_A(q, r)]^\alpha = \{qa + r : a \in [A]^\alpha\}, \alpha \in [0, 1].$$

For convenience, denote the fuzzy number $\psi_A(q, r)$ by $qA + r$ and the range of ψ_A by $\mathbb{R}_{\mathcal{F}(A)}$.

Definition 2.2 ([8]) A fuzzy number u is said to be symmetric with respect to $t \in \mathbb{R}$ if $u(t - y) = u(t + y)$ for all $y \in \mathbb{R}$. The fuzzy number u is non-symmetric if there doesn't exist $t \in \mathbb{R}$ such that u is symmetric.

Example 2.1 Consider a fuzzy number u defined by

$$u(t) = \begin{cases} 0 & \text{if } t < 1, \\ t - 1 & \text{if } 11 \le t < 2, \\ 1 & \text{if } 2 \le t \le 3, \\ 4 - t & \text{if } 3 < t \le 4, \\ 0 & \text{if } 4 < t. \end{cases}$$

According to Definition 2.2, this fuzzy number is a symmetric fuzzy number with respect to $\frac{5}{2}$ (Fig. 1).

Example 2.2 The fuzzy set $v : \mathbb{R} \to [0, 1]$, defined by

$$v(t) = \begin{cases} 0 & \text{if } t < 1, \\ \frac{t-1}{2} & \text{if } 1 \le t < 3, \\ \frac{7-t}{4} & \text{if } 3 < x \le 7, \\ 0 & \text{if } 7 < t, \end{cases}$$

is a fuzzy number on $\mathbb{R}$. According to Definition 2.2 and by similar arguments as in Example 2.2, we can see that the v is a non-symmetric fuzzy number (Fig. 2).

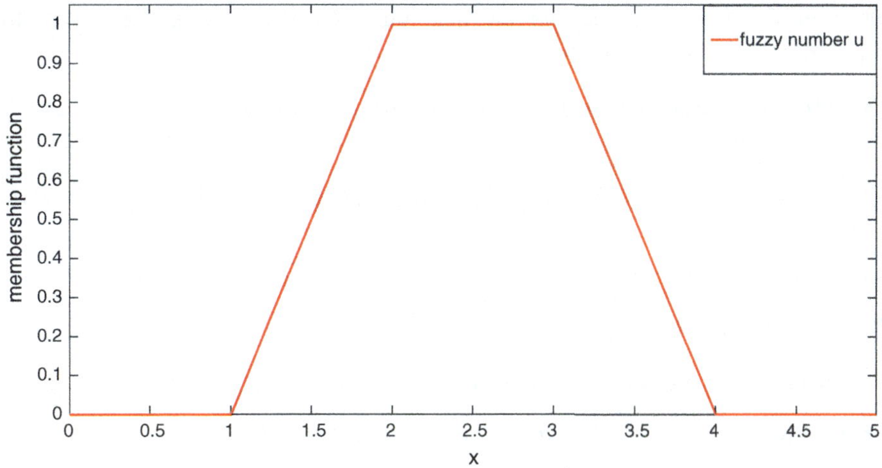

Fig. 1 The fuzzy number $u = (1, 2, 3, 4)$

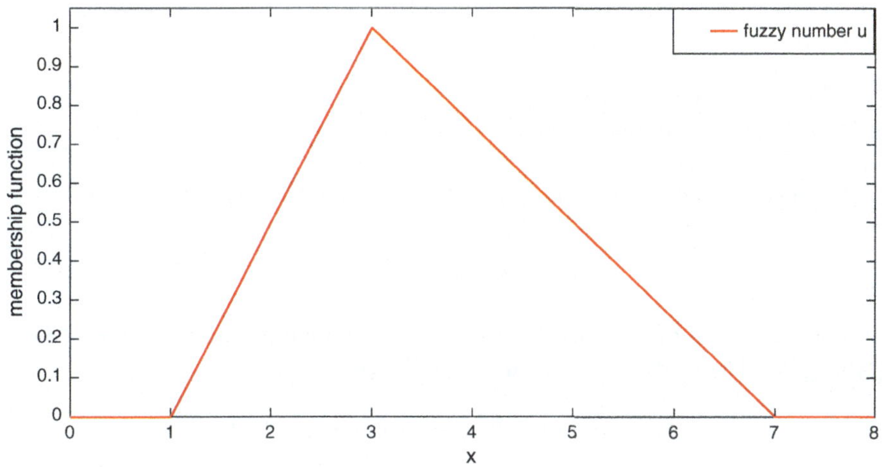

Fig. 2 The triangular fuzzy number $u = (1, 3, 7)$

For simply, we denote $n\mathbb{R}_{\mathcal{F}} = \{A \in \mathbb{R}_{\mathcal{F}}, A$ is a non-symmetric fuzzy number$\}$. It well-known result that ψ_A (Definition 2.2) is isomorphism when $A \in n\mathbb{R}_{\mathcal{F}}$.

Definition 2.3 ([8]) Let $A \in n\mathbb{R}_{\mathcal{F}}$. The arithmetic operations on the space $\mathbb{R}_{\mathcal{F}(A)}$ such as addition, scalar multiplication, subtraction are defined as follows

(i) $u +_A v = \psi_A \left(\psi_A^{-1}(u) + \psi_A^{-1}(v)\right)$,
(ii) $\lambda u = \psi_A \left(\lambda \psi_A^{-1}(u)\right)$,
(iii) $u -_A v = u +_A (-1)v = \psi_A \left(\psi_A^{-1}(u) + (-1)\psi_A^{-1}(v)\right)$,

where $u, v \in \mathbb{R}_{\mathcal{F}(A)}$, $\lambda \in \mathbb{R}$ and the mapping $\psi_A^{-1} : \mathbb{R}_{\mathcal{F}(A)} \to \mathbb{R} \times \mathbb{R}$, given by $\psi_A^{-1}(qA + r) = (q, r)$, is the inverse mapping of ψ_A.

Definition 2.4 Assume that $A \in n\mathbb{R}_{\mathcal{F}}$. For each $u = \psi_A(q_u, r_u)$, $v = \psi_A(q_v, r_v) \in \mathbb{R}_{\mathcal{F}(A)}$, we define

$$d_A(u, v) = |q_u - q_v| + |r_u - r_v|. \tag{2}$$

Denote $\|u\|_A = d_A(u, \hat{0})$, where $\hat{0} = \psi_A(0, 0)$ and $n\mathbb{R}_{\mathcal{F}(A)}$ by the range of ψ_A.

Proposition 2.1 *[8] $(n\mathbb{R}_{\mathcal{F}(A)}, d_A)$ is a Banach space where $A \in n\mathbb{R}_{\mathcal{F}}$.*

Following results are necessary in the calculus of $\mathbb{R}_{\mathcal{F}(A)}$.

Proposition 2.2 *If $A \in n\mathbb{R}_{\mathcal{F}}$, the following properties hold true*

(i) $d_A(u +_A w, v +_A w) = d_A(u, v), \forall u, v, w \in \mathbb{R}_{\mathcal{F}(A)}$,
(ii) $d_A(ku, kv) = |k| d_A(u, v), \forall u, v \in \mathbb{R}_{\mathcal{F}(A)}, k \in \mathbb{R}$,
(iii) $d_A(u +_A v, w +_A e) \le d_A(u, w) + d_A(v, e), \ \forall u, v, w, e \in \mathbb{R}_{\mathcal{F}(A)}$,
(iv) $d_A(u -_A v, \hat{0}) = d_A(u, v) \ \forall u, v \in \mathbb{R}_{\mathcal{F}(A)}$.

Proof (i) By the definition of the addition in $n\mathbb{R}_{\mathcal{F}(A)}$, we have $u +_A v = \psi_A(q_u + q_v, r_u + r_v)$. It implies that

$$\begin{aligned} d_A(u + v, v + w) &= |q_u + q_v - (q_v + q_w)| + |r_u + r_v - (r_v + r_w)| \\ &= |q_u - q_w| + |r_u - r_w| \\ &= d_A(u, w). \end{aligned}$$

(ii) For $k \in \mathbb{R}$, $u \in n\mathbb{R}_{\mathcal{F}(A)}$, by the definition of the multiplication in $n\mathbb{R}_{\mathcal{F}(A)}$, $ku = \psi_A(k\psi_A^{-1}u)$. Therefore $ku = \psi_A(kq_u, kr_u)$. We get

$$\begin{aligned} d_A(ku, kv) &= |kq_u - kq_v| + |kr_u - kr_v| \\ &= |k||q_u - q_v| + |k||r_u - r_v| \\ &= |k| d_A(u, v). \end{aligned}$$

(iii) For $u, v, w, e \in n\mathbb{R}_{\mathcal{F}(A)}$, we have

$$u +_A v = \psi_A(q_u + q_v, r_u + r_v), w +_A e = \psi_A(q_w + q_e, r_w + r_e).$$

It implies that

$$\begin{aligned} d_A(u +_A v, w +_A e) &= |q_u + q_v - q_w - q_e| + |r_u + r_v - r_w - r_e| \\ &\le |q_u - q_w| + |r_u - r_w| + |q_v - q_e| + |r_v + r_e| \\ &\le d_A(u, w) + d_A(v, e). \end{aligned}$$

(iv) For $u, v \in n\mathbb{R}_{\mathcal{F}(A)}$, from the subtraction in $n\mathbb{R}_{\mathcal{F}(A)}$, we have $u -_A v = \psi_A(q_u - q_v, r_u - r_v)$. It implies that

$$\begin{aligned} d(u -_A v, \hat{0}) &= |q_u - q_v| + |r_u - r_v| \\ &= d(u, v). \end{aligned}$$

□

Denote by $C([0, +\infty), n\mathbb{R}_{\mathcal{F}(A)})$ the space of all continuous functions $f : [0, +\infty) \to n\mathbb{R}_{\mathcal{F}(A)}$ and $\mathfrak{L}(n\mathbb{R}_{\mathcal{F}(A)})$ the space of all continuous linear operators $\mu : n\mathbb{R}_{\mathcal{F}(A)} \to n\mathbb{R}_{\mathcal{F}(A)}$, $A \in n\mathbb{R}_{\mathcal{F}}$.

Definition 2.5 [8] For each $g \in \mathscr{L}(J, \mathbb{R}_{\mathcal{F}(A)})$, we say that

(i) The mapping g is Fréchet differentiable at $t \in J$ if there is a linear operator $k_t \in \mathscr{L}(J, \mathbb{R}_{\mathcal{F}(A)})$ such that

$$g(t+h) = g(t) +_A k_t h +_A \mathcal{O}(h),$$

where $\mathcal{O}(h)$ approaches to $\hat{0}$ as $h \to 0$.

(ii) The mapping g is Fréchet differentiable on J if and only if it is Fréchet differentiable at every $t \in J$.

(iii) The Fréchet derivative of g is a linear operator $g'_{\mathcal{F}} : J \to \mathscr{L}(J, \mathbb{R}_{\mathcal{F}(A)})$, defined by $t \mapsto g'_{\mathcal{F}}(t) = k_t$.

Theorem 2.1 *[18] Let $A \in n\mathbb{R}_{\mathcal{F}}$, $q, r : \mathbb{R} \to \mathbb{R}$ and $f : \mathbb{R} \to n\mathbb{R}_{\mathcal{F}(A)}$ such that $f(t) = \psi_A(q(t, r(t)))$ for all $t \in \mathbb{R}$. The function f is Fréchet differentiable at $t \in \mathbb{R}$ if and only if q, r are differentiable at $t \in \mathbb{R}$. Additionally, the Fréchet derivative of f at t is given by $f'(t, h) = \psi_A(q'(t)h, r'(t)h)$, $\forall h \in \mathbb{R}$.*

Definition 2.6 [19] Let $A \in n\mathbb{R}_{\mathcal{F}}$, $f \in \mathscr{L}(J, \mathbb{R}_{\mathcal{F}(A)})$ and $f(t) = q(t)A + r(t)$, $t \in J$ with $q, r \in L^1(J, \mathbb{R}) \cap C(J, \mathbb{R})$. Then, the Riemann-Liouville fractional integral of order $p \in (0, 1]$ of the function f is defined by

$${}^{RL}_{\mathcal{F}}\mathcal{I}^p_{0^+} f(t) = \psi_A\left(I^p_{0^+} q(t), I^p_{0^+} r(t)\right), \ t \in J.$$

Definition 2.7 [19] Let $A \in n\mathbb{R}_{\mathcal{F}}$ and $f : J \subset \mathbb{R} \to \mathbb{R}_{\mathcal{F}(A)}$, $f(t) = q(t)A \oplus r(t)$ for each $t \in J$ and $q(\cdot), r(\cdot) \in L^1(J, \mathbb{R}) \cap C^1(J, \mathbb{R})$. For $p \in (0, 1]$, the Fréchet Caputo fractional derivative of order p of function f is defined by

$${}^{C}_{\mathcal{F}}\widetilde{D}^p_{0^+} f(t) = \psi_A\left(I^{1-p}_{0^+} q'(t), I^{1-p}_{0^+} r'(t)\right), \ t \in J. \tag{3}$$

2.2 Laplace Transformation for Functions in $\mathbb{R}_{\mathcal{F}(A)}$

Definition 2.8 [1, 19] Let $f : [0, +\infty) \to n\mathbb{R}_{\mathcal{F}(A)}$. Assume that the function $f(t)$ is given by $f(t) = \psi_A(q(t), r(t)), t \in J$, where $q(.), r(.)$ are continuous real-valued functions. Then, the fuzzy Laplace transform of the function $f(.)$ is defined by

$$\mathcal{L}[f(t)](s) = \psi_A\left(L[q(t)](s), L[r(t)](s)\right),$$

where $L[q(t)](s), L[r(t)](s)$ are classical Laplace transforms of real functions of $q, r : [0, +\infty) \to \mathbb{R}$ respectively.

Proposition 2.3 *Let $f : [0, +\infty) \to n\mathbb{R}_{\mathcal{F}(A)}$, $f(t) = \psi_A(q(t), r(t)), t \in [0, +\infty)$, where $q, r : [0, +\infty) \to \mathbb{R}$, $s \in \mathbb{R}$. Assume that $f'(.)$ is an Frechet derivative of f (see [8]). Then*

$$\mathcal{L}[f'(t)](s) = s\mathcal{L}[f(t)](s) -_A f(0).$$

Proof From [8], since $f'(.)$ be an integrable on $[0, \infty)$, $q'(t), r'(t)$ are integrable on $[0, +\infty)$ and $f'(t) = \psi_A(q'(t, r'(t))), t \in [0, +\infty)$. From Definition 2.8, we have

$$\mathcal{L}[f'(t)](s) = \psi_A(L[q'(t)](s), L[r'(t)](s)).$$

Since $q'(.), r'(.)$ are integrable on $[0, +\infty)$, we have

$$L[q'(t)] = sL[q(t)] - q(0) \text{ and } L[r'(t)](s) = sL[r(t)] - r(0).$$

It leads to

$$\begin{aligned}\mathcal{L}[f'(t)](s) &= \psi_A(sL[q(t)] - q(0), sL[r(t)] - r(0))\\ &= s\psi_A L[q(t)](s), L[r(t)](s)) -_A \psi_A(q(0), r(0))\\ &= s\mathcal{L}[f(t)](s) -_A f(0).\end{aligned}$$

It completes the proof. □

Proposition 2.4 *Let $f, g : [0, +\infty) \to n\mathbb{R}_{\mathcal{F}(A)}$ be continuous fuzzy-valued functions. Suppose that c_1, c_2 are real constants, then*

$$\mathcal{L}[c_1 f(t) +_A c_2 g(t)](s) = c_1\mathcal{L}[f(t)](s) +_A c_2\mathcal{L}[g(t)](s), s \in \mathbb{R}.$$

Proof According Definition 2.8 and the definition of addition operator in space $n\mathbb{R}_{\mathcal{F}(A)}$, we have

$$
\begin{aligned}
c_1\mathcal{L}[f(t)](s)+_A c_2\mathcal{L}[g(t)] &= c_1\psi_A(L[q_f(t)](s), L[r_f(t)](s)) +_A c_2\psi_A(L[q_g(t)](s), L[r_g(t)](s)) \\
&= \psi_A(c_1L[q_f(t)](s), c_1L[r_f(t)](s)) +_A \psi_A(c_2L[q_g(t)](s), c_2L[r_g(t)](s)) \\
&= \psi_A\left(c_1L[q_f(t)](s) + c_2L[q_g(t)](s), c_1L[r_f(t)](s) + c_2L[r_g(t)](s)\right) \\
&= \psi_A(c_1L[q_f(t)](s), c_1L[r_f(t)](s)) +_A \psi_A(c_2L[q_f(t)](s), c_2L[r_f(t)](s)) \\
&= c_1\mathcal{L}[f(t)](s) +_A c_2\mathcal{L}[g(t)](s)
\end{aligned}
$$

for all $s \in \mathbb{R}$. It ends the proof. □

Remark 2.1 Let $f : [0, +\infty) \rightarrow n\mathbb{R}_{\mathcal{F}(A)}$ be continuous fuzzy functions. Then

$$\mathcal{L}[\lambda f(t)](s) = \lambda\mathcal{L}[f(t)](s) \text{ for } \lambda \geq 0, s \in \mathbb{R}.$$

2.3 *The C_0—Semigroup of Fuzzy-Valued Mappings*

Applying the ideas in [?], we give some basic concepts on the semigroup of bounded linear mappings that take values on the space $n\mathbb{R}_{\mathcal{F}(A)}$.

Definition 2.9 A mapping $T : n\mathbb{R}_{\mathcal{F}(A)} \rightarrow n\mathbb{R}_{\mathcal{F}(A)}$ is called a bounded linear mapping if

(i) $T(u + v) = T(u) + T(v)$ and $T(\lambda u) = \lambda T(u)$ for all $u, v \in n\mathbb{R}_{\mathcal{F}(A)}$ and $\lambda \in \mathbb{R}$.
(ii) There exists $\mathsf{K} > 0$ such that $d_A(T(u), \hat{0}) \leq \mathsf{K} d_A(u, \hat{0})$ for all $u \in n\mathbb{R}_{\mathcal{F}(A)}$.

Definition 2.10 For each bounded operator $T : n\mathbb{R}_{\mathcal{F}(A)} \rightarrow n\mathbb{R}_{\mathcal{F}(A)}$, denote

$$\|T\|_{op} := \sup\{d_A(T(u), \hat{0}) | u \in n\mathbb{R}_{\mathcal{F}(A)}, d_A(u, \hat{0}) \leq 1\}.$$

Then, we have

(i) $d_A(T(u), \hat{0}) \leq \|T\|_{op} d_A(u, \hat{0})$ for all $u \in n\mathbb{R}_{\mathcal{F}(A)}$.
(ii) $d_A(T(u), T(v)) \leq \|T\|_{op} d_A(u, v)$ for all $u, v \in n\mathbb{R}_{\mathcal{F}(A)}$.

Definition 2.11 A family $\{T(t)\}_{t\geq 0}$ of bounded linear mappings on $n\mathbb{R}_{\mathcal{F}(A)}$ is called a strongly continuous semigroup (or C_0—semigroup) on $n\mathbb{R}_{\mathcal{F}(A)}$ if it satisfies

(i) $T(0) = \mathrm{Id}_A$, where Id_A is the identity mapping on $n\mathbb{R}_{\mathcal{F}(A)}$,
(ii) $T(t + s) = T(t)T(s)$ for all $t, s \geq 0$.
(iii) For each $w \in n\mathbb{R}_{\mathcal{F}(A)}$, the orbit maps $\xi_w : [0, \infty) \rightarrow n\mathbb{R}_{\mathcal{F}(A)}$, given by $\xi_w(t) = T(t)w$, are continuous.

Definition 2.12 A strongly continuous semigroup $\{T(t)\}_{t\geq 0}$ on $n\mathbb{R}_{\mathcal{F}(A)}$ is called a contraction if for each $t \geq 0$, there has $\mathsf{K}_1 \geq 1$ such that

$$\|T(t)\|_{op} \leq \mathsf{K}_1.$$

Proposition 2.5 *If* $\{T(t)\}_{t\geq 0} \subseteq \mathfrak{L}(n\mathbb{R}_{\mathcal{F}(A)})$ *is a strongly continuous semigroup on* $n\mathbb{R}_{\mathcal{F}(A)}$ *then*

$$\lim_{t\to 0} \|T(t) -_A Id_A\|_{op} = 0.$$

Proof We have

$$\begin{aligned}\|T(t) -_A \mathrm{Id}_A\|_{op} &= \sup_{x\in n\mathbb{R}_{\mathcal{F}(A)}} \left\{d_A\left([T(t) -_A \mathrm{Id}_A]x, \hat{0}\right) : d_A(x, \hat{0}) \leq 1\right\} \\ &= \sup_{x\in n\mathbb{R}_{\mathcal{F}(A)}} \left\{d_A\left(T(t)x -_A x, \hat{0}\right) : d_A(x, \hat{0}) \leq 1\right\} \\ &= \sup_{x\in n\mathbb{R}_{\mathcal{F}(A)}} \left\{d_A(T(t)x, x) : d_A(x, \hat{0}) \leq 1\right\}.\end{aligned}$$

Since $\{T(t)\}_{t\geq 0}$ is a strongly continuous semigroup on $n\mathbb{R}_{\mathcal{F}(A)}$, for each $x \in n\mathbb{R}_{\mathcal{F}(A)}$, the orbit maps $\xi_x(t) = T(t)x$ are continuous. Chosing $t \to 0$ then $T(t)x \to T(0)x$. Futhermore, $T(0) = Id_A$. It implies that $T(0)x = x$. There fore, $\lim_{t\to 0} d_A(T(t)x, x) = 0$ for all $x \in n\mathbb{R}_{\mathcal{F}(A)}$ then the proof is completed. □

From [26], for $p \in (0, 1)$ and $t \in (0, \infty)$, we define the probability density functions on $(0, \infty)$ as follows

$$S_p(t)z = \int_0^\infty \phi_p(\theta)T(t^p\theta)zd\theta, \quad T_p(t)z = p\int_0^\infty \theta\phi_p(\theta)T(t^p\theta)zd\theta.$$

where

$$\psi_p(\theta) = \frac{1}{\pi}\sum_{n=1}^{\infty}(-1)^{n-1}\theta^{-pn-1}\frac{\Gamma(np+1)}{n!}\sin(n\pi p), \theta \in [0; +\infty)$$

$$\phi_p(\theta) = \frac{1}{p}\theta^{-1-\frac{1}{p}}\psi_p(\theta^{-\frac{1}{p}}), \theta \in [0; +\infty).$$

Proposition 2.6 *For each* $t \geq 0$, *the mappings* $S_p(.)$ *and* $T_p(.)$ *satisfy the following assertions*

(i) $S_p(.)$ and $T_p(.)$ are bounded linear mappings, specifically for all $x \in n\mathbb{R}_{\mathcal{F}(A)}$, we have

$$\|S_p(t)x\|_A \leq \mathsf{K}_1\|x\|_A \text{ and } \|T_p(t)x\|_A \leq \frac{p\mathsf{K}_1}{\Gamma(p+1)}\|x\|_A,$$

(ii) the families $\{S_p(t)\}_{t\geq 0}$ and $\{T_p(t)\}_{t\geq 0}$ are strongly continuous, specifically, for each $x \in n\mathbb{R}_{\mathcal{F}(A)}$, we have

$$d_A\left(S_p(t)x, S_p(t')x\right) \to 0 \quad \text{and} \quad d_A\left(T_p(t)x, T_p(t')x\right) \to 0 \qquad \text{as } t \to t'.$$

(iii) $S_p(t)$ and $T_p(t)$ are compact operators if $T(t)$ compact.

Proof (i) Let $t \in J$ be fixed. Since $T(t)$ is a linear mapping, it implies that the mappings $S_p(t)$, $T_p(t)$ are also linear mappings. In addition, from [26], we get

$$\int_0^\infty \theta\phi_p(\theta)d\theta = \frac{1}{\Gamma(p+1)} \text{ and } \int_0^\infty \phi_p(\theta)d\theta = 1.$$

For each $x \in n\mathbb{R}_{\mathcal{F}(A)}$, we have

$$\begin{aligned}
\|S_p(t)x\|_A &\leq \int_0^\infty \phi_p(\theta)d_A(T(t^p\theta)x, \hat{0})d\theta \\
&\leq \mathsf{K}_1 d_A(x, \hat{0}) \int_0^\infty \phi_p(\theta)d\theta = \mathsf{K}_1 \|x\|_A, \\
\|T_p(t)x\|_A &\leq p \int_0^\infty \theta\phi_p(\theta)d_A(T(t^p\theta)x, \hat{0})d\theta \\
&\leq p\mathsf{K}_1 d_A(x, \hat{0}) \int_0^\infty \phi_p(\theta)d\theta \\
&= \frac{p\mathsf{K}_1}{\Gamma(p+1)}\|x\|_A.
\end{aligned}$$

(ii) Let $x \in n\mathbb{R}_{\mathcal{F}(A)}$ be arbitrary and $t, t' \in J$ such that $t \leq t'$. Then, we have

$$\begin{aligned}
d_A\left(T_p(t')x, T_p(t)x\right) &\leq p \int_0^\infty \theta\phi_p(\theta)d_A\left(T(t'^p\theta)x, T(t^p\theta)x\right)d\theta \\
&\leq p \int_0^\infty \theta\phi_p(\theta)d_A\left(T(t'^p\theta - t^p\theta + t^p\theta)x, T(t^p\theta)x\right)d\theta \\
&\leq p\mathsf{K}_1 \int_0^\infty \theta\phi_p(\theta)d_A\left(T(t'^p\theta - t^p\theta)x, x\right)d\theta \\
&\leq p\mathsf{K}_1 \int_0^\infty \theta\phi_p(\theta)\|T(t'^p\theta - t^p\theta) -_A \mathrm{Id}_A\|_{op}\, d_A\left(x, \hat{0}\right)d\theta \\
&\leq \frac{p\mathsf{K}_1}{\Gamma(p+1)}\|T(t'^p\theta - t^p\theta) -_A \mathrm{Id}_A\|_{op}\, d_A\left(x, \hat{0}\right).
\end{aligned}$$

From Proposition 2.5, it implies that $d_A\left(T_p(t')x, T_p(t)x\right)$ tends to 0 as $t' \to t$, which means that $\{T_p(t)\}_{t\geq 0}$ is strongly continuous on $n\mathbb{R}_{\mathcal{F}(A)}$. By using similar arguments, we also obtain that the family $\{S_q(t)\}_{t\geq 0}$ is strongly continuous on $n\mathbb{R}_{\mathcal{F}(A)}$.
(iii) For each positive constant k, set $D_k = \{x \in n\mathbb{R}_{\mathcal{F}(A)} : \|x\|_A \leq k\}$ and

$$T_1(t) = \left\{ \int_0^\infty \phi_p(\theta) T(t^p\theta) x d\theta, x \in D_k \right\},$$

$$T_2(t) = \left\{ p \int_0^\infty \theta \phi_p(\theta) T(t^p\theta) x d\theta, x \in D_k \right\}.$$

We need to prove that $T_1(t), T_2(t)$ are relatively compact in $n\mathbb{R}_{\mathcal{F}(A)}$.

Let $t > 0$ be fixed. For $\delta > 0$, define the subset in $n\mathbb{R}_{\mathcal{F}(A)}$ by

$$T_{1\delta} = \left\{ \int_\delta^\infty \phi_p(\theta) T(t^p\theta) x d\theta, x \in D_k \right\}.$$

Then for any $x \in D_k$, we have

$$\int_\delta^\infty \phi_p(\theta) T(t^p\theta) x d\theta = T(t^p\delta) \int_\delta^\infty \phi_p(\theta) T(t^p\theta - t^p\delta) x d\theta.$$

Since $\{T(t)\}(t \geq 0)$ is compact, $T(t^p\delta)$ is compact. For this reason, the set $T_{1\delta}$ is relatively compact in $n\mathbb{R}_{\mathcal{F}(A)}$ for $\delta > 0$. Moreover, for $x \in D_k$, we have

$$d_A\left(\int_0^\infty \phi_p(\theta) T(t^p\theta) x d\theta, \int_\delta^\infty \phi_p(\theta) T(t^p\theta) x d\theta \right)$$
$$= d_A\left(\int_0^\delta \phi_p(\theta) T(t^p\theta) x d\theta, \hat{0} \right) \leq \|T(t^p\theta)\|_{op}.\|x\|_A \int_0^\delta \phi_p(\theta) d\theta \leq \mathsf{K}_1 k \int_0^\delta \phi_p(\theta) d\theta.$$

Therefore, there are relatively compact sets arbitrarily close to the set $T_1(t), t > 0$. Hence $T_1(t), t > 0$ is relatively compact in $n\mathbb{R}_{\mathcal{F}(A)}$. Similarly, $T_2(t), t > 0$ is relatively compact in $n\mathbb{R}_{\mathcal{F}(A)}$. Therefore, $S_p(t)$ and $T_p(t)$ are also compact operators. □

Definition 2.13 Let $\{T(t)\}_{t\geq 0}$ be a strongly continuous semigroup on $n\mathbb{R}_{\mathcal{F}(A)}$. Then, we define the infinitesimal generator $B : D(B) \subset n\mathbb{R}_{\mathcal{F}(A)} \to n\mathbb{R}_{\mathcal{F}(A)}$ of semigroup $\{T(t)\}_{t\geq 0}$ by

$$Bx := \lim_{h\to 0^+} \frac{1}{h}(T(h)x -_A x), \quad x \in D(B),$$

where $D(B) = \left\{ x \in n\mathbb{R}_{\mathcal{F}(A)} : \lim_{h\to 0^+} \frac{1}{h}(T(h)x -_A x) \text{ exists} \right\}$.

For $x \in D(B)$, denote $\mathcal{M}_\lambda x = \lambda x -_A Bx$ and $\rho(B) := \{\lambda \in \mathbb{C} : \mathcal{M}_\lambda x$ is bijective and $\mathcal{M}_\lambda^{-1} \in \mathfrak{L}(n\mathbb{R}_{\mathcal{F}(A)})\}$. Then for all $\lambda \in \rho(B)$, the resolve operator is

$$R(\lambda, B) := \mathcal{M}_\lambda^{-1} \in \mathfrak{L}(n\mathbb{R}_{\mathcal{F}(A)}).$$

Theorem 2.2 *Assume that* $(B, D(B))$ *is the infinitesimal generator of a contraction semigroup* $\{T(t)\}_{t\geq 0}$ *on* $n\mathbb{R}_{\mathcal{F}(A)}$. *Then, we have*

(i) $(0, \infty) \subseteq \rho(B)$,

(ii) For every $\lambda > 0$, $R(\lambda, B)x = \int_0^\infty e^{\lambda s} T(s)xds$ for all $x \in D(B)$ and $\|R(\lambda, B)\|_{op} \leq \frac{1}{\lambda}$.

Proof For each $\lambda > 0$, denote $S(t) := e^{\lambda t} T(t)$, $t \geq 0$ and $Rx = \int_0^\infty S(t)xdt$ with $x \in D(B)$. Since $\{T(t)\}_{t\geq 0}$ is a contraction semigroup, using similar arguments of Lemma 5.4 in [21], we can prove that $\{S(t)\}_{t\geq 0}$ is also a contraction semigroup with the infinitesimal operator $\mathcal{M} = (-1)\mathcal{M}_\lambda$. For all $h > 0$, we have

$$S(h)Rx = S(h)\int_0^{+\infty} S(t)xdt = \int_h^{+\infty} S(t)xdt,$$

which implies that

$$Rx = \int_0^{+\infty} S(t)xdt = \int_0^h S(t)xdt + S(h)Rx.$$

Thus, we have

$$\lim_{h\to 0^+} \frac{1}{h}(S(h)Rx -_A Rx) = \lim_{h\to 0^+} \left(\frac{-1}{h}\right) \int_0^h S(t)xdt \to (-1)x \text{ when } h \to 0^+.$$

It means that $Rx \in D(\mathcal{M})$ and $\mathcal{M}Rx = (-1)x$.

Since $\mathcal{M}$ is a closed operator, for all $x \in n\mathbb{R}_{\mathcal{F}(A)}$ we have

$$R\mathcal{M}x = \int_0^\infty S(t)\mathcal{M}xdt = \lim_{s\to\infty} \int_0^s S(t)\mathcal{M}xdt = \lim_{s\to\infty} \mathcal{M} \int_0^s S(t)xdt = \mathcal{M}Rx.$$

This shows that $R((-1)\mathcal{M})x = ((-1)\mathcal{M})Rx = x$ for all $x \in D(\mathcal{M})$ or equivalently, the operator $\mathcal{M}$ is invertible and

$$d_A(R(\lambda, B)x, \hat{0}) \leq \int_0^\infty e^{-\lambda t} d_A(T(t)x, \hat{0})dt \leq \frac{1}{\lambda} d_A(x, \hat{0}).$$

It follows that $\|R(\lambda, B)x\|_{op} \leq \frac{1}{\lambda}$. □

2.4 Measure of Noncompactness and Condensing Mappings

Let β, β_C be the Hausdorff measures of noncompactness (MNC) in $n\mathbb{R}_{\mathcal{F}(A)}$ and $C(J, n\mathbb{R}_{\mathcal{F}(A)})$, respectively. By Theorem 2.1 in [12], the following assertion holds

Lemma 2.1 *[12] If $j : n\mathbb{R}_{\mathcal{F}(A)} \to n\mathbb{R}_{\mathcal{F}(A)}$ is an embedding map and G is a bounded subset in $n\mathbb{R}_{\mathcal{F}(A)}$, then $j(G)$ is a bounded subset in Banach space $n\mathbb{R}_{\mathcal{F}(A)}$ and*

$$\chi(j(G)) \leq \beta(G) \leq 2\chi(j(G)).$$

Lemma 2.2 *[7] Let $\{\mu_n\}_{n\geq 1}$ be an integrally bounded sequence of measurable continuous $n\mathbb{R}_{\mathcal{F}(A)}$−valued functions. Then, for each $t \in J$, mapping $t \mapsto \beta\left(\{\mu_n(t)\}_{n\geq 1}\right)$ is measurable and*

$$\beta\left(\left\{\int_0^t \mu_n(s)ds\right\}_{n\geq 1}\right) \leq 2\int_0^t \beta\left(\{\mu_n(s)\}_{n\geq 1}\right) ds.$$

Proposition 2.7 *[2] Let $Y \subset C(J, n\mathbb{R}_{\mathcal{F}(A)})$ be a bounded set. Then, we have*

(i) $\beta(Y(t)) \leq \beta_C(Y)$ for all $t \in J$ and $Y(t) = \{\mu(t) : \mu \in Y\}$.
(ii) If Y is an equicontinuous set then $\beta_C(Y) = \sup_{t\in J} \beta(Y(t))$.

Definition 2.14 A continuous mapping $\mathcal{F} : n\mathbb{R}_{\mathcal{F}(A)} \to n\mathbb{R}_{\mathcal{F}_{(A)}}$ is called condensing with respect to the MNC β (or β −condensing for short) if for an arbitrary bounded set $\Omega \subset \mathbb{R}_{\mathcal{F}(A)}$, It implies from the relation $\beta(\mathcal{F}(\Omega)) \geq \beta(\Omega)$ that the relative compactness of Ω.

To end this section, we recall an extension version of Krasnoselskii's fixed point theorem, which plays the key role in the proof of controllability.

Theorem 2.3 (Krasnoselskii's fixed point theorem)*[13] Assume that X is a complete semilinear metric space having cancellation property and Ω is a nonempty, bounded, closed, convex subset of X. Let $\mathcal{P} : X \to X$ be an operator satisfying*

(i) $\mathcal{P} = \mathcal{P}_1 + \mathcal{P}_2$, where $\mathcal{P}_1$ is a completely continuous operator and $\mathcal{P}_2$ is a contraction,
(ii) $\mathcal{P}_1(\Omega) + \mathcal{P}_2(\Omega) \subset \Omega$.

Then $\mathcal{P}$ has at least a fixed point in Ω.

3 Complete Controllability

Consider the control problem under the Fréchet–Caputo differentiability.

$$\begin{cases} {}_F^C\mathcal{D}_{0+}^p \mu(t) = B\mu(t) +_A f(t, \mu(t)) +_A Eu(t), & t \in J = [0, b], \\ \mu(0) = \mu_0, \end{cases} \tag{4}$$

where

- ${}_F^C\mathcal{D}_{0+}^p \mu(t)$ is Fréchet Caputo of function $\mu : J \to \mathbb{R}^n_{\mathcal{F}_A}$,
- B is the generator of a C_0 –semigroup $\{T(t)\}_{t\geq 0}$ on $n\mathbb{R}_{\mathcal{F}(A)}$,
- the fuzzy-valued function $u \in L^1\left(J, n\mathbb{R}_{\mathcal{F}(A)}\right)$ is called control input where $L^1\left(J, \mathbb{R}_{\mathcal{F}(A)}\right)$ is space of integrable function in J.
- The linear operator E and the function $f : J \times n\mathbb{R}_{\mathcal{F}(A)} \to n\mathbb{R}_{\mathcal{F}(A)}$ satisfy some given conditions mentioned later.

Lemma 3.1 *Assume that the function $\mu : J \times n\mathbb{R}_{\mathcal{F}(A)} \to n\mathbb{R}_{\mathcal{F}(A)}$ is Fréchet differentiable on J and satisfies the problem (4). Then, for each $t \in J$, we have*

$$\mu(t) = S_p(t)\mu_0 +_A \int_0^t (t-s)^{p-1} T_p(t-s)(f(s, \mu(s)) +_A Eu(s))ds, \quad 0 \leq t \leq b. \tag{5}$$

Proof For each $t \in J$, denote

$$V(\lambda) := \mathcal{L}[\mu(t)] = \int_0^\infty e^{-\lambda t}\mu(t)dt,$$

$$\mathrm{G}(\lambda) := \mathcal{L}\left[f(t, \mu(t)) +_A Eu(t)\right] = \int_0^\infty e^{-\lambda t}\left[f(t, \mu(t)) +_A Eu(s)\right]dt.$$

By the assumption that μ is Fréchet differentiable on J and by using the property of Fréchet–Caputo fractional derivative, (4) can be transformed into the following form

$$\mu(t) = \mu_0 +_A \frac{1}{\Gamma(p)} \int_0^t (t-s)^{p-1}\left[B\mu(s) +_A f(s, \mu(s) +_A Eu(s))\right]ds.$$

By using fuzzy Laplace transform, we receive

$$\begin{aligned} V(\lambda) &= \int_0^\infty e^{-\lambda t}\mu_0 dt +_A \frac{1}{\Gamma(p)} \int_0^t (t-s)^{p-1}\left[BV(\lambda) +_A \mathrm{G}(\lambda)\right]ds \\ &= \frac{1}{\lambda}\mu_0 +_A \frac{1}{\lambda^p} BV(\lambda) +_A \frac{1}{\lambda^p}\mathrm{G}(\lambda), \end{aligned} \tag{6}$$

or equivalently,

$$\lambda^p V(\lambda) = \lambda^{p-1}\mu_0 +_A BV(\lambda) +_A \mathrm{G}(\lambda).$$

In addition, since B is an infinitesimal generator of the C_0 −semigroup $\{T(t)\}_{t\geq 0}$, we have

$$\left(\lambda^p \mathrm{Id}_A -_A B\right)^{-1} \mu(t) = R(\lambda^p, B)\mu(t) = \int_0^\infty e^{-\lambda^p \tau} T(\tau)\mu(t) d\tau.$$

Thus, the expression (6) becomes

$$\begin{aligned} V(\lambda) &= \lambda^{p-1}\int_0^\infty e^{-\lambda^p \nu} T(\nu)\mu_0 d\nu +_A \int_0^\infty e^{-s^p \nu} T(\nu)\mathrm{G}(\lambda) d\nu \\ &= V_1(\lambda) +_A V_2(\lambda). \end{aligned}$$

By changing variables $\nu = t'^p$, we have

$$V_1(\lambda) = \int_0^\infty p(\lambda t')^{p-1} e^{-(\lambda t')^p} T\left(t'^p\right)\mu_0 dt'.$$

Note that $e^{-(st')^p} = \int_0^\infty e^{-\lambda t'\theta}\psi_p(\theta) d\theta$. Thus,

$$\frac{d}{dt'}\left(e^{-(\lambda t')^p}\right) = \frac{d}{dt'}\left(\int_0^\infty e^{-\lambda t'\theta}\psi_p(\theta) d\theta\right) = -\lambda \int_0^\infty \theta e^{-\lambda t'\theta}\psi_p(\theta) d\theta.$$

Then, the integral $V_1(\lambda)$ can be rewritten as follows

$$V_1(\lambda) = \int_0^\infty -\frac{1}{\lambda}\frac{d}{dt'}\left(e^{-(\lambda t')^p}\right) T\left(t'^p\right)\mu_0 dt' = \int_0^\infty \int_0^\infty \theta\psi_p(\theta) e^{-\lambda t'\theta} T\left(t'^p\right)\mu_0 d\theta dt'.$$

Finally, by changing variable $t = t'\theta$, we obtain

$$V_1(\lambda) = \int_0^\infty e^{-\lambda t}\left(\int_0^\infty \psi_p(\theta) T\left(\frac{t^p}{\theta^p}\right)\mu_0 d\theta\right) dt. \tag{7}$$

Then we obtain

$$\begin{aligned} V_2(\lambda) &= \int_0^\infty \int_0^\infty p t^{p-1} e^{-(\lambda t)^p} T\left(t^p\right) e^{-\lambda\nu}[f(\nu, \mu(\nu))] d\nu dt \\ &= \int_0^\infty \int_0^\infty \int_0^\infty p\psi_p(\theta) e^{-\lambda(t^p+\nu^p)} T\left(\frac{t^p}{\theta^p}\right)\frac{t^{p-1}}{\theta^p}[f(\nu, \mu(\nu))] d\theta d\nu dt \\ &= \int_0^\infty p\psi_p(\theta)\left[\int_0^\infty e^{-\lambda t} T\left(\frac{t^p}{\theta^p}\right)\frac{t^{p-1}}{\theta^p} dt\right]\left[\int_0^\infty e^{\lambda\nu}[f(\nu, \mu(\nu))] d\nu\right] d\theta. \end{aligned}$$

Applying Laplace transform of correlated fuzzy functions, we obtain

$$V_2(\lambda) = \int_0^\infty p\psi_p(\theta)\mathcal{L}\left[T\left(\frac{t^p}{\theta^p}\right)\frac{t^{p-1}}{\theta^p}\right](\lambda)\mathcal{L}f(\lambda)dtd\theta$$
$$= \int_0^\infty p\psi_p(\theta)\mathcal{L}\left[\left(T\left(\frac{t^p}{\theta^p}\right)\frac{t^{p-1}}{\theta^p}\right) * f\right](\lambda)dtd\theta$$
$$= \int_0^\infty p\psi_p(\theta)\int_0^t\int_0^\infty e^{-\lambda t}T\left(\frac{(t-\nu)^p}{\theta^p}\right)\frac{(t-\nu)^{p-1}}{\theta^p}[f(\nu,\mu(\nu))]d\theta d\nu dt. \tag{8}$$

By combing (7) and (8), it yields

$$V(\lambda) = \int_0^\infty e^{-\lambda t}\left[\int_0^\infty \psi_p(\theta)T\left(\frac{t^p}{\theta^p}\right)\mu_0 d\theta\right]dt$$
$$+_A \int_0^\infty e^{-\lambda t}\left[p\int_0^t\int_0^\infty \psi_p(\theta)T\left(\frac{(t-\nu)^p}{\theta^p}\right)\frac{(t-\nu)^{p-1}}{\theta^p}[f(\nu,\mu(\nu))]d\theta d\nu\right]dt.$$

Finally, applying the inverse Laplace transform, we obtain

$$\mu(t) = \int_0^\infty \psi_p(\theta)T\left(\frac{t^p}{\theta^p}\right)\mu_0 d\theta +_A p\int_0^t\int_0^\infty \psi_p(\theta)T\left(\frac{(t-\nu)^p}{\theta^p}\right)\frac{(t-\nu)^{p-1}}{\theta^p}$$
$$[f(\nu,\mu(\nu))]d\theta d\nu$$
$$= \int_0^\infty \psi_p(\theta)T\left(t^p\theta\right)\mu_0 d\theta +_A p\int_0^t\int_0^\infty \theta(t-\nu)^{p-1}\psi_p(\theta)T\left((t-\nu)^p\theta\right)$$
$$[f(\nu,\mu(\nu))]d\theta d\nu$$
$$= S_p(t)\mu_0 +_A \int_0^t (t-s)^{p-1}T_p(t-s)\left[f(s,\mu(s) +_A Eu(s)\right]ds.$$

The proof is completed. □

Definition 3.1 A continuous function $\mu : J \times C\left(J, n\mathbb{R}_{\mathcal{F}(A)}\right) \to n\mathbb{R}_{\mathcal{F}(A)}$ satisfying the integral equation (5) is said to be an integral fuzzy solution of the problem (4).

For $\mu, \overline{\mu} \in C(J, n\mathbb{R}_{\mathcal{F}(A)})$, we denote $\mathcal{H}(\mu,\overline{\mu}) = \sup_{t\in J} d_A(\mu(t),\overline{\mu}(t))$.

Firstly, we need the following hypothese

(P) The operator $\Pi : L^1\left(J, n\mathbb{R}_{\mathcal{F}(A)}\right) \to n\mathbb{R}_{\mathcal{F}(A)}$, given by

$$\Pi(u) = \int_0^b (b-s)^{p-1}T_p(b-s)Eu(s)ds,$$

is bijective and there exists $\mathsf{K}_3 \in \left(0, \frac{\mathsf{K}_1\mathsf{K}_2 b^p}{\Gamma(p+1)}\right)$ such that $\|\Pi(u)\|_A \geq \mathsf{K}_3\mathcal{H}(\mu,\hat{0})$ for all $u \in L^1\left(J, n\mathbb{R}_{\mathcal{F}(A)}\right)$.

Remark 3.1 Since E is a linear mapping, we get the linearity of the operator Π. In addition, the boundedness of the operator Π is implied from following estimation

$$\begin{aligned}\|\Pi(u)\|_A &= \left\| \int_0^b (b-s)^{p-1} T_p(t-s) E\mu(s) ds \right\|_A \\ &\le \int_0^b (b-s)^{p-1} d_A\left(T_p(t-s)Eu(s), \hat{0}\right) ds \\ &\le \int_0^b (b-s)^{p-1} \frac{\mathsf{K}_1\mathsf{K}_2 p}{\Gamma(p+1)} d_A\left(u(s), \hat{0}\right) ds \\ &\le \frac{\mathsf{K}_1\mathsf{K}_2 b^p}{\Gamma(p+1)} \mathcal{H}(u, \hat{0}).\end{aligned}$$

Moreover, since the assumption that Π is a bijection satisfying $\|\Pi(u)\|_A \ge \mathsf{K}_3\mathcal{H}(u, \hat{0})$ for all $\mu \in L^1\left(J, \mathbb{R}_{\mathcal{F}(A)}\right)$, the inverse operator Π^{-1} exists and $\|\Pi^{-1}\|_{op} \le \mathsf{K}_3$.

Definition 3.2 The fuzzy fractional evolution problem (4) is said to be completely controllable on J if for any initial state μ_0 and final state μ_1, there exists a control input $u \in L^1\left(J, n\mathbb{R}_{\mathcal{F}(A)}\right)$ such that $\mu(b) = \mu_1$, where $\mu(\cdot)$ is a integral solution of (4).

We define an operator $\mathscr{P} : C(J, n\mathbb{R}_{\mathcal{F}(A)}) \to C(J, n\mathbb{R}_{\mathcal{F}(A)})$ by

$$\begin{aligned}\mathscr{P}[\mu](t) = S_p(t)\mu_0 &+_A \int_0^t (t-s)^{p-1} T_p(t-s) f(s, \mu(s)) ds \\ &+_A \int_0^t (t-s)^{p-1} T_p(t-s) E u_\mu(s) ds,\end{aligned}$$

where the control input $u_\mu(.)$ is given by

$$u_\mu(t) = \Pi^{-1}\left[\mu_1 -_A \left(S_p(b)\mu_0 +_A \int_0^b (b-s)^{p-1} T_p(b-s) f(s, \mu(s)) ds\right)\right](t).$$

Now, we will prove that the operator $\mathscr{P}$ has a unique fixed point $\mu \in C\left(J, n\mathbb{R}_{\mathcal{F}(A)}\right)$.

Theorem 3.1 *Assume that following conditions are satisfied.*

(F1) *For each $t \in J$, the mapping $f(t, \cdot) : J \times n\mathbb{R}_{\mathcal{F}(A)} \to n\mathbb{R}_{\mathcal{F}(A)}$ is a continuous fuzzy-valued function, $f(t, \hat{0}) = 0$ and for each $\mu \in n\mathbb{R}_{\mathcal{F}(A)}$, the mapping $f(., \mu) : J \to n\mathbb{R}_{\mathcal{F}(A)}$ is measurable.*

(F2) *For arbitrary $\mu, \overline{\mu} \in B_\rho := \{\mu \in C(J, n\mathbb{R}_{\mathcal{F}(A)}) : \mathcal{H}(\mu, \hat{0}) \le \rho\}$, then there exists a constant $a_f > 0$ such that*

$$d_A(f(t, \mu(t)), f(t, \overline{\mu}(t))) \le a_f d_A(\mu, \overline{\mu}).$$

(C) There exists a constant $\mathsf{K}_2 > 0$ *such that* $\|E\|_{op} \leq \mathsf{K}_2$.

Then the problem (4) is completely controllable on J*. Furthermore, the control input is unique.*

Proof Let $T_1 > 0$, we define

$$\mathcal{S}(1, T_1) = \{\mu \in C(J, n\mathbb{R}_{\mathcal{F}(A)}) : \mathcal{H}(\mu, \mu_0) \leq 1 \text{ for } t \in [0, T_1]\}.$$

Then $\mathcal{S}(1, T_1) \subseteq C(J, n\mathbb{R}_{\mathcal{F}(A)})$ is a closed convex subset of $C(J, \mathbb{R}^n_{\mathcal{F}_A})$.

According to (**F1**) and (**F2**), it implies that $f(t, \mu(t))$ is a measurable function on $[0, T_1]$. Let $\mu \in \mathcal{S}(1, T_1)$. Consider $\mathscr{P} : \mathcal{S}(1, T_1) \to C(J, n\mathbb{R}_{\mathcal{F}(A)})$

$$\begin{aligned}\mathscr{P}[\mu](t) = S_p(t)\mu_0 +_A \int_0^t (t-s)^{p-1} T_p(t-s) f(s, \mu(s)) ds \\ +_A \int_0^t (t-s)^{p-1} T_p(t-s) E u_\mu(s) ds.\end{aligned}$$

For $t \in [0, T_1]$, from Proposition 2.6, we have the following inequality

$$\int_0^t (t-s)^{p-1} \|T_p(t-s) f(s, \mu(s))\|_A ds \leq \frac{p\mathsf{K}_1}{\Gamma(p+1)} \int_0^t (t-s)^p \|f(s, \mu(s))\|_A ds.$$

We have $\mathcal{H}(\mu, \hat{0}) \leq \mathcal{H}(\mu, \mu_0) + \mathcal{H}(\mu_0, \hat{0})$. Since $\mu \in \mathcal{S}(1, T_1), \mathcal{H}(\mu, \mu_0) \leq 1$. It leads to $\mathcal{H}(\mu, \hat{0}) \leq \rho = 1 + \mathcal{H}(\mu_0, \hat{0})$. Using (**F2**), we get

$$\|f(s, \mu(s))\|_A \leq a_f \|\mu(s)\|_A.$$

Therefore,

$$\begin{aligned}\int_0^t (t-s)^{p-1} \|T_p(t-s) f(s, \mu(s))\|_A ds &\leq \int_0^t (t-s)^{p-1} \frac{\mathsf{K}_1 p}{\Gamma(1+p)} \|f(s, \mu(s))\|_A ds \\ &\leq \frac{\mathsf{K}_1 p}{\Gamma(1+p)} a_f \|\mu(s)\|_A \int_0^t (t-s)^{p-1} ds \\ &\leq \frac{\mathsf{K}_1 t^p a_f}{\Gamma(1+p)} \|\mu(t)\|_A.\end{aligned}$$

By using the hypothesis (**F2**), we have

$$\begin{aligned}\|u_\mu(t)\|_A &\leq \|\Pi^{-1}\|_{op} \left(\|\mu_1\|_A + d_A\left(S_p(b)\mu_0, \hat{0}\right) + \int_0^b (b-s)^{p-1} \|T_q(b-s) f(s, \mu(s))\|_A ds \right) \\ &\leq \mathsf{K}_3 \left(\|\mu_1\|_A + \mathsf{K}_1 \|\mu_0\|_A + \frac{\mathsf{K}_1 b^p a_f}{\Gamma(1+p)} \|\mu(t)\|_A \right).\end{aligned}$$

This follows

$$\mathcal{H}(u_\mu, \hat{0}) \le \mathsf{K}_3 \left(\|\mu_1\|_A + \mathsf{K}_1 \|\mu_0\|_A + \frac{\mathsf{K}_1 a_f b^p}{\Gamma(p+1)} \mathcal{H}(\mu, \hat{0}) \right).$$

Since $\mathcal{H}(\mu, \hat{0}) \le \rho$, it leads to

$$\mathcal{H}(u_\mu, \hat{0}) \le \mathsf{K}_3 \left(\|\mu_1\|_A + \mathsf{K}_1 \|\mu_0\|_A + \frac{\mathsf{K}_1 a_f^p \rho}{\Gamma(p+1)} \right). \tag{9}$$

From Remark 3.1, we get

$$\begin{aligned} \int_0^t (t-s)^{p-1} \|T_p(t-s) E u_\mu(s)\|_A ds &\le \int_0^t (t-s)^{p-1} \frac{p\mathsf{K}_1}{\Gamma(p+1)} \|E u_\mu(s)\|_A \\ &\le \frac{p\mathsf{K}_1\mathsf{K}_2}{\Gamma(p+1)} \|u_\mu\|_A \frac{t^p}{p} \le \frac{\mathsf{K}_1\mathsf{K}_2 t^p}{\Gamma(p+1)} \mathcal{H}(u_\mu, \hat{0}). \end{aligned} \tag{10}$$

From (9) and (10), we have

$$\begin{aligned} \int_0^t (t-s)^{p-1} \|T_p(t-s) E u_\mu(s)\|_A ds \le & \frac{\mathsf{K}_1\mathsf{K}_2 t^p}{\Gamma(p+1)} \\ & \left[\mathsf{K}_3 \left(\|\mu_1\|_A + \mathsf{K}_1 \|\mu_0\|_A + \frac{\mathsf{K}_1 a_f b^p}{\Gamma(p+1)} \rho \right) \right]. \end{aligned}$$

For $t \in [0, T_1]$, let $a = \frac{\mathsf{K}_1\mathsf{K}_2\mathsf{K}_3}{\Gamma(p+1)}$. It is easy to obtain the following inequality

$$\begin{aligned} & d_A(\mathscr{P}[\mu](t), \mu_0) \\ & \le d_A(S_p(t)\mu_0, \mu_0) + \int_0^t (t-s)^{p-1} \|T_p(t-s) f(s, \mu(s))\|_A \\ & \quad + \int_0^t (t-s)^{p-1} \|T_p(t-s) E u_\mu(s)\|_A ds \\ & \le d_A(S_p(t)\mu_0, \mu_0) + \frac{t^p \mathsf{K}_1 a_f}{\Gamma(p+1)} + a t^p \left(\|\mu_1\|_A + \mathsf{K}_1 \|\mu_0\|_A + \frac{\mathsf{K}_1 a_f b^p \rho}{\Gamma(p+1)} \right). \end{aligned} \tag{11}$$

Since $\{S_p(t), t \ge 0\}$ is a continuous operator in $\mathbb{R}_{\mathcal{F}(A)}$, we can choose $\epsilon = \frac{1}{3}$ such that

$$\|S_p(t)\mu_0 - \mu_0\|_A \le \frac{1}{3} \tag{12}$$

Let

$$T_{11} = \min\left\{\frac{1}{3}, \left(\frac{\mathsf{K}_1 a_f}{\Gamma(p+1)}\right)^{\frac{1}{p}}, \left[a\left(\|\mu_1\|_A + \mathsf{K}_1\|\mu_0\|_A + \frac{\mathsf{K}_1 a_f b^p \rho}{\Gamma(p+1)}\right)\right]^{\frac{1}{p}}\right\}.$$

Then for all $t \leq T_{11}$, from (11) and (12) we obtain

$$d_A(\mathscr{P}[\mu](t), \mu_0) \leq 1.$$

It implies $\mathscr{P}(S(1, T_1)) \subseteq \mathscr{P}(1, T_1)$. Let $\mu, \overline{\mu} \in \mathcal{S}(1, T_1)$ and **(F2)**, we get

$$\begin{aligned} d_A\left(u_\mu(t), u_{\overline{\mu}}(t)\right) &\leq \frac{p\mathsf{K}_1\mathsf{K}_3}{\Gamma(p+1)} \int_0^b (b-s)^{p-1} a_f d_A(\mu(s), \overline{\mu}(s))ds \\ &\leq \frac{\mathsf{K}_1\mathsf{K}_3 b^p a_f}{\Gamma(p+1)} d_A(\mu(s), \overline{\mu}(s)). \end{aligned}$$

Taking supremum both sides, we obtain

$$\mathcal{H}(\left(u_\mu(t), u_{\overline{\mu}}(t)\right) \leq \frac{\mathsf{K}_1\mathsf{K}_3 b^p a_f}{\Gamma(p+1)} \mathcal{H}(\mu, \overline{\mu}).$$

It implies

$$\begin{aligned} \int_0^t (t-s)^{p-1} \|T_p(t-s) d_A(E\mu_1(s), Eu_{\mu_2}(s))ds &\leq \int_0^t (t-s)^{p-1} d_A\left(T_p(t-s)Eu_\mu(s),\right. \\ &\qquad \left. T_p(t-s)Eu_{\overline{\mu}}(s)\right) ds \\ &\leq \frac{\mathsf{K}_1\mathsf{K}_2 p}{\Gamma(p+1)} \int_0^t (t-s)^{p-1} d_A(u_\mu(s), u_{\overline{\mu}}(s))ds \\ &\leq \frac{\mathsf{K}_1\mathsf{K}_2}{\Gamma(p+1)} \frac{\mathsf{K}_1\mathsf{K}_3 b^p a_f}{\Gamma(p+1)} \mathcal{H}(\mu, \overline{\mu}) t^p \\ &\leq a \frac{\mathsf{K}_1 t^p a_f b^p}{\Gamma(p+1)} \mathcal{H}(\mu, \overline{\mu}). \end{aligned}$$

On the other hand, we have

$$\begin{aligned} \int_0^t (t-s)^{p-1} \|T_p(t-s) d_A(f(s, \mu(s)), f(s, \overline{\mu}(s)))\|_A ds &\leq \frac{p\mathsf{K}_1}{\Gamma(p+1)} \int_0^t (t-s)^p d_A(f(s, \mu(s)), \\ &\qquad f(s, \overline{\mu}(s)))ds \\ &\leq \frac{p\mathsf{K}_1}{\Gamma(p+1)} \mathcal{H}(\mu, \overline{\mu}) \int_0^t (t-s)^p a_f(t) ds \\ &\leq \frac{t^p \mathsf{K}_1 a_f}{\Gamma(p+1)} \mathcal{H}(\mu, \overline{\mu}). \end{aligned}$$

For $t \in [0, T_1]$, let $\mu, \overline{\mu} \in \Omega(1, T_1)$. Using Proposition 2.6(ii) and **(F2)**, we get

$$\begin{aligned}
& d_A(\mathscr{P}[\mu_1](t), \mathscr{P}[\mu_2](t)) \\
& \leq \int_0^t (t-s)^{p-1} \| T_p(t-s) d_A(f(s, \mu_1(s)), f(s, \mu_2(s))) \|_A ds \\
& \quad + \int_0^t (t-s)^{p-1} \| T_p(t-s) d_A(Cu_{\mu_1}(s), Cu_{\mu_2}(s)) \|_A ds \\
& \leq a \frac{\mathsf{K}_1 t^p a_f b^p}{\Gamma(p+1)} \mathcal{H}(\mu, \overline{\mu}) + \frac{t^p \mathsf{K}_1 a_f}{\Gamma(p+1)} \mathcal{H}(\mu, \overline{\mu}).
\end{aligned}$$

Let $T_{12} = \frac{1}{2}\left(a\frac{\mathsf{K}_1 a_f b^p}{\Gamma(p+1)} + \frac{\mathsf{K}_1 a_f}{\Gamma(p+1)}\right)^{\frac{1}{p}}$ and $T_1 = \min\{T_{11}, T_{12}\}$. Then $\mathscr{P}$ is a contraction mapping on $S(1, T_1)$. It follows from the contraction mapping principle that $\mathscr{P}$ has a unique fixed point $\mu \in \mathcal{S}(1, T_1)$. Let

$$T_{21} = T_1 + T_{11}, \qquad T_{22} = T_1 + T_{12}, \qquad \Delta T = \min\{T_{21} - T_1, T_{12} > 0\}.$$

Similarly, one can point out that problem (4) has a unique control u on $[0, \Delta T]$. Repeating the above procedures on each interval $[\Delta T, 2\Delta T], [2\Delta T, 3\Delta T], ..$ we immediately obtain the unique existence of control u for problem (4) on J. □

Example 3.1 Consider the following problem

$$\begin{cases} {}^{C}_{F}\mathcal{D}^{1/3}_{0^+} \mu(t) = \frac{e^{-\frac{t}{3}}}{3} \mu(t) +_A \frac{2t+1}{3} \mu(t) +_A \frac{1}{\pi} u \\ \mu(0) = 0 \end{cases} \tag{13}$$

where A is a given non-symmetric fuzzy number, $t \in I = [0, 1], 0 < s < t$. We can see that

$$f(t, \mu(s)) = \frac{2t+1}{3} \mu(t).$$

Firstly, the $B = e^{-t/3} I_{d_A}$ generates a semigroup $\{T(t)\}_{t \geq 0}$ and $\|T(t)\|_{op} \leq 1$. Therefore $\mathsf{K}_1 = \frac{1}{3}$ is the number satisfying hypothesis **(F1)**. For $t \in [0, 1]$, we have

$$\|f(t, \mu(t))\|_A \leq \frac{2t+1}{3} \|\mu\|_A \leq \|\mu\|_A$$

and

$$d_A(f(t, \mu(t)), f(t, \overline{\mu}(t)) = \frac{2t+1}{3} d_A(\mu(t), \overline{\mu}(t)) \leq d_A(\mu(t), \overline{\mu}(t)),$$

for all $\mu, \overline{\mu} \in C(J, n\mathbb{R}_{\mathcal{F}(A)})$. It implies that (**F2**) is satisfied with $a_f = 1$. Moreover, one gets $E = \frac{1}{\pi} I_{d_A}$ then

$$\|E\|_{op} \leq 1.$$

And so (**C**) are satisfied with $K_2 = 1$. For $u \in L^1(J, \mathbb{R}^n_{\mathcal{F}}(A))$, the linear operator Π is given by

$$\Pi u = \frac{1}{\pi} \int_0^t (t-s)^{\frac{2}{3}} T_p(t-s) u(s) ds.$$

Let $\mathrm{K}_3 = \frac{1}{4} \in \left(0, \frac{1}{\Gamma(\frac{2}{3})}\right)$, we have $\|\Pi u\|_A \leq \frac{1}{4}\mathcal{H}(u, \hat{0})$. By applying Theorem 3.1, we can conclude that the fractional differential system (13) is completely controllable on $I = [0, 1]$.

Remark 3.2 In the Theorem 3.1, we have used the fixed point theorem in Bannach space to prove the existence of only a control function u to ensure that the given problem iscompletely controllable. However, in reality, the problem does not always need a unique control function. By replacing the Lipschitz condition of the right-hand side by the sublinear condition of f, the complete control of the problem remains. However, this control variable is not unique.

We consider following conditions

(F3) There exist $p_1 \in (0, p)$ and $g_f(\cdot) \in L^{\frac{1}{p_1}}(J, \mathbb{R}^+)$ such that

$$d_A(f(t, \mu(t)), f(t, \overline{\mu}(t))) \leq g_f(t)\mathcal{H}(\mu, \overline{\mu})$$

for all $\mu, \overline{\mu} \in C(J, n\mathbb{R}_{\mathcal{F}(A)})$ and $t \in J$.

(F4) For each bounded set $\Omega \subset n\mathbb{R}_{\mathcal{F}(A)}$, there is a $p_1 \in (0, p)$ and $\mathfrak{X} : J \times J \to \mathbb{R}^+$ such that $\mathfrak{X}(t, \cdot) \in L^{\frac{1}{p_1}}([0, t), \mathbb{R}^+)$ and

$$\beta(T_p(t-s) f(s, \Omega)) \leq \mathfrak{X}(t, s)\beta(\Omega)$$

and

$$\|\mathfrak{X}(t, \cdot)\|_{L^{\frac{1}{p_1}}(J, \mathbb{R}^+)} < \frac{1}{8}\left[\left(\frac{p-p_1}{1-p_1}\right) b^{\frac{1-p_1}{p-p_1}}\right]^{1-p_1} \tag{14}$$

for all $t \in [0, b]$, $s \in [0, t]$, where $\beta(\cdot)$ is a measure of noncompactness on $n\mathbb{R}_{\mathcal{F}(A)}$. We recall that the operator $\mathscr{P} : C(J, n\mathbb{R}_{\mathcal{F}(A)}) \to C(J, n\mathbb{R}_{\mathcal{F}(A)})$ by

$$\mathscr{P}[\mu](t) = S_p(t)\mu_0 +_A \int_0^t (t-s)^{p-1} T_p(t-s) f(s,\mu(s))ds$$
$$+_A \int_0^t (t-s)^{p-1} T_p(t-s) E u_\mu(s)ds,$$

where the control input $u_\mu(t)$ is given by

$$u_\mu(t) = \Pi^{-1}\left[\mu_1 -_A \left(S_p(b)\mu_0 +_A \int_0^b (b-s)^{p-1} T_p(b-s) f(s,\mu(s))ds\right)\right]$$
$$(t)\,,\, t \in [0,b].$$

Now, we will prove that the operator $\mathscr{P}$ has at least one fixed point $\overline{\mu} \in C\left(J, \mathbb{R}_{\mathcal{F}(A)}\right)$ by the virtue of Krasnoselskii's fixed point theorem. To do this, we represent $\mathscr{P}$ as a sum of two following operators

$$\mathscr{P}_1[\mu](t) = S_p(t)\mu_0 +_A \int_0^t (t-s)^{p-1} T_p(t-s) E u_\mu(s)ds$$
$$\mathscr{P}_2[\mu](t) = \int_0^t (t-s)^{p-1} T_p(t-s) f(s,\mu(s))ds. \tag{15}$$

For $r > 0$, $\Omega_r = \{\mu \in C\left(J, n\mathbb{R}_{\mathcal{F}(A)}\right) : \mathcal{H}(\mu, \hat{0}) \le r\}$.

Lemma 3.2 *The operator $\mathscr{P}_2$ defined by (15) is continuous on Ω_r* □

Proof To prove the continuity of $\mathscr{P}_2$, we consider a sequence $\{\mu_n\} \subset \Omega_r$ satisfying $\mu_n \to \mu \in \Omega_r$. In addition, for simplicity in representation, we denote

$$\mathsf{F}_n(t) := f(t,\mu_n(t)) \quad \text{and} \quad \mathsf{F}(t) := f(t,\mu(t)).$$

It is inferred from the hypothesis **(F1)** that for all $0 \le s \le t \le b$, we have

$$(t-s)^{p-1}\mathsf{F}_n(s) \to (t-s)^{p-1}\mathsf{F}(s) \quad \text{as } n \to \infty.$$

Next, by using the hypotheses **(F1)**, **(F3)** and Lebesgue's dominated convergence theorem, we obtain

$$d_A\left(\mathscr{P}_2[\mu_n](t), \mathscr{P}_2[\mu](t)\right) \le \int_0^t (t-s)^{p-1} d_A\left(T_p(t-s)\mathsf{F}_n(s), T_p(t-s)\mathsf{F}(s)\right)ds$$
$$\le \frac{p\mathsf{K}_1}{\Gamma(p+1)} \int_0^t (t-s)^{p-1} d_A(\mathsf{F}_n(s), \mathsf{F}(s))ds \to 0$$
$$\text{as } n \to \infty.$$

It means that the operator $\mathscr{P}_2$ is continuous on Ω_r. □

Lemma 3.3 *Under assumption that the hypotheses* ***(F1)***,***(F3)***,***(F4)*** *and* ***(P)*** *hold then $\mathscr{P}_2$ is compact, provided that*

$$\|\mathfrak{X}(t,\cdot)\|_{L^{\frac{1}{p_1}}(J,\mathbb{R}^+)} < \frac{1}{8}\left[\left(\frac{p-p_1}{1-p_1}\right)b^{\frac{1-p_1}{p-p_1}}\right]^{1-p_1}. \tag{16}$$

Proof To prove that $\mathscr{P}_2$ is compact, we prove that $\mathscr{P}_2$ satisfies two following conditions

(i) $\mathscr{P}_2$ is equicontinuous on Ω_r,
(ii) for each $t \in J$, $\{\mathscr{P}_2[\mu](t) : \mu \in\}$ is relatively compact in $n\mathbb{R}_{\mathcal{F}(A)}$.

Step 1: To prove $\mathscr{P}_2$ is equicontinuous on Ω_r, we prove that for each $\mu \in \Omega_r, t \in J$, $d_A(\mathscr{P}_2[\mu](t+h), \mathscr{P}_2[\mu](t)) \to 0$ as $h \to 0$. Let $\mu \in \Omega_r$ be arbitrary and denote $f(t, \mu_t, \mu(t))$ by $\mathsf{F}(t)$ for simplicity. Then, for each $t \in J$ and $0 < \varepsilon < h$ such that $t-\varepsilon, t+h \in J$, we have

$$\begin{aligned}
d_A(\mathscr{P}_2[\mu](t+h), \mathscr{P}_2[\mu](t)) \le& \int_t^{t+h} (t+h-s)^{p-1} d_A\left(T_p(t+h-s)\mathsf{F}(s), \hat{0}\right) ds \\
&+ \int_{t-\varepsilon}^{t} (t+h-s)^{p-1} d_A\left(T_p(t+h-s)\mathsf{F}(s), T_p(t-s)\mathsf{F}(s)\right) ds \\
&+ \int_{t-\varepsilon}^{t} \left[(t-s)^{p-1} - (t+h-s)^{p-1}\right] d_A\left(T_p(t-s)\mathsf{F}(s), \hat{0}\right) ds \\
&+ \int_0^{t-\varepsilon} \left[(t-s)^{p-1} - (t+h-s)^{p-1}\right] d_A\left(T_p(t-s)\mathsf{F}(s), \hat{0}\right) ds \\
=& \mathrm{J}_1 + \mathrm{J}_2 + \mathrm{J}_3 + \mathrm{J}_4 + \mathrm{J}_5.
\end{aligned}$$

Now, we will prove that for all $i = \overline{1,5}$, the integrals J_i tends to 0 as $h \to 0$.

Indeed, we have

$$\begin{aligned}
\mathrm{J}_1 =& \int_t^{t+h} (t+h-s)^{p-1} d_A\left(T_p(t+h-s)\mathsf{F}(s), \hat{0}\right) ds \\
\le& \frac{\mathsf{K}_1 p}{\Gamma(p+1)} \int_t^{t+h} (t+h-s)^{p-1} g_f(s)\mathcal{H}(\mu, \hat{0}) ds \\
\le& \frac{\mathsf{K}_1 pR}{\Gamma(p+1)} \left(\int_t^{t+h} (t+h-s)^{p-1} g_f(s) ds\right) \\
\le& \frac{\mathsf{K}_1 pR}{\Gamma(p+1)} \left[\left(\frac{1-p_1}{p-p_1}\right) h^{\frac{p-p_1}{1-p_1}}\right]^{1-p_1} \|g_f\|_{L^{\frac{1}{p_1}}(J,\mathbb{R}^+)} \to 0 \quad \text{as } h \to 0.
\end{aligned}$$

Then

$$
\begin{aligned}
d_A\left(T_p(t+h-s)\mathsf{F}(s), T_p(t-s)\mathsf{F}(s)\right) &\le d_A\left(T_p(t+h-s)\mathsf{F}(s), \hat{0}\right) \\
&\quad + d_A\left(T_p(t-s)\mathsf{F}(s), \hat{0}\right) \\
&\le \frac{2\mathsf{K}_1 p}{\Gamma(p+1)} g_f(s)\mathcal{H}(\mu, \hat{0}) \\
&\le \frac{2\mathsf{K}_1 pr}{\Gamma(p+1)} g_f.
\end{aligned}
$$

By using the inequality $|a^\eta - b^\eta| \le (b-a)^\eta$ for $\eta \in (0, 1]$ and $0 < a \le b$. We deduce that

$$
\begin{aligned}
\mathrm{J}_2 &= \int_{t-\varepsilon}^{t} (t+h-s)^{p-1} d_A\left(T_p(t+h-s)\mathsf{F}(s), T_p(t-s)\mathsf{F}(s)\right) ds \\
&\le \frac{2\mathsf{K}_1 pR}{\Gamma(p+1)} \left(\int_{t-\varepsilon}^{t} (t+h-s)^{p-1} g_f(s) ds\right) \\
&\le \frac{2\mathsf{K}_1 pR}{\Gamma(p+1)} \left[\left(\frac{1-p_1}{p-p_1}\right) \left|(h+\varepsilon)^p - h^p\right|\right]^{1-p_1} \|g_f\|_{L^{\frac{1}{p_1}}(J,\mathbb{R}^+)} \\
&\le \frac{2\mathsf{K}_1 pR}{\Gamma(p+1)} \left[\left(\frac{1-p_1}{p-p_1}\right) \varepsilon^p\right]^{1-p_1} \|g_f\|_{L^{\frac{1}{p_1}}(J,\mathbb{R}^+)} \to 0 \quad \text{as } h \to 0.
\end{aligned}
$$

For $0 < p_1 < p < 1$, set $\Upsilon = \frac{p-1}{1-p_1} \in (-1, 0)$. Then, by applying the following estimation

$$
\begin{aligned}
\int_{t-\varepsilon}^{t} \left[(t-s)^{p-1} - (t+h-s)^{p-1}\right] g_f(s) ds &\le \left(\int_{t-\varepsilon}^{t} \left[(t-s)^{\Upsilon} - (t+h-s)^{\Upsilon}\right] ds\right)^{1-p_1} \\
&\quad \|g_f\|_{L^{\frac{1}{p_1}}(J,\mathbb{R}^+)} \\
&\le \frac{\left[\eta^{\Upsilon+1} + h^{\Upsilon+1} - (h+\varepsilon)^{\Upsilon+1}\right]^{1-p_1}}{(\Upsilon+1)^{1-p_1}} \|g_f\|_{L^{\frac{1}{p_1}}(J,\mathbb{R}^+)} \\
&\le \left(\frac{(2h)^{(1+\Upsilon)}}{(\Upsilon+1)}\right)^{1-p_1} \|g_f\|_{L^{\frac{1}{p_1}}(J,\mathbb{R}^+)}.
\end{aligned}
$$

We receive the following estimation for J_3 as follows

$$
\begin{aligned}
\mathrm{J}_3 &= \int_0^{t-\varepsilon} \left[(t-s)^{p-1} - (t+h-s)^{p-1}\right] d_A\left(T_p(t-s)\mathsf{F}(s), \hat{0}\right) ds \\
&\le \frac{\mathsf{K}_1 p}{\Gamma(p+1)} \left(\int_{t-\eta}^{t} \left[(t-s)^{p-1} - (t+h-s)^{p-1}\right] g_f(s) ds\right) \mathcal{H}(\mu, \hat{0}) \\
&\le \frac{\mathsf{K}_1 pR}{\Gamma(p+1)} \left(\frac{(2h)^{(1+\Upsilon)}}{\Upsilon+1}\right)^{1-p_1} \|g_f\|_{L^{\frac{1}{p_1}}(J,\mathbb{R}^+)} \to 0 \quad \text{as } h \to 0.
\end{aligned}
$$

From Proposition 2.6 (ii), one has

$$\begin{aligned}
J_4 &\le \int_0^{t-\varepsilon} (t+h-s)^{p-1} \frac{K_1 p}{\Gamma(p+1)} \left\| T\left((t+h-s)^p\varepsilon - (t-s)^p\varepsilon\right) -_A \mathrm{Id}_A \right\|_{op} d_A\left(F(s), \hat{0}\right) ds \\
&\le \frac{K_1 p}{\Gamma(p+1)} \left\| T\left((t+h-s)^p\varepsilon - (t-s)^p\varepsilon\right) -_A \mathrm{Id}_A \right\|_{op} \int_0^{t-\varepsilon} (t+h-s)^{p-1} g_f(s)\mathcal{H}_{0,b}(\mu, \hat{0}) ds \\
&\le \frac{K_1 pR}{\Gamma(p+1)} \| T\left((t+h-s)^p\varepsilon - (t-s)^p\varepsilon\right) -_A \mathrm{Id}_A \|_{op} \left(\frac{(t+h)^{\Upsilon+1} - (h+\varepsilon)^{\Upsilon+1}}{\Upsilon+1} \right)^{1-p_1} \|g_f\|_{L^{\frac{1}{p_1}}(J,\mathbb{R}^+)} \\
&\le \frac{K_1 pR}{\Gamma(p+1)} \|g_f\|_{L^{\frac{1}{p_1}}(J,\mathbb{R}^+)} \left(\frac{(b+h)^{\Upsilon+1}}{\Upsilon+1} \right)^{1-p_1} \left\| T\left((t+\varepsilon-\gamma)^p\varepsilon - (t-\gamma)^p\varepsilon\right) -_A \mathrm{Id}_A \right\|_{op}
\end{aligned}$$

tends to 0 as $h \to 0$. Similarly, we also have

$$\int_0^{t-\varepsilon} \left[(t-s)^{p-1} - (t+h-s)^{p-1}\right] g_f(s) ds \le \left(\frac{(2h)^{1+\Upsilon}}{(\Upsilon+1)} \right)^{1-p_1} \|g_f\|_{L^{\frac{1}{p_1}}(J,\mathbb{R}^+)} \to 0 \quad \text{as } h \to 0.$$

Then

$$\begin{aligned}
J_5 &\le \frac{K_1 pr}{\Gamma(p+1)} \left(\int_0^{t-\varepsilon} \left[(t-s)^{p-1} - (t+h-s)^{p-1}\right] g_f(s) ds \right) \\
&\le \frac{K_1 pr}{\Gamma(p+1)} \left(\frac{2h^{1+\Upsilon}}{(\Upsilon+1)} \right)^{1-p_1} \|g_f\|_{L^{\frac{1}{p_1}}(J,\mathbb{R}^+)} \to 0 \quad \text{as } h \to 0.
\end{aligned}$$

As a consequence, these estimations imply that for all $\mu \in \Omega_r$, $d_A\left(\mathscr{P}_2[\mu](t+h), \mathscr{P}_2[\mu](t)\right) \to 0$ as $h \to 0$. Hence, we can conclude that the set $\mathscr{P}_2(\Omega_r)$ is equicontinuous on Ω_r.

Step 2: Recall that β, β_C are the Hausdorff measures of noncompactness in $n\mathbb{R}_{\mathcal{F}(A)}$ and $C(J, nn\mathbb{R}_{\mathcal{F}(A)})$, respectively. We will point out that $\mathscr{P}_2$ is condensing operator. Indeed, let $\Omega \subset \Omega_r$ be a bounded, equicontinuous set such that

$$\beta_C(\Omega) \le \beta_C\left(\mathscr{P}(\Omega)\right).$$

By applying Proposition 2.7 in [3] and Theorem 2.2 in [7], it follows that for each $t \in J$, for all $\epsilon > 0$, there has a sequence $\{\mu_n\}_{n \ge 1} \subset \Omega$ such that

$$\begin{aligned}\beta(\mathscr{P}_2(\Omega)(t)) &\le 2\chi(j(\mathscr{P}_2(\Omega)(t))) \le 4\chi\left(j(\{\mathscr{P}_2[\mu_n](t)\}_{n\ge1})\right)+\epsilon\\ &\le 4\beta\left(\left\{\int_0^t (t-s)^{p-1}T_p(t-s)f(s,\mu_n(s))ds\right\}_{n\ge1}\right)+\epsilon\\ &\le 8\int_0^t (t-s)^{p-1}\beta\left(\{T_p(t-s)f(s,\mu_n(s))\}_{n\ge1}\right)ds+\epsilon\\ &\le 8\int_0^t (t-s)^{p-1}\mathfrak{X}(t,s)\beta\{\mu_n(s)\}_{n\ge1}ds+\epsilon\\ &\le 8\int_0^t (t-s)^{p-1}\mathfrak{X}(t,s)\beta\left(\Omega(s)\right)ds+\epsilon\\ &\le 8\left[\left(\frac{1-p_1}{p-p_1}\right)b^{\frac{p-p_1}{1-p_1}}\right]^{1-p_1}\|\mathfrak{X}(t,\cdot)\|_{L^{\frac{1}{p_1}}(J,\mathbb{R}^+)}\beta_C\left(\Omega\right)+\epsilon.\end{aligned}$$

Then, by letting $\epsilon \to 0$ and taking supremum for $t \in J$, we obtain

$$\beta_C\left(\mathscr{P}_2(\Omega)\right) \le 8\left[\left(\frac{1-p_1}{p-p_1}\right)b^{\frac{p-p_1}{1-p_1}}\right]^{1-p_1}\|\mathfrak{X}(t,\cdot)\|_{L^{\frac{1}{p_1}}(J,\mathbb{R}^+)}\beta_C\left(\Omega\right).$$

Finally, by employing the condition (16), we immediately get that $\beta_C\left(\Omega\right) \le \beta_C\left(\Omega_r\right) < \beta_C(\Omega)$. As a result, $\beta_C\left(\Omega\right) = 0$. Consequently, the operator $\mathscr{P}_2$ is β_C- condensing. Futhermore since Ω_r is a bounded subset of $C\left(J, n\mathbb{R}_{\mathcal{F}(A)}\right)$, we can conclude that the set $\mathscr{P}_2(\Omega_r)$ is a relatively compact subset. □

Theorem 3.2 *If the hypotheses* ***(F1)****-* ***(F3)****-****(F4)****-* ***(C)*** *and* ***(P)*** *are fulfilled, then the problem (4) is completely controllable on* $J = [0, b]$, *provided that*

$$\left[(\sigma-1)\Gamma(p+1)+\sigma \mathsf{K}_1\mathsf{K}_2\mathsf{K}_3 b^p\right] \ge 0, \tag{17}$$

where $\mathsf{K}_4 = \left(\frac{1-p_1}{p-p_1}\right)^{1-p_1} b^{p-p_1}\|g_f\|_{L^{\frac{1}{p_1}}(J,\mathbb{R}^+)}$ *and* $\sigma = \frac{\mathsf{K}_1\mathsf{K}_4 p}{\Gamma(p+1)}$.

Proof Combining the Arzela-Ascoli Theorem with Lemma 3.2 and Lemma 3.3, we can see that $\mathscr{P}_2$ is completely continuous. Therefore, by virtue of Krasnoselskii's fixed point theorem, we will show that

(i) the operator $\mathscr{P}$ maps each closed, bounded subset Ω_r of the functional space $C\left(J, n\mathbb{R}_{\mathcal{F}(A)}\right)$ into itself,
(ii) the operator $\mathscr{P}_1$ is a contraction mapping.

Firstly, we prove that the operator $\mathscr{P}$ maps each closed, bounded subset Ω_r of the functional space $C\left(J, n\mathbb{R}_{\mathcal{F}(A)}\right)$ into itself. Indeed, assume by contrary that there exists an element $\mu \in \Omega_r$, but $\mathscr{P}[\mu] \notin \Omega_r$. We have

$$
\begin{aligned}
r &< d_A\left(\mathscr{P}[\mu](t), \hat{0}\right) \\
&\le \mathsf{K}_1\|\mu_0\|_A + \int_0^t (t-s)^{p-1}\frac{\mathsf{K}_1\mathsf{K}_2 p}{\Gamma(p+1)} d_A(u_\mu(s), \hat{0})ds + \int_0^t (t-s)^{p-1}\frac{\mathsf{K}_1 p}{\Gamma(p+1)} g_f(s)\mathcal{H}(\mu, \hat{0})ds \\
&\le \mathsf{K}_1\|\mu_0\|_A + \frac{\mathsf{K}_1\mathsf{K}_2 p}{\Gamma(p+1)}\frac{b^p}{p}\mathcal{H}(u_\mu, \hat{0}) + \frac{\mathsf{K}_1 p}{\Gamma(p+1)}\mathcal{H}(\mu, \hat{0})\left(\int_0^t (t-s)^{p-1} g_f ds\right).
\end{aligned}
$$

By applying Hölder inequality and taking supremum both sides, we have

$$
\begin{aligned}
\int_0^t (t-s)^{p-1} g_f(t)ds &\le \left(\int_0^t (t-s)^{\frac{p-1}{1-p_1}} ds\right)^{1-p_1} \|g_f\|_{L^{\frac{1}{p_1}}(J,\mathbb{R}^+)} \\
&\le \left[\left(\frac{1-p_1}{p-p_1} b^{\frac{p-p_1}{1-p_1}}\right)^{1-p_1}\right] \|g_f\|_{L^{\frac{1}{p_1}}(J,\mathbb{R}^+)} \\
&\le \mathsf{K}_4. \qquad (18)
\end{aligned}
$$

As a result, one has

$$
r \le \mathsf{K}_1\|\mu_0\|_A + \frac{\mathsf{K}_1\mathsf{K}_2 b^p}{\Gamma(p+1)}\mathcal{H}(u_\mu, \hat{0}) + \frac{\mathsf{K}_1\mathsf{K}_4 p}{\Gamma(p+1)}\mathcal{H}(\mu, \hat{0}). \qquad (19)
$$

Furthermore, from **(F3)** and (18), we can prove

$$
\mathcal{H}(u_\mu, \hat{0}) \le \mathsf{K}_3\left(\|\mu_1\|_A + \mathsf{K}_1\|\mu_0\|_A + \frac{\mathsf{K}_1\mathsf{K}_4 p}{\Gamma(p+1)}\mathcal{H}(\mu, \hat{0})\right). \qquad (20)
$$

From (19) and (20), we receive

$$
\begin{aligned}
r &\le \mathsf{K}_1\|\mu_0\|_A + \frac{\mathsf{K}_1\mathsf{K}_2 b^p}{\Gamma(p+1)}\mathsf{K}_3\left(\|\mu_1\|_A + \mathsf{K}_1\|\mu_0\|_A + \frac{\mathsf{K}_1\mathsf{K}_4 p}{\Gamma(p+1)}\mathcal{H}(\mu, \hat{0})\right) + \frac{\mathsf{K}_1\mathsf{K}_4 p}{\Gamma(p+1)}\mathcal{H}(\mu, \hat{0}) \\
&\le \mathsf{K}_1\|\mu_0\|_A\left(1 + \frac{\mathsf{K}_1\mathsf{K}_2\mathsf{K}_3}{\Gamma(p+1)}\right) + \frac{\mathsf{K}_1\mathsf{K}_2\mathsf{K}_3 b^p}{\Gamma(p+1)}\|\mu_1\|_A + \frac{\mathsf{K}_1\mathsf{K}_4 p}{\Gamma(p+1)}\left[1 + \frac{\mathsf{K}_1\mathsf{K}_2\mathsf{K}_3}{\Gamma(p+1)}\right] r \\
&\le \mathsf{K}_1\|\mu_0\|_A\left(1 + \frac{\mathsf{K}_1\mathsf{K}_2\mathsf{K}_3}{\Gamma(p+1)}\right) + \frac{\mathsf{K}_1\mathsf{K}_2\mathsf{K}_3 b^p}{\Gamma(p+1)}\|\mu_1\|_A + \sigma\left[1 + \frac{\mathsf{K}_1\mathsf{K}_2\mathsf{K}_3}{\Gamma(p+1)}\right] r.
\end{aligned}
$$

Combining with (9) and dividing both sides of the above inequality by r and taking the limit as $r \to \infty$, we directly obtain

$$
(\sigma - 1)\Gamma(p+1) + \sigma\mathsf{K}_1\mathsf{K}_2\mathsf{K}_3 b^p \ge 0.
$$

It implies that $(\sigma - 1)\Gamma(p+1) + \sigma\mathsf{K}_1\mathsf{K}_2\mathsf{K}_3 b^p \ge 0$. That contradicts to (17). Hence, for $r > 0$, it implies that the operator $\mathscr{P}$ maps Ω_r into itself.

Secondary, we prove that the operator $\mathscr{P}_1$ is a contraction on Ω_r. For each $\mu, \overline{\mu} \in \Omega_r$ and $t \in J$, by employing Proposition 2.6 (i) and hypotheses **(F2)**–**(P)**, for all $t \in J$, we have

$$\begin{aligned}
d_A\left(u_\mu(t), u_{\overline{\mu}}(t)\right) &\le \frac{p\mathsf{K}_1\mathsf{K}_3}{\Gamma(p+1)}\int_0^b (b-s)^{p-1} g_f(s) d_A(\mu(s), \overline{\mu}(s))ds \\
&\le \frac{p\mathsf{K}_1\mathsf{K}_3}{\Gamma(p+1)}\left(\int_0^b (b-s)^{p-1} g_f(s)ds\right)\mathcal{H}(\mu,\overline{\mu}) \\
&\le \frac{p\mathsf{K}_1\mathsf{K}_3\mathsf{K}_4}{\Gamma(p+1)}\mathcal{H}(\mu,\overline{\mu}) \\
&= \sigma\mathsf{K}_3\mathcal{H}(\mu,\overline{\mu}).
\end{aligned}$$

Next, by applying hypothesis **(C)**, for all $t \in J$, we receive

$$\begin{aligned}
d_A\left(\mathscr{P}_1[\mu](t), \mathscr{P}_1[\overline{\mu}](t)\right) &\le \int_0^t (t-s)^{p-1} d_A\left(T_p(t-s)Eu_\mu(s), T_p(t-s)Eu_{\overline{\mu}}(s)\right)ds \\
&\le \frac{\mathsf{K}_1\mathsf{K}_2 p}{\Gamma(p+1)}\int_0^t (t-s)^{p-1} d_A(u_\mu(s), u_{\overline{\mu}}(s))ds \\
&\le \frac{\sigma\mathsf{K}_1\mathsf{K}_2\mathsf{K}_3 p}{\Gamma(p+1)}\frac{t^p}{p}\mathcal{H}(\mu,\overline{\mu}) \\
&\le \frac{\sigma\mathsf{K}_1\mathsf{K}_2\mathsf{K}_3 b^p}{\Gamma(q+1)}\mathcal{H}(\mu,\overline{\mu})
\end{aligned}$$

Then, by taking supremum for $t \in J$, one gets

$$\mathcal{H}\left(\mathscr{P}_1[\mu], \mathscr{P}_1[\overline{\mu}]\right) \le \frac{\sigma\mathsf{K}_1\mathsf{K}_2\mathsf{K}_3 b^p}{\Gamma(p+1)}\mathcal{H}(\mu,\overline{\mu}).$$

It can be deduced from (16) that the constant $\frac{a\mathsf{K}_1\mathsf{K}_2\mathsf{K}_3 b^p}{\Gamma(p+1)} < 1$. Hence, we can conclude that $\mathscr{P}_1$ is a contraction.

Consequently, $\mathscr{P}$ is a continuous function from Ω_r to itself and $\mathscr{P} = \mathscr{P}_1 + \mathscr{P}_2$ where $\mathscr{P}_1$ is a contraction. Recall that $\mathscr{P}_2$ is a completely continuous. By the Kranoselskii's fixed point theorem, it follows that $\mathscr{P} = \mathscr{P}_1 + \mathscr{P}_2$ has at least a fixed point in Ω_r. This completes the proof. □

Example 3.2 This example is devoted to investigate the controllability of the following fuzzy fractional differential system

$$\begin{cases} {}^C_F\mathcal{D}^{1/3}_{0+}\mu_1(t) = \frac{e^{-\frac{t}{3}}}{3}\mu_1(t) +_A \frac{2t+1}{3}\mu_2(t) +_A \frac{2}{\pi}u_1(t) \\ {}^C_F\mathcal{D}^{1/3}_{0+}\mu_2(t) = (-1)\mu_1(t) +_A \frac{e^{-\frac{t}{3}}}{3}\mu_2(t) +_A \frac{1}{\pi}u_2(t), \end{cases} \tag{21}$$

that satisfies the initial conditions

$$\begin{cases} \mu_1(0) = \mu_1^0 \\ \mu_2(0) = \mu_2^0, \end{cases}$$

where A is a given non-symmetric fuzzy number, $t \in I = [0, 1]$ and the input controls $u_1, u_2 : [0, 1] \to n\mathbb{R}_{\mathcal{F}(A)}$ are continuous fuzzy-valued functions. For simplicity, for each $\omega(t) = \left[\mu_1(t)\ \mu_2(t)\right]^T$, we define

$$
{}_F^C\mathcal{D}_{0+}^{1/3}\mu(t) = \begin{bmatrix} {}_F^C\mathcal{D}_{0+}^{1/3}\mu_1(t) \\ {}_F^C\mathcal{D}_{0+}^{1/3}\mu_2(t) \end{bmatrix}, \quad B = \begin{bmatrix} 0 & 1 \\ -1 & 0 \end{bmatrix}, \quad E = \begin{bmatrix} \frac{2}{\pi} & 0 \\ 0 & \frac{1}{\pi} \end{bmatrix},
$$

$$
f(t, \omega(t)) = \begin{bmatrix} \frac{e^{-\frac{t}{3}}}{3}\mu_1(t) + \frac{2t-2}{3}\mu_2(t) \\ \frac{e^{-\frac{t}{3}}}{3}\mu_2(t) \end{bmatrix}.
$$

Then, (21) can be rewritten in following fractional evolution equation:

$$
{}_F^C\mathcal{D}_{0+}^{1/3}\omega(t) = B\mu(t) + f(t, \mu(t)) + Eu(t).
$$

Firstly, the matrix $B \in \mathrm{Mat}_{2\times 2}(\mathbb{R})$ generates a semigroup $\{T(t)\}_{t\geq 0}$ given by

$$
T(t)\mu = \begin{bmatrix} \cos t & \sin t \\ -\sin t & \cos t \end{bmatrix} \begin{bmatrix} \omega_1 \\ \omega_2 \end{bmatrix},
$$

where $\mu = \left[\mu_1\ \mu_2\right]^T \in n\mathbb{R}_{\mathcal{F}(A)} \times n\mathbb{R}_{\mathcal{F}(A)}$.

Since the orbit maps ξ_ω, given by $t \mapsto T(t)\mu$, are continuous for each $\mu \in n\mathbb{R}_{\mathcal{F}(A)} \times n\mathbb{R}_{\mathcal{F}(A)}$, it implies that $\{T(t)\}_{t\geq 0}$ is a strongly continuous semigroup on $n\mathbb{R}_{\mathcal{F}(A)} \times n\mathbb{R}_{\mathcal{F}(A)}$.

Moreover, for each $\mu \in n\mathbb{R}_{\mathcal{F}(A)} \times n\mathbb{R}_{\mathcal{F}(A)}$, we also have

$$
\|T(t)\|_{op} \leq 2,
$$

where $\mathsf{K}_1 = 2$ is the number satisfying hypothesis **(F1)**. Next, we denote

$$
\Omega_r = \left\{ \omega = \begin{bmatrix} \omega_1 \\ \omega_2 \end{bmatrix} \in C(I, n\mathbb{R}_{\mathcal{F}(A)}) \times C(I, n\mathbb{R}_{\mathcal{F}(A)}) : \begin{bmatrix} \mathcal{H}(\mu_1, \hat{0}) \\ \mathcal{H}(\mu_2, \hat{0}) \end{bmatrix} \preceq_{\mathbb{R}^2} \begin{bmatrix} r_1 \\ r_2 \end{bmatrix}, \ r_1, r_2 > 0 \right\},
$$

where $\preceq_{\mathbb{R}^2}$ is an ordered relation on $\mathbb{R}^2$, defined by

$$
y \preceq z \Leftrightarrow \begin{cases} y_1 \leq z_1 \\ y_2 \leq z_2. \end{cases}
$$

Then, for each $\mu, \overline{\mu} \in \Omega_r$ and $t \in I$, we have

$$
d_A(f(t, \mu(t)), \hat{0}) \leq \begin{bmatrix} \frac{1}{3} & 0 \\ 0 & \frac{1}{3} \end{bmatrix} \begin{bmatrix} d_A(\mu_1(t), \hat{0}) \\ d_A(\mu_2(t), \hat{0}) \end{bmatrix} = g_f \begin{bmatrix} d_A(\mu_1(t), \hat{0}) \\ d_A(\mu_2(t), \hat{0}) \end{bmatrix},
$$

and

$$d_A(f(t,\mu(t)), f(t,\overline{\mu}(t)) = \begin{bmatrix} \frac{e^{-\frac{t}{3}}}{3} & \frac{2t-1}{3} \\ 0 & \frac{e^{-\frac{t}{3}}}{3} \end{bmatrix} \begin{bmatrix} d_A(\mu_1(t), \overline{\mu}_1(t)) \\ d_A(\mu_2(t), \overline{\mu}_2(t)) \end{bmatrix} = \hat{D}_f(t) \begin{bmatrix} d_A(\mu_1(t), \overline{\mu}_1(t)) \\ d_A(\mu_2(t), \overline{\mu}_2(t)) \end{bmatrix},$$

where $g_f \in \text{Mat}_{2\times 2}(\mathbb{R}^+)$ and $g_f(t) \in L^{\frac{1}{p_1}}(I, \mathbb{R}^+)$, $p_1 = \frac{1}{4}$ satisfy hypotheses **(F3)** and **(F4)**, respectively. Moreover, by using Hölder inequality, one gets

$$\|g_f\|_{L^{\frac{1}{p_1}}(I,\mathbb{R}^+)} = 2\sqrt{\frac{1}{108}\left(1 - \frac{1}{\sqrt[3]{e^4}}\right)} + \frac{1}{\sqrt{162}}.$$

For each $u_1, u_2 \in L^1(J, \mathbb{R}_{\mathcal{F}(A)})$, the linear operator Π is given by

$$\Pi(u) = \frac{1}{\pi} \begin{bmatrix} \int_0^b (b-s)^{-\frac{2}{3}} T_{\frac{2}{3}}(b-s) u_1(s) ds \\ 2\int_0^b (b-s)^{-\frac{2}{3}} T_{\frac{2}{3}}(b-s) u_2(s) ds \end{bmatrix}.$$

Then, by applying Theorem 3.2, we can conclude that the fractional differential system (21) is completely controllable on I, provided that there exists a constant $\mathsf{K}_3 \in \left(0, \frac{2\sqrt{\pi}}{5\Gamma(\frac{2}{3})}\right)$ such that

$$\left[(\sigma - 1)\Gamma(p+1) + \sigma \mathsf{K}_1 \mathsf{K}_2 \mathsf{K}_3\right] \geq 0,$$

where $\mathsf{K}_1 = 2, \mathsf{K}_2 = \frac{\sqrt{5}}{\pi}$ and $\sigma \approx \frac{0.88}{\Gamma(\frac{3}{2})}, \mathsf{K}_4 = \sqrt[4]{729}\left(2\sqrt{\frac{1}{108}\left(1 - \frac{1}{\sqrt[3]{e^4}}\right)} + \frac{1}{\sqrt{162}}\right) \approx 0.44$.

4 Conclusions

In this paper, the complete controllability of fuzzy solutions of the correlated fuzzy fractional evolution equations was presented. Some conditions that ensure the problem is completely controllable are given. The control variable u may be not unique.

References

1. Allahviranloo, T., Ahmadi, M.B.: Fuzzy Laplace transforms. Soft Comput. **14**, 235–243 (2010)
2. Akhmerov, R.R., Kamenskii, M.I., Potapov, A.S., Rodkina, A.E., Sadovskii, B.N.: Measures of Noncompactness and Condensing Operators. Operator Theory, Advances and Applications, Birkhäuser Basel (1992)
3. Anh, C.T., Ke, T.D.: On nonlocal problems for retarded fractional differential equations in Banach spaces. Fixed Point Theory **15**(2), 373–392 (2014)
4. Barros, L.C., Gomes, L.T., Tonelli, P.A.: Fuzzy differential equations: an approach via fuzzification of the derivative operator. Fuzzy Sets Syst. **230**, 39–52 (2013)
5. Bede, B., Gal, S.G.: Generalizations of the differentiability of fuzzy number valued functions with applications to fuzzy differential equations. Fuzzy Sets Syst. **151**, 581–599 (2005)
6. Bede, B., Stefanini, L.: Generalized differentiability of fuzzy-valued functions. Fuzzy Sets Syst. **230**, 119–141 (2013)
7. Donchev, T., Nosheen, A., Lupulescu, V.: Fuzzy integro-differential equations with compactness type conditions. Hacettepe J. Maths Stats. **43**(2), 259–267 (2014)
8. Esmi, E., Pedro, F.S., Barros, L.C.: Fréchet derivative for linearly correclated fuzzy function. Inf. Sci. **435**, 150–160 (2018)
9. Ignatius, O.I., Oviawe, C.I.: On the observability and controllability of active suspension system. Netw. Complex Syst. **6**(1), (2016)
10. Gal, C.G., Gal, S.G.: Semigroups of operators on spaces of fuzzy-number-valued functions with applications to fuzzy differential equations. J. Fuzzy Math. 13(3)(2005) 647–682. (arXiv:1306.3928) Klamka
11. Klamka, J.: Controllability and Minimum Energy Control. Springer, Berlin (2018)
12. Kaleva, O.: The Cauchy problem for fuzzy differential equations. Fuzzy Sets Syst. **420, 35**(3), 389–396 (1990)
13. Long, H.V., Dong, N.P.: An extension of Krasnoselskii's fixed point theorem and its application to nonlocal problems for implicit fractional differential systems with uncertainty. J. Fixed Point Theory Appl. **20**(1), 1–27 (2018)
14. Mazandarani, M., Pariz, N., Kamyad, A.V.: Granular differentiability of fuzzy-number-valued functions. IEEE Tran. Fuzzy Syst. **26**(1), 310–323 (2018)
15. Najariyan, M., Zhao, Y.: Fuzzy fractional quadratic regulator problem under granular fuzzy fractional derivatives. IEEE Tran. Fuzzy Syst. **26**(4), 2273–2288 (2018)
16. Pedro, F.S., Barros, L.C., Esmi, E.: Population growth model via interactive fuzzy differential equation. Inf. Sci. **481**, 160–173 (2019)
17. Puri, M.L., Ralescu, D.A.: Differentials of fuzzy functions. J. Math. Anal. Appl. **91**(2), 552–558 (1983)
18. Pedroa, F.S., Esmi, E., Barros, L.C.: Calculus for linearly correlated fuzzy function using Fréchet derivative and Riemann integral **512**, 219–237 (2020)
19. Son, N.T.K., Thao, H.T.P., Dong, N.P., Long, H.V.: Fréchet fractional order derivative of linear correlated fuzzy functions with application to fuzzy fractional differential equations. Fuzzy Sets Syst. https://doi.org/10.1016/j.fss.2020.10.019
20. Son, N.T.K.: A foundation on semigroup of operators defined on the set of triangular fuzzy numbers and its application to fuzzy fractional evolution equations. Fuzzy Sets Syst. **347**, 1–28 (2018)
21. Son, N.T.K., Dong, N.P.: Asymptotic behavior of C^0-solutions of evolution equations with uncertainties. J. Fixed Point Theory Appl. **20**(1), 1–27 (2018)
22. Son, N.T.K., Dong, N.P., Long, H.V.: Fuzzy delay differential equations under granular differentiability with applications. Comp. Appl. Math. **38**(3), 107–136 (2019)
23. Son, N.T.K., Dong, N.P., Long, H.V., Son, L.H.: Towards granular calculus of single-valued neutrosophic functions under granular computing. Multimed. Tools Appl. 1–37 (2019)

24. Son, N.T.K., Dong, N.P., Long, H.V., Son, L.H., Khastan, A.: Linear quadratic regulator problem governed by granular neutrosophic fractional differential equations. ISA Trans. **97**, 296–316 (2020)
25. Shukla, A., Sukavanam, N., Pandey, D.N.: Approximate controllability of semi-linear system with state delay using sequence method. J. Franklin Inst. **352**(1), 5380–5392 (2015)
26. Zhou, Yong, Jiao, Feng: Existence of mild solutions for fractional neutral evolution equations. Comput. Math. Appl. **59**, 1063–1077 (2010)

An Estimation of the Solution of First Order Fuzzy Differential Equations

N. Ahmady, T. Allahviranloo, and E. Ahmady

Abstract The aim of this paper is to suggest two numerical methods for solving first order fuzzy differential equation under generalized differentiability. Modified Euler method and Piecewise approximate method are two methods that we introduce in this work.

Keywords Fuzzy differential equations · Generalized Hukuhara differentiability · Numerical method

MR Subject Classification 34A07

1 Introduction

The topic of fuzzy differential equations (FDEs) are utilized for modeling problems in science and engineering and it was first introduced by Kandel and Byatt [29]. In recent years, FDEs has attracted widespread attention and has been rapidly growing. Fuzzy differential equations base on the Hukuhara derivative introduced by Hukuhara [26], and developed by many authors [28, 36]. Numerical methods for solving FDEs under Hukuhara differentiability concept such as Euler method, Predictor-Corrector and Improved Predictor-Corrector methods and Taylor method were introduced in [1, 12, 13, 30]. Some other papers on fuzzy differential equations and fuzzy integral equations that can be mentioned were presented in [2–4, 8–10].

It is well-known that the solution of fuzzy differential equations under Hukuhara derivatives has the property that the diameter is non-decreasing as t increase, in order

N. Ahmady (✉)
Department of Mathematics, Islamic Azad University, Varamin-Pishva Branch, Varamin, Iran
e-mail: n.ahmadi@iauvaramin.ac.ir

T. Allahviranloo
Faculty of Engineering and Natural Sciences, Bahcesehir University, Istanbul, Turkey

E. Ahmady
Department of Mathematics, Islamic Azad University, Shahr-e-Qods Branch, Tehran, Iran

T. Allahviranloo and S. Salahshour (eds.), *Advances in Fuzzy Integral and Differential Equations*, Studies in Fuzziness and Soft Computing 412,
https://doi.org/10.1007/978-3-030-73711-5_4

to overcome this short-coming, Bede and Gal introduced the strongly generalized differential of set value fuzzy function [16]. Moreover general properties studied in [17–19].

Many researches have been suggested some numerical methods for fuzzy differential equations and fuzzy integral equations under generalized differentiability (see, e.g. [11, 14, 15]). In [11], authors proposed Taylor series expansion for fuzzy valued function based on the concept of generalized Hukuhara differentiability, moreover, they introduced Euler method for solving fuzzy differential equations. In the present work we propose two numerical methods for solving FDEs. First one, is Modified Euler method and the second one is Piecewise Approximation method (PWA method). Modified Euler method estimated FDEs by using a two-stage predictor-corrector algorithm with local truncation error of order two and PWA method interpolated the solution of FDE by three order fuzzy piecewise polynomial in each subintervals of solution. The existence, uniqueness and convergence of the proposed methods are investigated in details. method are compared with Euler method [?], and we show these methods are more accurate than Euler method. The paper is organized as follows:

In Sect. 2, some basic definitions are brought. The numerical methods for solving FDE are introduced in Sect. 3. The numerical examples are presented in Sect. 4 and finally conclusion is drawn.

2 Preliminaries

In this section the most basic notations and definitions used are introduced.

The set of fuzzy numbers, that is, normal, fuzzy convex, upper semi-continuous and compactly supported fuzzy sets which are defined over the real line and denoted by $\mathbb{R}_{\mathcal{F}}$. For $0 < \alpha \leq 1$, set $[u]^\alpha = \{t \in \mathbb{R} | u(t) \geq \alpha\}$, and $[u]^0 = cl\{t \in \mathbb{R} | u(t) > 0\}$. We represent $[u]^\alpha = [u^-(\alpha),\ u^+(\alpha)]$, so if $u \in \mathbb{R}_{\mathcal{F}}$, the α-level set $[u]^\alpha$ is a closed interval for all $\alpha \in [0,\ 1]$. For arbitrary $u,\ v \in \mathbb{R}_{\mathcal{F}}$ and $k \in \mathbb{R}$, the addition and scalar multiplication are defined by $[u + v]^\alpha = [u]^\alpha + [v]^\alpha$, $[ku]^\alpha = k[u]^\alpha$ respectively.

Definition 2.1 *(see* [28]*)* The Hausdorff distance between fuzzy numbers is given by $D : \mathbb{R}_{\mathcal{F}} \times \mathbb{R}_{\mathcal{F}} \longrightarrow \mathbb{R}^+ \cup \{0\}$ as

$$D(u,\ v) = \sup_{\alpha \in [0,\ 1]} \max \Big\{ |u^-(\alpha) - v^-(\alpha)|,\ |u^+(\alpha) - v^+(\alpha)| \Big\}.$$

Consider $u, v, w, z \in \mathbb{R}_{\mathcal{F}}$ and $\lambda \in \mathbb{R}$, then the following properties are well-known for metric D: (see [31])

1. $D(u \oplus w,\ v \oplus w) = D(u,\ v)$;
2. $D(\lambda u,\ \lambda v) = |\lambda| D(u,\ v)$;

3. $D(u \oplus v,\ w \oplus z) \leq D(u,\ w) + D(v,\ z)$;
4. $D(u \ominus v,\ w \ominus z) \leq D(u,\ w) + D(v,\ z)$, as long as $u \ominus v$ and $w \ominus z$ exist, where $u,\ v,\ w,\ z \in \mathbb{R}_{\mathcal{F}}$.

where, $\ominus$ is the Hukuhara difference (H-difference), it means that $w \ominus v = u$ if and only if $u \oplus v = w$.

Definition 2.2 *(see* [35].*)* The generalized Hukuhara difference of two fuzzy numbers $u,\ v \in \mathbb{R}_{\mathcal{F}}$ is defined as follows

$$u \ominus v = w \Longleftrightarrow \begin{cases} (a).\ u = v + w; \\ or\ (b).\ v = u + (-1)w. \end{cases}$$

Proposition 2.1 (see [35].) *Let $A,\ B \in K_n^C$ be two compact convex set; then*

1. *if the gH-difference exists, it is unique and it is a generalization of the usual Hukuhara difference since $A \ominus B = A \sim_h B$, whenever $A \sim_h B$ exists,*
2. $A \ominus A = 0$,
3. *if $A \ominus B$ exists in the sence (a), then $B \ominus A$ exists in sense (b) and vice versa,*
4. $(A \oplus B) \ominus B = A$,
5. $\{0\} \ominus (A \ominus B) = (-B) \ominus (-A)$,
6. *we have $(A \ominus B) = (B \ominus A) = C$ if and only if $C = \{0\}$ and $A = B$.*

Definition 2.3 *(see* [23].*)* A fuzzy valued function $f : [a, b] \to \mathbb{R}_{\mathcal{F}}$ is said to be continuous at $t_0 \in [a, b]$ if for each $\epsilon > 0$ there is $\delta > 0$ such that $D(f(t),\ f(t_0)) < \epsilon$, whenever $t \in [a, b]$ and $|t - t_0| < \delta$. We say that f is fuzzy continuous on $[a, b]$ if f is continuous at each $t_0 \in [a,\ b]$.

Definition 2.4 *(see* [16]*)* Let $F : I \to \mathbb{R}$. Fix $t_0 \in I$. We say F is strongly generalized differentiable at t_0, if there exists an element $F'(t_0) \in \mathbb{R}$ such that either

1. foe all $h > 0$ sufficiently closed to 0, the H-differences $F(t_0 + h) \ominus F(t_0)$, $F(t_0) \ominus F(t_0 - h)$ exist and limits (in the metric D)

$$\lim_{h \to 0^+} \frac{F(t_0 + h) \ominus F(t_0)}{h} = \lim_{h \to 0^+} \frac{F(t_0) \ominus F(t_0 - h)}{h} = F'(t_0) \tag{2.1}$$

or

2. for all $h > 0$ sufficiently closed to 0, the H-differences $F(t_0) \ominus F(t_0 + h)$, $F(t_0 - h) \ominus F(t_0)$ exist and limits (in the metric D)

$$\lim_{h \to 0^+} \frac{F(t_0) \ominus F(t_0 + h)}{-h} = \lim_{h \to 0^+} \frac{F(t_0 - h) \ominus F(t_0)}{-h} = F'(t_0) \tag{2.2}$$

or

3. for all $h > 0$ sufficiently closed to 0, the H-differences $F(t_0 + h) \ominus F(t_0)$, $F(t_0 - h) \ominus F(t_0)$ exist and limits (in the metric D)

$$\lim_{h\to 0^+} \frac{F(t_0+h)\ominus F(t_0)}{h} = \lim_{h\to 0^+} \frac{F(t_0-h)\ominus F(t_0)}{-h} = F'(t_0) \tag{2.3}$$

or

4. for all $h > 0$ sufficiently closed to 0, the H-differences $F(t_0) \ominus F(t_0 + h)$, $F(t_0) \ominus F(t_0 - h)$ exist and limits (in the metric D)

$$\lim_{h\to 0^+} \frac{F(t_0)\ominus F(t_0+h)}{-h} = \lim_{h\to 0^+} \frac{F(t_0)\ominus F(t_0-h)}{h} = F'(t_0) \tag{2.4}$$

Definition 2.5 Let $f : I \to \mathbb{R}_F$. we say f is $[(i) - gH]$-differentiable on I if f is differentiable in the sense (1) and its derivative is denoted $f_{i.gH}$ and similarly for $[(ii) - gH]$-differentiable we have $f_{ii.gH}$.

Theorem 2.1 (see [21]) *Let* $f : I \to \mathbb{R}_F$ *and put* $[f(t)]^\alpha = [(f^+)(t;\alpha), (f^-)(t;\alpha)]$ *for each* $\alpha \in [0, 1]$.

1. *If f is $[(i) - gH]$ differentiable the f^+ and f^- are differentiable functions and*

$$f'_{i.gH}(t_0;\alpha) = [(f^-)'(t;\alpha)\ ,\ (f^+)'(t;\alpha)], \tag{2.5}$$

2. *If f is $[(ii) - gH]$ differentiable the f^+ and f^- are differentiable functions and*

$$f'_{ii.gH}(t_0;\alpha) = [(f^+)'(t;\alpha)\ ,\ (f^-)'(t;\alpha)], \tag{2.6}$$

Definition 2.6 *(see* [21]*.)* We say that a point $t_0 \in (a, b)$, is a switching point for the differentiability of f, if in any neighborhood V of t_0 there exist points $t_1 < t_0 < t_2$ such that

type(I) at t_1 (2.5) holds while (2.6) does not hold and at t_2 (2.6) holds and (2.5) does not hold, or

type(II) at t_1 (2.6) holds while (2.5) does not hold and at t_2 (2.5) holds and (2.6) does not hold.

The following condition and notations are used in the reminder of paper.

In all of the paper the generalized Hukuhara difference of two fuzzy numbers is exists.

$C_{\mathcal{F}}([a, b], \mathbb{R}_{\mathcal{F}})$ is the set of fuzzy valued function f which are defined on $[a, b]$ and be fuzzy continuous from the interior points of $[a, b]$ such that the continuity is one-sided at endpoints a, b.

$C^k_{gH}([a, b], \mathbb{R}_{\mathcal{F}})$, is the space of functions f such that f and it's first k, gH-derivatives are all in $C_{\mathcal{F}}([a, b], \mathbb{R}_{\mathcal{F}})$.

Theorem 2.2 (see [11].) *Consider $f : [a, b] \longrightarrow \mathbb{R}_{\mathcal{F}}$ is gH-differentiable such that type of differentiability f in $[a, b]$ don't change. Then for $a \leq s \leq b$*

(i) If $f(t)$ is $[(i) - gH]$-differentiable then $f'_{i.gH}(t)$ is (FR)-integrable over $[a,\ b]$ and

$$f(s) = f(a) \oplus \int_a^s f'_{i.gH}(t)dt. \tag{2.7}$$

(ii) If $f(t)$ is $[(ii) - gH]$-differentiable then $f'_{ii.gH}(t)$ is (FR)-integrable over $[a,\ b]$ and

$$f(a) = f(s) \oplus (-1) \int_a^s f'_{ii.gH}(t)dt. \tag{2.8}$$

Theorem 2.3 (see [11].) *Let $f^{(i)} : [a,\ b] \rightarrow \mathbb{R}_{\mathcal{F}}$ and $f \in C^4_{gH}([a, b],\ \mathbb{R}_{\mathcal{F}})$. For all $s \in [a,\ b]$*

(i) Consider $f^{(i)}_{gH}$, $i = 1, ..., n$ are $[(i) - gH]$-differentiable and type of differentiability don't change in interval $[a,\ b]$, then

$$f^{(i-1)}_{i.gH}(s) = f^{(i-1)}_{i.gH}(a) \oplus \int_a^s f^{(i)}_{i.gH}(t)dt. \tag{2.9}$$

(ii) If $f^{(i)}_{gH}$, $i = 1, ..., n$ are $[(ii) - gH]$-differentiable and type of differentiability don't change in interval $[a,\ b]$, then

$$f^{(i-1)}_{ii.gH}(s) = f^{(i-1)}_{ii.gH}(a) \oplus \int_a^s f^{(i)}_{ii.gH}(t)dt. \tag{2.10}$$

(iii) Suppose that $f^{(i)}$, $i = 2k - 1,\ k \in \mathbb{N}$ are $[(i) - gH]$-differentiable and $f(t)$, $f^{(i)}$, $i = 2k,\ k \in \mathbb{N}$ are $[(ii) - gH]$-differentiable, so

$$f^{(i-1)}_{i.gH}(s) = f^{(i-1)}_{i.gH}(a) \ominus (-1) \int_a^s f^{(i)}_{ii.gH}(t)dt. \tag{2.11}$$

(iv) Consider for $i = 2k - 1, k \in \mathbb{N}$, $f^{(i)}$ are $[(ii) - gH]$-differentiable and $f(t)$, $f^{(i)}$ are $[(i) - gH]$-differentiable for $i = 2k,\ k \in \mathbb{N}$, then

$$f^{(i-1)}_{ii.gH}(s) = f^{(i-1)}_{ii.gH}(a) \ominus (-1) \int_a^s f^{(i)}_{i.gH}(t)dt. \tag{2.12}$$

3 Numerical Methods

Now, let us consider the following fuzzy differential equation (FDE),

$$\begin{cases} y'(t) = f(t, y(t)),\ t \in [0, T]; \\ y(0) = y_0 \in \mathbb{R}_{\mathcal{F}}, \end{cases} \tag{3.1}$$

here $y(t)$ is an unknown fuzzy function of crisp variable t and $f : [0,\ T] \times \mathbb{R}_{\mathcal{F}} \rightarrow \mathbb{R}_{\mathcal{F}}$ is continuous, also $y'(t)$ is the generalized Hukuhara derivative of $y(t)$ such that the set of switching points is finite.

For solving FDE (3.1) by numerical method, first to integrate the system given Eq. (3.1), we replace the interval $[0, T]$ by a set of discrete equally spaced grid points, $0 = t_0 < t_1 < ... < t_N = T$, where $t_n = nh,\ h = \frac{T}{N}$.

Now we introduce two numerical methods for solving FDE (3.1).

3.1 Modified Euler Method

Let $y(t) \in C^4_{gH}([0, T],\ \mathbb{R}_{\mathcal{F}})$ is solution of problem (3.1) and $y^{(i)}(t),\quad i = 1, ..., 4$ are gH-differentiable. The basic idea for this method is applying the Taylor series expansion of $y(t_{k+1})$ at the point t_k, for each $k = 0, 1, ..., N$, such as

$$y(t_{k+1}) = y(t_k) \oplus (t_{k+1} - t_k) \odot y'(t_k) \oplus \frac{(t_{k+1} - t_k)^2}{2} \odot y''(t_k) \oplus \frac{(t_{k+1} - t_k)^3}{3!} \odot y'''(\eta_k),$$

for some points η_k lie between t_k and t_{k+1}. For this purpose, by using the following theorem, we are going to approximate $y''(t)$, by using concept of generalized differentiability four different cases are considered.

Theorem 3.1 *Let* $f \in C^4_{gH}([a, b],\ \mathbb{R}_{\mathcal{F}})$. *Then*

(i) *if* $f^{(i)}_{gH}$, $i = 1, ..., 4$ *are* $[(i) - gH]$*-differentiable and type of differentiability don't change in interval* $[a,\ b]$, *then*

$$f''_{i.gH}(a) \approx \frac{f'_{i.gH}(s) \ominus f'_{i.gH}(a)}{s - a}, \tag{3.2}$$

(ii) *let* $f^{(i)}_{gH}$, $i = 1, ..., 4$ *are* $[(ii) - gH]$*-differentiable and type of differentiability don't change in interval* $[a,\ b]$, *then*

$$f''_{ii.gH}(a) \approx \frac{f'_{ii.gH}(s) \ominus f'_{ii.gH}(a)}{s - a}, \tag{3.3}$$

(iii) *suppose that* $f^{(i)}$, $i = 1, 3$ *are* $[(i) - gH]$*-differentiable and* $f(t)$, $f^{(i)}$, $i = 2, 4$ *are* $[(ii) - gH]$*-differentiable, so*

$$(-1) f''_{i.gH}(a) \approx \frac{f'_{ii.gH}(s) \ominus f'_{ii.gH}(a)}{s - a}, \tag{3.4}$$

(iv) *consider for* $i = 1, 3$, $f^{(i)}$ *are* $[(ii) - gH]$*-differentiable and* $f(t)$, $f^{(i)}$ *are* $[(i) - gH]$*-differentiable for* $i = 2, 4$, *then*

$$(-1) f''_{ii.gH}(a) \approx \frac{f'_{i.gH}(s) \ominus f'_{i.gH}(a)}{s - a}. \tag{3.5}$$

Proof Since $f \in C^4_{gH}([a, b], \mathbb{R}_{\mathcal{F}})$, so $f^{(i)}_{gH}$, $i = 0,\ 1,\ 2,\ 3,\ 4$, are (FR)-integrable on T.

We will prove Parts (i) and (iii); the other parts are similar, and we omit the details.
(i) Let $f^{(i)}$ is $[(i) - gH]$-differentiable, by Theorem 2.3, we can write

$$f'_{i.gH}(s) = f'_{i.gH}(a) \oplus \int_a^s f''_{i.gH}(s_1) ds_1,$$

where

$$f''_{i.gH}(s_1) = f''_{i.gH}(a) \oplus \int_a^{s_1} f'''_{i.gH}(s_2) ds_2. \tag{3.6}$$

by integration of Eq. (3.6), we get

$$\begin{aligned}
&\int_a^s f''_{i.gH}(s_1) ds_1 = \\
&\quad = \int_a^s f''_{i.gH}(a) ds_1 \oplus \int_a^s \Big(\int_a^{s_1} f'''_{i.gH}(s_2) ds_2 \Big) ds_1, \\
&\quad = f''_{i.gH}(a) \odot (s - a) \oplus \int_a^s \Big(\int_a^{s_1} f'''_{i.gH}(s_2) ds_2 \Big) ds_1,
\end{aligned}$$

where the last double (FR)-integral belongs to $\mathbb{R}_{\mathcal{F}}$. So

$$\begin{aligned}
f'_{i.gH}(s) &= f'_{i.gH}(a) \oplus f''_{i.gH}(a) \odot (s - a) \\
&\quad \oplus \int_a^s \Big(\int_a^{s_1} f'''_{i.gH}(s_2) ds_2 \Big) ds_1.
\end{aligned}$$

Consequently,

$$f'''_{i.gH}(s_2) = f'''_{i.gH}(a) \oplus \int_a^{s_2} f^{(4)}_{i.gH}(s_3) ds_3,$$

applying the (FR)-integral operator to $f'''_{i.gH}(s_2)$, gives

$$\int_a^{s_1} f'''_{i.gH}(s_2)ds_2 =$$

$$f'''_{i.gH}(a) \odot (s_1 - a) \oplus \int_a^{s_1} \left(\int_a^{s_2} f^{(4)}_{i.gH}(s_3)ds_3 \right) ds_2,$$

furthermore,

$$\int_a^{s} \left(\int_a^{s_1} f'''_{i.gH}(s_2)ds_2 \right) ds_1 = f'''_{i.gH}(a) \odot \int_a^{s} (s_1 - a)ds_1$$
$$\oplus \int_a^{s} \left(\int_a^{s_1} \left(\int_a^{s_2} f^{(4)}_{i.gH}(s_3)ds_3 \right) ds_2 \right) ds_1,$$

where the last triple integral belongs to $\mathbb{R}_{\mathcal{F}}$. Hence

$$f'_{i.gH}(s) = f'_{i.gH}(a) \oplus f''_{i.gH}(a) \odot (s - a) \oplus f'''_{i.gH}(a)$$
$$\odot \frac{(s-a)^2}{2!} \oplus \int_a^{s} \left(\int_a^{s_1} \left(\int_a^{s_2} f^{(4)}_{i.gH}(s_3)ds_3 \right) ds_2 \right) ds_1.$$

If s is very closed to a and $(s - a) \to 0$,

$$D(f'_{i.gH}(s), f'_{i.gH}(a) \oplus f''_{i.gH}(a) \odot (s - a)) \to 0,$$

and

$$D(f'_{i.gH}(s) \ominus f'_{i.gH}(a), f''_{i.gH}(a) \odot (s - a)) \to 0,$$

this means where $(s - a) \to 0$, we obtain:

$$f''_{i.gH}(a) \approx \frac{f'_{i.gH}(s) \ominus f'_{i.gH}(a)}{s - a}.$$

(iii) Suppose that $f^{(i)}$, $i = 1, 3$, are $[(i) - gH]$-differentiable and $f^{(i)}$, $i = 2, 4$, are $[(ii) - gH]$-differentiable, so

$$f'_{ii.gH}(s) = f'_{ii.gH}(a) \ominus (-1) \odot \int_a^{s} f''_{i.gH}(s_1)ds_1,$$

As it is known, $f''_{i.gH}$ is obtain from

$$f''_{i.gH}(s_1) = f''_{i.gH}(a) \ominus (-1) \odot \int_a^{s_1} f'''_{ii.gH}(s_2)ds_2,$$

by integration we obtain

$$\int_a^s f''_{i.gH}(s_1)ds_1 = f''_{i.gH}(a) \odot (s-a) \ominus (-1) \odot \int_a^s \left(\int_a^{s_1} f'''_{ii.gH}(s_2)ds_2 \right) ds_1,$$

therefore

$$f'_{ii.gH}(s) = f'_{ii.gH}(a) \oplus (-1) f''_{i.gH}(a) \odot (s-a) \ominus \int_a^s \left(\int_a^{s_1} f'''_{ii.gH}(s_2)ds_2 \right),$$

similarly, in order to find $f'''_{ii.gH}$ by Theorem 2.3, we have

$$f'''_{ii.gH}(s_2) = f'''_{ii.gH}(a) \ominus (-1) \int_a^{s_2} f^{(4)}_{i.gH}(s_3)ds_3,$$

and

$$\int_a^s \left(\int_a^{s_1} f'''_{ii.gH}(s_2)ds_2 \right) ds_1 = \frac{(s-a)^2}{2} \odot f'''_{ii.gH}(a)$$
$$\ominus(-1) \int_a^s \left(\int_a^{s_1} \left(\int_a^{s_2} f^{(4)}_{ii.gH}(s_3)ds_3)ds_2 \right) \right) ds_1,$$

thus, we conclude:

$$f'_{ii.gH}(s) = f'_{ii.gH}(a) \oplus (-1) f''_{i.gH}(a) \odot (s-a) \ominus (-1) f'''_{ii.gH}(a) \odot \frac{(s-a)^2}{2}$$
$$\oplus \int_a^s \left(\int_a^{s_1} \left(\int_a^{s_2} f^{(4)}_{i.gH}(s_3)ds_3)ds_2 \right) \right) ds_1,$$

now, if $(s-a) \to 0$, we get

$$D(f'_{ii.gH}(s) \ominus f'_{ii.gH}(a), (-1) f''_{i.gH}(a) \odot (s-a)) \to 0,$$

this means

$$(-1) f''_{i.gH}(a) \approx \frac{f'_{ii.gH}(s) \ominus f'_{ii.gH}(a)}{s-a}.$$ ■

Now in order to obtain the Modified Euler method for solving FDE (3.1), we consider four cases:

Case 1.
Let $y(t) \in C^4_{gH}([0, T], \mathbb{R}_{\mathcal{F}})$ is $[(i)-gH]$-differentiable solution of problem (3.39) and $y^{(i)}(t), \quad i = 1, ..., 4$ are $[(i)-gH]$-differentiable. Now by using the Taylor series expansion of $y(t_{k+1})$ at the point t_k, for each $k = 0, 1, ..., N$, we get

$$y(t_{k+1}) = y(t_k) \oplus (t_{k+1} - t_k) \odot y'_{i.gH}(t_k)$$
$$\oplus \frac{(t_{k+1} - t_k)^2}{2} \odot y''_{i.gH}(t_k) \oplus \frac{(t_{k+1} - t_k)^3}{3!} \odot y'''_{i.gH}(\eta_k),$$

for some points η_k lie between t_k and t_{k+1}. Since $h = t_{k+1} - t_k$, gives

$$y(t_{k+1}) = y(t_k) \oplus h \odot y'_{i.gH}(t_k) \oplus \frac{h^2}{2} \odot y''_{i.gH}(t_k) \oplus \frac{h^3}{3!} \odot y'''_{i.gH}(\eta_k). \tag{3.7}$$

Consequently, by taking Eq. (3.2) into Eq. (3.7) we obtain:

$$y(t_{k+1}) = y(t_k) \oplus \frac{h}{2} \odot \left(\frac{y'_{i.gH}(t_{k+1}) \ominus y'_{i.gH}(t_k)}{h} \right) \oplus \frac{h^3}{3!} \odot y'''_{i.gH}(\eta_k).$$

Fuzzy differential equation (3.39) implies that $y'_{gH}(t) = f(t, y(t))$, so we have

$$y(t_{k+1}) = y(t_k) \oplus \frac{h}{2} \odot \left(\frac{f(t_{k+1}, y(t_{k+1})) \oplus f(t_k, y(t_k))}{h} \right) \oplus \frac{h^3}{3!} \odot y'''_{i.gH}(\eta_k). \tag{3.8}$$

In order to obtain a numerical method, the value of $y(t_{k+1})$ appearing on the RHS is not known. To handel this, the value of $y(t_{k+1})$ is first predicted by Euler method [11], and then the predicted value is used in Eq. (3.8).

Thus, the Modified Euler method can be written as:

$$\begin{cases} y^*_{k+1} = y_k \oplus h \odot f(t_k, y_k), \\ y_{k+1} = y_k \oplus \frac{h}{2} \odot \Big(f(t_{k+1}, y^*(t_{k+1})) \oplus f(t_k, y(t_k)) \Big), \ k = 0, 1, ..., N-1. \end{cases} \tag{3.9}$$

Case 2.

Now, consider $y(t)$ is $[(ii) - gH]$-differentiable and belongs to $C^3_{gH}([0, T], \mathbb{R}_{\mathcal{F}})$ and $y^{(i)}(t),\ i = 1, ..., 4$ are $[(ii) - gH]$-differentiable. So the Taylor's series expansion of $y(t)$ about the point t_k at t_{k+1} is

$$\begin{aligned} y(t_{k+1}) &= y(t_k) \ominus (-1)h \odot y'_{ii.gH}(t_k) \\ &\ominus (-1)\frac{h^2}{2} \odot y''_{ii.gH} \\ &\ominus (-1)\frac{h^3}{3!} \odot y'''_{ii.gH}(\eta_k), \end{aligned} \tag{3.10}$$

Now, by taking Eq. (3.3) into Eq. (3.10) we obtain

$$y(t_{k+1}) = y(t_k) \ominus (-1)\frac{h}{2} \odot \Big(y'_{ii.gH}(t_k) \oplus y'_{ii.gH}(t_{k+1}) \Big),$$

As described in Case 1, the Modified Euler method takes the following form:

$$\begin{cases} y_{k+1}^* = y_k \ominus (-1)h \odot f(t_k, y_k), \\ y_{k+1} = y_k \ominus (-1)\frac{h}{2} \odot \Big(f(t_{k+1}, y^*(t_{k+1})) \oplus f(t_k, y(t_k)) \Big), k = 0, 1, ..., N-1. \end{cases} \tag{3.11}$$

Case 3.
We Consider partition of $[0, T]$ as follows

$$t_0 = 0, t_1, ..., t_j, \gamma, t_{j+1}, ..., t_N = T. \tag{3.12}$$

If the assumptions of case 1 are true for $y(t),\ t \in [0, t_j]$ and case 2 are true for $y(t),\ t \in [t_{j+1}, T]$ ($\gamma \in [0, T]$ is a switching point type I), the Modified Euler method can be written as follows:

$$\begin{cases} y_{k+1}^* = y_k \oplus h \odot f(t_k, y_k), \\ y_{k+1} = y_k \oplus \frac{h}{2} \odot \Big(f(t_{k+1}, y^*(t_{k+1})) \oplus f(t_k, y(t_k)) \Big), \qquad k = 0, 1, ..., j. \\ \\ y_{k+1}^* = y_k \ominus (-1)h \odot f(t_k, y_k), \\ y_{k+1} = y_k \ominus (-1)\frac{h}{2} \odot \Big(f(t_{k+1}, y^*(t_{k+1})) \oplus f(t_k, y(t_k)) \Big), k = j+1, 1, ..., N-1. \end{cases} \tag{3.13}$$

Case 4.
If the assumptions of case 2, are true for $y(t),\ t \in [0, t_j]$ and case 1 are true for $y(t),\ t \in [t_{j+1}, T]$ ($\gamma \in [0, T]$ is a switching point type II), hence the Modified Euler method can be written as:

$$\begin{cases} y_{k+1}^* = y_k \ominus (-1)h \odot f(t_k, y_k), \\ y_{k+1} = y_k \ominus (-1)\frac{h}{2} \odot \Big(f(t_{k+1}, y^*(t_{k+1})) \oplus f(t_k, y(t_k)) \Big), k = 0, 1, ..., j, \\ \\ y_{k+1}^* = y_k \oplus h \odot f(t_k, y_k), \\ y_{k+1} = y_k \oplus \frac{h}{2} \odot \Big(f(t_{k+1}, y^*(t_{k+1})) \oplus f(t_k, y(t_k)) \Big), \qquad k = j+1, 1, ..., N-1. \end{cases} \tag{3.14}$$

Theorem 3.2 *The Modified Euler method is consistent.*

Proof First, let $y(t)$ is $[(i) - gH]$-differentiable, by Eq. (3.9) we can write

$$\phi_k = y_{k+1} \ominus \Big(y_k \oplus \frac{h}{2} \odot \Big(f(t_{k+1}, y_k \oplus f(t_k, y_k)) \oplus f(t_k, y_k) \Big) \Big).$$

The local truncation error is defined as

$$\tau_k = \frac{1}{h} \odot \phi_k,$$

now, it is sufficient to show

$$\lim_{h \to 0} \max D(\tau_k, 0) = 0,$$

where $\tau_k = \frac{h^2}{3!} y'''_{i.gH}(\xi_k)$ and $D(y'''_{i.gH}, 0) \leq M_1$.
Consequently,

$$\lim_{h \to 0} \max D(\tau_k, 0) = \lim_{h \to 0} \frac{h^2}{3!} \odot \max D(y'''_{i.gH}(\xi_k), 0) \leq \frac{h^2}{3!} M_1 = 0.$$

If $y(t)$ is $[(ii) - gH]$-differentiable,

$$\tau_k = \ominus(-1)\frac{h^2}{3!} \odot y'''_{ii.gH}(\xi_k),$$

and $D(y'''_{ii.gH}, 0) \leq M_2$,
therefore

$$\lim_{h \to 0} \max D(\tau_k, 0) = \lim_{h \to 0} |\frac{-h^2}{3!}| \odot \max D(\ominus y'''_{i.gH}(\xi_k), 0) \leq \frac{h^2}{3!} M_2 = 0,$$

so, the proof of the theorem is complete.

Lemma 3.1 (see. [23]) *For all real z,*

$$1 + z \leq e^z. \tag{3.15}$$

Theorem 3.3 *Let $y'''_{gH}(t)$ exists and $f(t, y(t))$ satisfies in Lipschitz condition on the $\{(t, y(t)) | t \in [0, p], y \in \overline{B}(y_0, q), \ p, q > 0\}$, then the Modified Euler Method converges to the solution of fuzzy differential equation (3.1).*

Proof Let $y(t)$ is $[(i) - gH]$-differentiable, the Modified Euler method may be written in the form

$$y_{k+1} = y_k \oplus h \odot \phi(t_n, y_n; h), \tag{3.16}$$

where

$$\phi(t_n, y_n; h) = \frac{1}{2}[f(t_n, y_n) \oplus f(t_n + h, y_n \oplus h \odot f(t_n, y_n))],$$

and $\phi(., .; .)$ is continuous function of its variables. First we want to verify the Lipschitz condition of the function ϕ for the Modified Euler method.

From Lipschitz condition, we obtain

$$\begin{aligned} D(\phi(t,u;h),\phi(t,v;h)) &\le \frac{1}{2}D(f(t,u),f(t,v)) \\ &+ \frac{1}{2}D(f(t+h,u\oplus h\odot f(t,u)),f(t+h,v\oplus h\odot f(t,v))), \\ &\le \frac{L}{2}D(u,v)+\frac{L}{2}D(u\oplus h\odot f(t,u),v\oplus h\odot f(t,v)), \\ &\le L(1+\frac{hL}{2})D(u,v)=L_\phi D(u,v), \end{aligned}$$

i.e. ϕ satisfies a Lipschitz condition with constant $(L+\frac{1}{2}hL^2)$.

Now the exact solution of Eq. (3.39) is satisfies:

$$y(t_{k+1}) = y(t_k)\oplus h\odot \phi(t_k, y(t_k); h)\oplus R_k, \tag{3.17}$$

where $R_k = \frac{h^3}{3!}\odot y'''_{i.gH}(t_k)$.

Subtracting Eq. (3.17) from Eq. (3.16), and by using Lipschitz condition, the following equation is obtained

$$D(y(t_{k+1}), y_{k+1}) \le (1+hL_{\phi_k})D(y(t_k), y_k)\oplus D(R_k, 0), \tag{3.18}$$

we put

$$L_\phi = \max_{0\le k\le N}\{L_{\phi_k}\},\quad R = \max_{0\le k\le N} D(R_k, 0),$$

and by substituting in Eq. (3.18), get

$$D(y(t_{k+1}), y_{k+1}) \le (1+hL_\phi)D(y(t_k), y_k)\oplus R,$$

it is easy to see that

$$D(y(t_{k+1}), y_{k+1}) \le (1+hL_\phi)^{k+1}D(y(t_0), y_0)\oplus[1+(1+hL_\phi)+\cdots+(1+hL_\phi)^k]\odot R.$$

On the other hand,

$$\sum_{i=0}^{k}(1+hL_\phi)^i = \frac{(1+hL_\phi)^{k+1}-1}{hL_\phi},$$

Now for $0\le (k+1)h\le T$, $(k+1)\le(N-1)$, and by using Lemma 3.15, we obtain

$$D(y(t_{k+1}), y_{k+1}) \le e^{TL_{\phi_k}}D(y(t_0), y_0)+\frac{R}{hL_\phi}[e^{TL_\phi}-1],$$

where $R = \max_{0\le k\le N} D(\frac{h^3}{3!}\odot y'''_{i.gH}(\xi_k), 0)$.

It is clear that $D(y(t_0), y_0) = 0$, so

$$D(y(t_{k+1}), y_{k+1}) \leq \frac{h^2}{6L_\phi}[e^{TL_\phi} - 1] \max_{0\leq t\leq T} D(y'''_{i.gH}(t), 0).$$

Thus, $\lim_{h\to 0} D(y(t_{k+1}), y_{k+1}) \to 0$ and in this case, Modified Euler method is converges.

Consider $y(t)$ is $[(ii) - gH]$-differentiable, the Modified Euler method is written as

$$y_{k+1} = y_k \ominus (-1)h \odot \varphi(t_n, y_n; h), \tag{3.19}$$

where

$$\varphi(t_n, y_n; h) = \frac{1}{2}[f(t_n, y_n) \oplus f(t_n + h, y_n \ominus (-1)h \odot f(t_n, y_n))],$$

and $\varphi(.,.;.)$ is continuous function of its variables. The Lipschitz condition of the function φ for the Modified Euler method is as follows

$$\begin{aligned} D(\varphi(t, u; h), \varphi(t, v; h)) &\leq \frac{1}{2}D(f(t, u), f(t, v)) \\ &+ \frac{1}{2}D(f(t+h, u \ominus (-1)hf(t, u)), f(t+h, v \ominus (-1)hf(t, v))), \\ &\leq \frac{L}{2}D(u, v) + \frac{L}{2}D(u \ominus (-1)h \odot f(t, u), v \ominus (-1)h \odot f(t, v)), \\ &\leq L(1 - \frac{hL}{2})D(u, v) = L_\varphi D(u, v), \end{aligned}$$

i.e. φ satisfies a Lipschitz condition with constant $(L - \frac{1}{2}hL^2)$.

Now the exact solution of Eq. (3.1) is satisfies:

$$y(t_{k+1}) = y(t_k) \ominus (-1)h \odot \varphi(t_k, y(t_k); h) \oplus J_k, \tag{3.20}$$

where $J_k = -\frac{h^3}{3!} \odot y'''_{ii.gH}(t_k)$.

Subtracting Eq. (3.20) from Eq. (3.19), and by using Lipschitz condition, the following equation is obtained

$$D(y(t_{k+1}), y_{k+1}) \leq (1 - hL_{\phi_k})D(y(t_k), y_k) \oplus D(J_k, 0), \tag{3.21}$$

now by putting

$$L_\varphi = \max_{0\leq k\leq N}\{L_{\varphi_k}\}, \quad J = \max_{0\leq k\leq N} D(J_k, 0),$$

and by similar procedure we obtain

$$D(y(t_{k+1}), y_{k+1}) \leq \frac{-h^2}{6L_\phi}[\frac{1}{e^{TL_\phi}} - 1] \max_{0\leq t\leq T} D(y'''_{ii.gH}(t), 0).$$

Thus, $\lim_{h\to 0} D(y(t_{k+1}), y_{k+1}) \to 0$.

Theorem 3.4 *The Modified Euler method is stable.*

Proof Let $y_{k+1}, k+1 \geq 0$ be the solution of the Modified Euler method with initial condition $y_0 \in \mathbb{R}_F$ and let τ_{k+1} be the solution of the Modified Euler method with perturbed fuzzy initial condition $\tau_0 = y_0 \oplus \epsilon \in \mathbb{R}_F$, therefore if $y(t)$ is $[(i) - gH]$-differentiable

$$\tau_{k+1} = \tau_k \oplus \frac{h}{2} \odot \left(f(t_{k+1}, \tau_k \oplus h \odot f(t_k, \tau_k)) \oplus f(t_k, \tau_k) \right), \quad \tau_0 = y_0 \oplus \epsilon,$$

by using Lipschitz condition and properties of Hausdorff distance, we have

$$D(y_{k+1}, \tau_{k+1}) \leq D(y_k, \tau_k) + \frac{h}{2}L\left(D(y_k, \tau_k) + hLD(y_k, \tau_k) \right) + \frac{h}{2}LD(y_k, \tau_k),$$

Thus

$$D(y_{k+1}, \tau_{k+1}) \leq (1 + hL + \frac{(hL)^2}{2})D(y_k, \tau_k).$$

Now by iterating the above inequality, we can write

$$D(y_{k+1}, \tau_{k+1}) \leq (1 + hL + \frac{(hL)^2}{2})^k D(y_0, \tau_0) \leq (e^{hL})^k D(y_0, \tau_0) \leq e^{LT} D(y_0, \tau_0) \leq \kappa D(y_0, \tau_0),$$

where $\kappa = e^{TL}$, and $kh \leq (k+1)h \leq T$. So, if $y(t)$ is $[(i) - gH]$-differentiable this method is stable.

Now, we suppose that $y(t)$ is $[(ii) - gH]$-differentiable, it is easy to see that

$$D(y_{k+1}, \tau_{k+1}) \leq D(y_k, \tau_k) - \frac{h}{2}L\left(D(y_k, \tau_k) - hLD(y_k, \tau_k) \right) - \frac{h}{2}LD(y_k, \tau_k),$$

therefor

$$D(y_{k+1}, \tau_{k+1}) \leq (1 - hL + \frac{(hL)^2}{2})D(y_k, \tau_k),$$

by repeating the earlier procedure we obtain

$$D(y_{k+1}, \tau_{k+1}) \leq (1 - hL + \frac{(hL)^2}{2})^k D(y_0, \tau_0) \leq (e^{-hL})^k D(y_0, \tau_0) \leq e^{-LT} D(y_0, \tau_0) \leq \kappa D(y_0, \tau_0),$$

where $\kappa = e^{-TL}$, and $kh \leq (k+1)h \leq T$, therefore the Modified Euler method is stable.

3.2 Piecewise Approximation Method

The second method is Piecewise Approximation method (PWA method), In this method we are interested in finding solution of FDE (3.1), in the form of fuzzy piecewise polynomial in each subintervals of solution. Without loss of generality, let $y(t) \in C^2_{gH}([0, T], \mathbb{R}_{\mathcal{F}})$. We construct fuzzy piecewise polynomial $s(t)$ of order 3, in four cases as follows:

Case(i): Let $y(t) \in C^2_{gH}([0, T], \mathbb{R}_{\mathcal{F}})$ is $[(i) - gH]$-differentiable solution of problem (3.1), and $y^{(i)}(t),\ \ i = 1, 2$ are $[(i) - gH]$-differentiable, then

$$s_{i.gH,0}(t) = y(0) \oplus y'_{i.gH}(0) \odot t \oplus y''_{i.gH}(0) \odot \frac{t^2}{2!} \oplus \alpha_0 \odot \frac{t^3}{3!}, \tag{3.22}$$

where $0 \le t \le h$, with the last coefficient α_0 as yet undetermine.

Then approximate solution $s_{i.gH,k}(t)$ in interval $[kh, (k+1)h]$ is obtained as follows:

$$s_{i.gH,k}(t) = \sum_{\ell=0}^{2} s^{(\ell)}_{i.gH,k-1}(kh) \odot \frac{(t-kh)^i}{i!} \oplus \alpha_k \odot \frac{(t-kh)^3}{3!}, \quad k = 1, \cdots, n-1. \tag{3.23}$$

Now we determine $\alpha_k = (\alpha_k^-(r), \alpha_k^+(r))$ for $k = 0, \cdots, n$ by requiring the $s_{i.gH,k}(t)$ should be satisfy Eq. (3.1) for $t = kh$. Therefore $s_{i.gH}(t)$ is obtained as follows

$$s_{i.gH}(t) = \begin{cases} s_{i.gH,0}(t), \ t \in [0, h] \\ s_{i.gH,k}(t), \ t \in [kh, (k+1)h] \end{cases}$$

Case(ii): Now, consider $y(t) \in C^2_{gH}([0, T], \mathbb{R}_{\mathcal{F}})$ is $[(ii) - gH]$-differentiable solution of problem (3.1), and $y^{(i)}(t),\ \ i = 1, 2$ are $[(ii) - gH]$-differentiable, then the first component of $s_{ii.gH}(t)$ denoted by $s_{ii.gH,0}(t)$ is obtained as follows:

$$s_{ii.gH,0}(t) = y(0) \ominus (-1)t \odot y'_{ii.gH}(0) \ominus (-1)\frac{t^2}{2} \odot y''_{ii.gH}(0) \ominus (-1)\frac{t^3}{3!} \odot \beta_0, \tag{3.24}$$

for $0 \le t \le h$, with the last coefficient β_0 as yet undetermine.
Also approximate solution $s_{ii.gH,k}(t)$ in interval $[kh, (k+1)h]$ is obtained as follows:

$$s_{ii.gH,k}(t) = \sum_{\ell=0}^{2} s^{(\ell)}_{ii.gH,k-1}(kh) \odot \frac{(t-kH)^i}{i!} \oplus \beta_k \odot \frac{(t-kH)^3}{3!}, \quad k = 1, \cdots, n-1, \tag{3.25}$$

Now we determine $\beta_k = (\beta_k^-(r), \beta_k^+(r))$ for $k = 0, \cdots, n$, by requiring the $s_{ii.gH,k}(t)$ should be satisfy equation (3.1) for $t = kh$.

Therefore $s_{ii.gH}(t)$ is obtained as follows

$$s_{ii.gH}(t) = \begin{cases} s_{ii.gH,0}(t), \ t \in [0, h] \\ s_{ii.gH,k}(t), \ t \in [kh, (k+1)h] \end{cases}$$

Case(iii): If the assumptions of case (i), are true for $y(t), \ t \in [0, t_j]$ and case(ii) are true for $y(t), \ t \in [t_{j+1}, T]$ ($\gamma \in [0, T]$ is a switching point type I), the PWA method can be written as follows:

$$s_{sw_I}(t) = \begin{cases} s_{i.gH}(t), \ \ t \in [0, t_j] \\ s_{ii.gH}(t), \ t \in [t_{j+1}, T] \end{cases}$$

Case(iv): If the assumptions of case(ii), are true for $y(t), \ t \in [0, t_j]$ and case(i), are true for $y(t), \ t \in [t_{j+1}, T]$ ($\gamma \in [0, T]$ is a switching point type II), the PWA method can be written as follows:

$$s_{sw_{II}}(t) = \begin{cases} s_{ii.gH}(t), \ t \in [0, t_j] \\ s_{i.gH}(t), \ \ t \in [t_{j+1}, T] \end{cases}$$

Now we prove that there exists a unique solution $s(t)$ approximating the solution $y(t)$ of fuzzy initial value problem (3.1), provided that the size of the subinterval h satisfies $h < \frac{3}{L}$.

Theorem 3.5 *If* $h < \frac{3}{L}$, *then the piecewise approximate solution* $s(t)$, *exists and is unique.*

Proof Without loss of generality, we suppose that $y(t)$ is $[(i) - gH]$-differentiable solution on the interval $[kh, (k+1)h]$ we define:

$$\begin{aligned} s_k^-(t,r) &= \sum_{i=0}^{2} (s_{k-1}^-)^{(i)}(kh,r) \odot \frac{(t-kh)^i}{i!} \oplus \alpha_k^-(r) \odot \frac{(t-kh)^3}{3!}, \qquad (3.26) \\ &= A_k^-(t,r) \oplus \alpha_k^-(r) \odot \frac{(t-kh)^3}{3!}, \end{aligned}$$

and

$$\begin{aligned} s_k^+(t,r) &= \sum_{i=0}^{2} (s_{k-1}^+)^{(i)}(kh,r) \odot \frac{(t-kh)^i}{i!} \oplus \alpha_k^+(r) \odot \frac{(t-kh)^3}{3!}, \qquad (3.27) \\ &= A_k^+(t,r) \oplus \alpha_k^-(r) \odot \frac{(t-kh)^3}{3!}, \end{aligned}$$

$A_k^-(t,r)$ and $A_k^+(t,r)$ are known by continuity conditions. Let us prove that $\alpha_k^-(r)$ and $\alpha_k^+(r)$ may be uniquely determined from

$$(s^-)'((k+1)h, r) = f^-((k+1)h, s((k+1)h), r), \tag{3.28}$$

$$(s^+)'((k+1)h, r) = f^+((k+1)h, s((k+1)h), r). \tag{3.29}$$

Replacing s_k^-, s_k^+ in (3.28)and (3.29)and we obtain:

$$\alpha_k^-(r) = \frac{2}{h^2}\{f^-((k+1)h, A_k^-((k+1)h) + \alpha_k^-(r)\frac{h^3}{3!}) - (A_k^-)'((k+1)h, r)\}, \tag{3.30}$$

$$\alpha_k^+(r) = \frac{2}{h^2}\{f^+((k+1)h, A_k^+((k+1)h, r) + \alpha_k^+(r)\frac{h^3}{3!}) - (A_k^+)'((k+1)h, r)\}, \tag{3.31}$$

for the unknown $\alpha_k^-(r)$ and $\alpha_k^+(r)$.
we define

$$\begin{aligned} \alpha_k^- &= g_{k,1}(\alpha_k^-), \\ \alpha_k^+ &= g_{k,2}(\alpha_k^+), \end{aligned}$$

that is

$$\begin{aligned} \alpha_k &= g(\alpha_k), \\ \alpha_k &= (\alpha_k^-, \alpha_k^+)^T \\ g(\alpha_k) &= (g_{k,1}, g_{k,2})^T, \end{aligned}$$

therefore

$$\|g(\alpha_k^1) - g(\alpha_k^2)\| \leq \frac{hL}{3}\|\alpha_k^1 - \alpha_k^2\|, \tag{3.32}$$

where L is the Lipschitz constant. Therefore $g(\alpha_k)$ is a strong contraction mapping for $h < \frac{3}{L}$. ∎

We are now interested in the order of convergence. We shall give a result which shows that our constructed method is of order 4.

Theorem 3.6 *There exists a constant K such that $D(s(h), y(h)) < Kh^4$.*

Proof Let $y(t)$ is $[(i) - gH]$-differentiable solution, therefore

$$s^-(h, r) = y^-(0, r) + h(y^-)'(0, r) + \frac{h^2}{2!}(y^-)''(0, r) + \frac{h^3}{3!}\alpha_0^-, \tag{3.33}$$

$$s^+(h, r) = y^+(0, r) + h(y^-)'(0, r) + \frac{h^2}{2!}(y^+)''(0, r) + \frac{h^3}{3!}\alpha_0^+, \tag{3.34}$$

also for $y(t)$, we have:

$$y^-(h,r) = y^-(0,r) + h(y^-)'(0,r) + \frac{h^2}{2!}(y^-)''(0,r) + \frac{h^3}{3!}(y^-)'''(0,r) + \frac{h^4}{4!}(y^-)^{(4)}(\xi,r),$$

$$y^+(h,r) = y^+(0,r) + h(y^+)'(0,r) + \frac{h^2}{2!}(y^+)''(0,r) + \frac{h^3}{3!}(y^+)'''(0,r) + \frac{h^4}{4!}(y^+)^{(4)}(\xi,r),$$

where $0 < \xi < h$.

Therefore by using Hausdorff distance D, we obtain:

$$D(s(h),\ y(h)) = \frac{h^3}{3!} \sup_{r\in[0,\ 1]} \max\Big\{|\alpha_0^- - (y^-)'''(0,r) - \frac{1}{4}(y^-)^{(4)}(\xi,r)|,$$
$$|\alpha_0^+ - (y^+)'''(0,r) - \frac{1}{4}(y^+)^{(4)}(\xi,r)|\Big\},$$

then

$$D(s(h),\ y(h)) = \frac{h^3}{3!} D(\alpha_0,\ y'''(0) + \frac{1}{4}y^{(4)}(\xi)). \tag{3.35}$$

The proof of theorem is reduced to showing that α_0 is uniformly bounded as function of h. For this propose by using (3.33) and (3.34), we have:

$$\alpha_0^- = \frac{2!}{h^2}\Big\{f(h,\ y^-(0,r) + h(y^-)'(0,r) + \frac{h^2}{2!}(y^-)''(0,r) + \frac{h^3}{3!}\alpha_0^-) - (y^-)'(0,r) - h(y^-)''(0,r),\Big\},$$
$$\alpha_0^+ = \frac{2!}{h^2}\Big\{f(h,\ y^+(0,r) + h(y^+)'(0,r) + \frac{h^2}{2!}(y^+)''(0,r) + \frac{h^3}{3!}\alpha_0^+) - (y^+)'(0,r) - h(y^+)''(0,r),\Big\},$$

we put

$$g_0^-(u) = \frac{2!}{h}\Big\{f(h,\ y^-(0,r) + h(y^-)'(0,r) + \frac{h^2}{2!}(y^-)''(0,r) + \frac{h^3}{3!}u^-) - (y^-)'(0,r) - h(y^-)''(0,r),\Big\},$$
$$g_0^+(u) = \frac{2!}{h}\Big\{f(h,\ y^+(0,r) + h(y^+)'(0,r) + \frac{h^2}{2!}(y^+)''(0,r) + \frac{h^3}{3!}u^+) - (y^+)'(0,r) - h(y^+)''(0,r),\Big\},$$

The fuzzy functions $g_0(u)$ is a strong contraction for all $h < 3/L$, because:

$$D(g_0(u), g_0(u^*)) \le \frac{hL}{3} D(u, u^*). \tag{3.36}$$

In particular for $h \le 1/L$, we have

$$D(g_0(u_1), g_0(u_2)) \le \frac{1}{3} D(u_1, u_2),$$

taking $u_1 = \alpha_0$, $u_2 = 0$,

$$D(g_0(\alpha_0), g_0(0)) \leq \frac{1}{3} D(\alpha_0, 0),$$

$$D(g_0(\alpha_0), 0) + D(g_0(0), 0) \leq \frac{1}{3} D(\alpha_0, 0),$$

$$D(\alpha_0, 0) + D(g_0(0), 0) \leq \frac{1}{3} D(\alpha_0, 0),$$

thus:

$$D(\alpha_0, 0) \leq \frac{3}{2} D(g_0(0), 0). \tag{3.37}$$

Now, we estimate $g_0^-(0)$ and $g_0^+(0)$, as follows:

$$\begin{aligned} g_0^-(0) &= \frac{2!}{h}\Big\{ f(h, y^-(0,r) + h(y^-)'(0,r) + \frac{h^2}{2!}(y^-)''(0,r)) - (y^-)'(0,r) - h(y^-)''(0,r), \Big\}, \\ &= \frac{2!}{h^2}\{(y^-)'(h,r) + O(h^3) - (y^-)'(0,r) - h(y^-)''(0,r))\}, \\ &= \frac{2!}{h^2}\{(y^-)'(0,r) + h(y^-)''(0,r) + o(h^2) - (y^-)'(0,r) - h(y^-)''(0,r))\} \leq M_1, \end{aligned}$$

and similarly

$$g_0^+(u) \leq M_2, \tag{3.38}$$

for some constants M_1 and M_2. We consider $M = \max\{M_1, M_2\}$, so that $D(\alpha_0, 0) < \frac{3}{2}M$, and α_0 is uniformly bounded for all $h \leq 1/L$.

Clearly we have

$$D(s(h), y(h)) = o(h^3)$$

also $D(s'(h), y'(h)) = o(h^3)$ because

$$D(s'(h), y'(h)) = D(f(h, s(h)), f(h, y(h))) \leq LD(s(h), y(h)) < kLh^3 = K^* h^3,$$

where let $k^* = \max\{k, Lk\}$. Therefore

$$(s^+)'(h,r) = (y')^+(h,r) + o(h^3) = (y^+)'(0,r) + h(y^+)''(0,r) + \frac{h^2}{2!}(y^+)'''(0,r) + o(h^3), \tag{3.39}$$

$$(s^-)'(h,r) = (y')^-(h,r) + o(h^3) = (y^-)'(0,r) + h(y^-)''(0,r) + \frac{h^2}{2!}(y^-)'''(0,r) + o(h^3), \tag{3.40}$$

by definitions of $(s')^+(h,r)$ and $(s')^-(h,r)$, we have

$$s^{+}(h,r) = y^{+}(0,r) + h(y^{-})'(0,r) + \frac{h^2}{2!}(y^{+})''(0,r) + \frac{h^3}{3!}\alpha_0^{+}, \tag{3.41}$$

$$s^{-}(h,r) = y^{-}(0,r) + h(y^{-})'(0,r) + \frac{h^2}{2!}(y^{-})''(0,r) + \frac{h^3}{3!}\alpha_0^{-}, \tag{3.42}$$

combining two equations (3.39), (3.41) and (3.40), (3.41) and solving for α_0^{+} and α_0^{-}, we obtain

$$\alpha_0^{-} = (y_0^{-})''' + o(h),$$
$$\alpha_0^{+} = (y_0^{+})''' + o(h),$$

Finally

$$D(s(h), y(h)) = o(h^4).$$

4 Numerical Example

In this section we will provide two examples to emphasize acceptable accuracy of Modified Euler Method and PWA method. All numerical computations performed by using Maple 13 software package.

Example 4.1 [11] The ordinary fuzzy differential equation

$$\begin{cases} y'_{ii}(t) = -y(t) \oplus t \odot (0.7, 1, 1.8), \quad 0 \le t \le 1, \\ y(0) = (r, 2.2 - 1.2r), \\ y'(0) = (-2.2 + 1.2r, -r), \\ y''(0) = (0.2r + 1.8, 2.9 - 0.9r), \end{cases} \tag{4.1}$$

is consider.

The global truncation error of Modified Euler and Euler method have been reported for $h = 0.025$ and 0.005 in Table 1.

Now applying the PWA method by $h = 0.25$ and reported in Table 2. The exact solution and approximated solution $s(t)$ are plotted in Figs. 1, 2, 3 and 4.

Example 4.2 [27] A tank with a heating system is displayed in Fig. 5, where $\tilde{R} = 0.5$, the thermal capacitance is $\tilde{C} = 2$ also the temperature is ψ. The model is formulated as follows,

$$\begin{cases} \phi'(t) = -\frac{1}{RC}\phi(t), \quad 0 \le t \le T, \\ \phi(0) = (\phi^{-}(0,\alpha), \phi^{+}(0,\alpha)), \end{cases} \tag{4.2}$$

Table 1 The global truncation errors for Modified Euler method and Euler method for Example 4.1

t	h = 0.025		h = 0.005	
	Modified Euler	Euler	Modified Euler	Euler
0	0	0	0	0
0.1	2.7852×10^{-5}	3.33362×10^{-3}	1.098×10^{-6}	6.58119×10^{-4}
0.2	5.0403×10^{-5}	6.02895×10^{-3}	1.984×10^{-6}	1.19083×10^{-3}
0.3	6.8410×10^{-5}	8.17763×10^{-3}	2.695×10^{-6}	1.61606×10^{-3}
0.4	8.2535×10^{-5}	9.85964×10^{-3}	3.252×10^{-6}	1.94945×10^{-3}
0.5	9.3351×10^{-5}	1.11446×10^{-2}	3.679×10^{-6}	2.20464×10^{-3}
0.6	1.01361×10^{-4}	1.20932×10^{-2}	3.994×10^{-6}	2.39351×10^{-3}
0.7	1.07002×10^{-4}	1.27580×10^{-2}	4.217×10^{-6}	2.52638×10^{-3}
0.8	1.10650×10^{-4}	1.31847×10^{-2}	4.360×10^{-6}	2.61220×10^{-3}
0.9	1.12636×10^{-4}	1.34127×10^{-2}	4.438×10^{-6}	2.65874×10^{-3}
1.0	1.13242×10^{-4}	1.34763×10^{-2}	4.461×10^{-6}	2.67269×10^{-3}

Table 2 Error of PWA method for h = 0.25 in Example 4.1

t	Error
0	0
0.1	2.5334×10^{-5}
0.2	1.11585×10^{-4}
0.3	9.8167×10^{-5}
0.4	1.8666×10^{-5}
0.5	6.3174×10^{-5}
0.6	1.0778×10^{-5}
0.7	9.2513×10^{-5}
0.8	8.1174×10^{-5}
0.9	4.3971×10^{-5}
1.0	1.02016×10^{-4}

where the initial condition be a symmetric triangular fuzzy number as $\phi(0) = (-a(1-\alpha), a(1-\alpha))$.

The solution of FDE (4.2), by using Modified Euler method, for $a = 2, h = 0.005$ is shown in Fig. 6.

By applying PWA method for $h = 0.25$, Error of this method reported in Table 3.

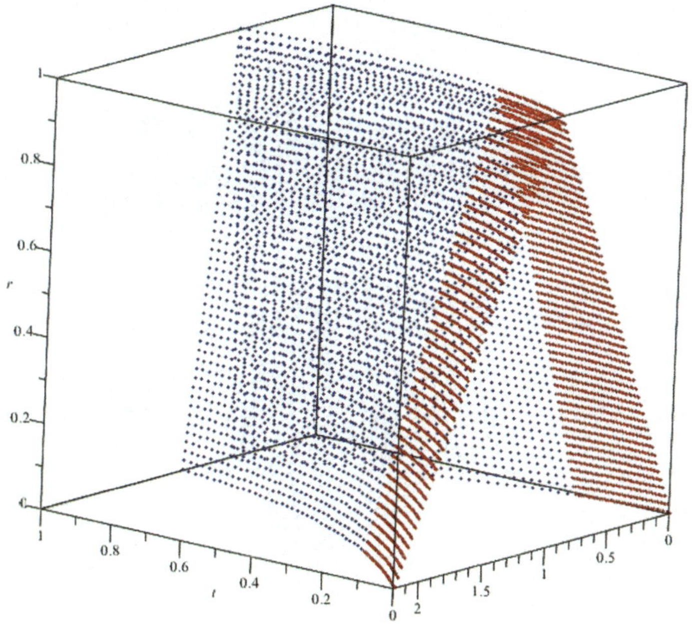

Fig. 1 Comparison of exact and approximated solution in Example 4.1. Red points: $s(t)$ for $t \in [0, 0.25]$; blue points: real solution

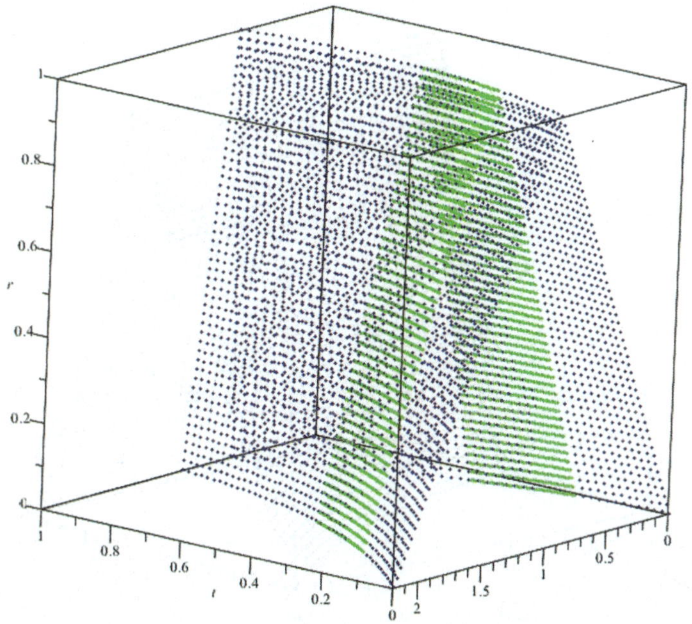

Fig. 2 Comparison of exact and approximated solution in Example 4.1. Green points: $s(t)$ for $t \in [0.25, 0.5]$; blue points: real solution

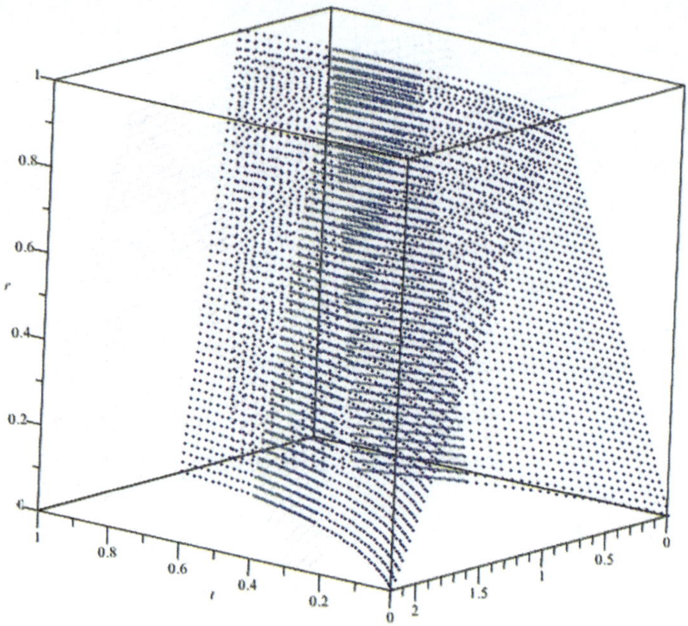

Fig. 3 Comparison of exact and approximated solution in Example 4.1. Dark blue points: $s(t)$ for $t \in [0.5, 0.75]$; blue points: real solution

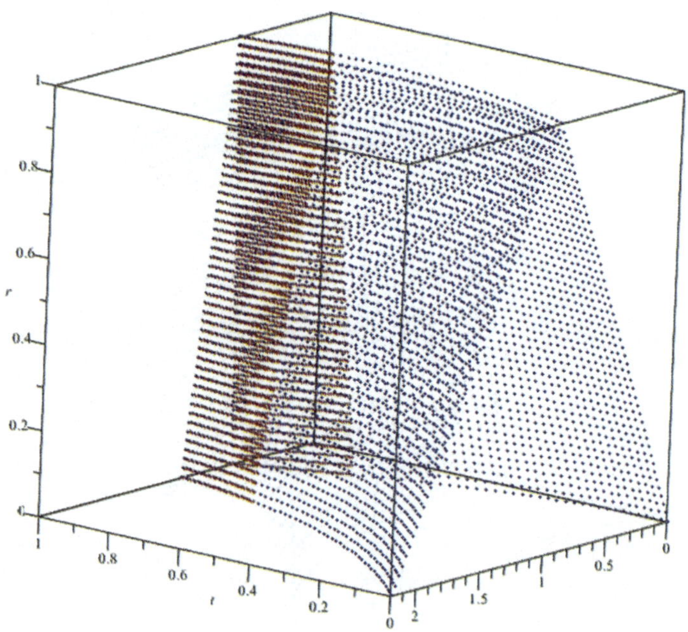

Fig. 4 Comparison of exact and approximated solution in Example 4.1. Brown points: $s(t)$ for $t \in [0.75, 1]$; blue points: real solution

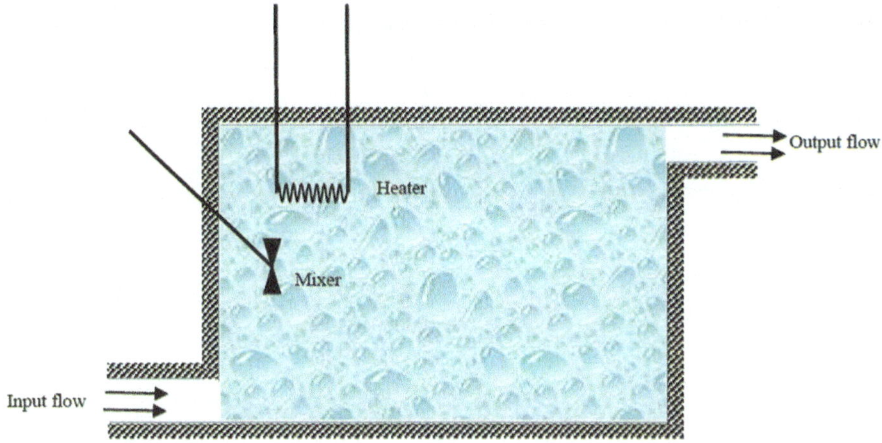

Fig. 5 Thermal system

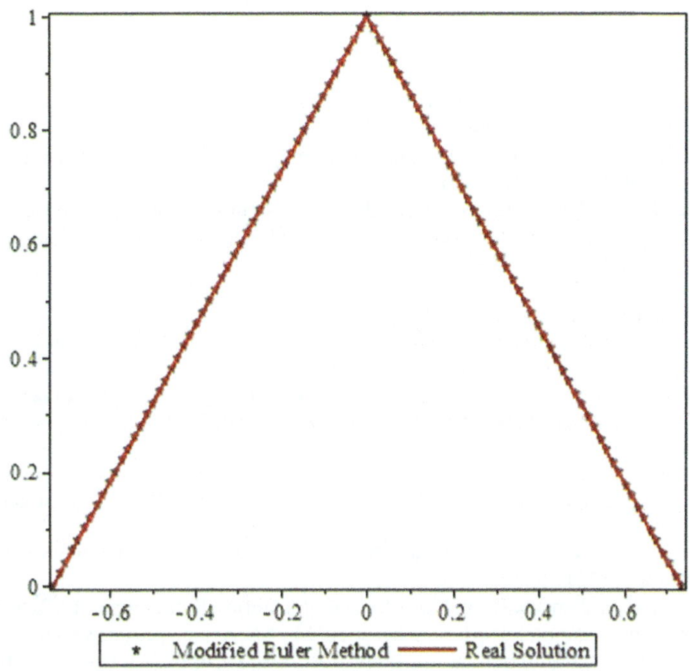

Fig. 6 Approximated and Real solution for Example 4.2

Table 3 Error of PWA method for h = 0.25 in Example 4.2

t	Error
0	0
0.1	0.000017472
0.2	0.000076956
0.3	0.000067702
0.4	0.000012873
0.5	0.000043567
0.6	0.000007431
0.7	0.0000638033
0.8	0.0006830269
0.9	0.0014116710
1.0	0.00168156

References

1. Abbasbandy, S., Allahviranloo, T.: Numerical solution of fuzzy differential equation by Taylor method. J. Comput. Method Appl. Math. **2**, 113–124 (2002)
2. Abbasbandy, S., Allahviranloo, T.: Numerical solution of fuzzy differential equation. Math. Comput. Appl. **7**(1), 41–52 (2002)
3. Abbasbandy, S., Allahviranloo, T.: The Adomian decomposition method applied to the Fuzzy system of Fredholm integral equations of the second kind. Int. J. Uncertain. Fuzziness Knowl. Based Syst. **14**(1), 101–110 (2006)
4. Abbasbandy, S., Allahviranloo, T., Lopez-Pouso, O., Nieto, J.J.: Numerical methods for fuzzy differential inclusions. Comput. Math. Appl. **48**, 1633–1641 (2004)
5. Abbasbandy, S., Allahviranloo, T.: Numerical solution of fuzzy differential equation by Runge-Kutta method. Nonlinear Stud. **11**(1), 117–129 (2004)
6. Ahmadian, A., Salahshour, S., Chan, C., Baleanu, D.: Numerical solutions of fuzzy differential equations by an efficient Runge-Kutta method with generalized differentiability. Fuzzy Sets Syst. **331**, 47–67 (2018)
7. Ahmadian, A., Suleiman, M., Ismail, F.: An improved Runge-Kutta method for solving fuzzy differential equations under generalized differentiability. AIP Conf. Proc. **1482**, 325–330 (2012)
8. Allahviranloo, T., Ahmady, E., Ahmady, N.: Nth order fuzzy linear differential equations. Inf. Sci. **178**: 1309–1324 (2008)
9. Allahviranloo, T., Ahmadi, M.B.: Fuzzy Laplace transforms. Soft Comput. **14**, 235–243 (2010)
10. Allahviranloo, T., Amirteimoori, A., Khezerloo, M., Khezerloo, S.: A new method for solving fuzzy volterra integro-differential equations. Aust. J. Basic Appl. Sci. **5**(4), 154–164 (2011)
11. Allahviranloo, T., Gouyandeh, Z., Armand, A.: A full fuzzy method for solving differential equation based on Taylor expansion. J. Intell. Fuzzy Syst. **29**, 1039–1055 (2015)
12. Allahviranloo, T., Ahmady, N., Ahmady, E.: Numerical solution of fuzzy differential equations by predictor-corrector method. Inf. Sci. **177**(7), 1633–1647 (2007)
13. Allahviranloo, T., Abbasbandy, S., Ahmady, N., Ahmady, E.: Improved predictor-corrector method for solving fuzzy initial value problems. Inf. Sci. **179**, 945–955 (2009)
14. BaloochShahryari, M.R., Salashour, S.: Improved predictor—corrector method for solving fuzzy differential equations under generalized differentiability. J. Fuzzy Set Valued Anal. **2012**, 1–16 (2012)

15. Allahviranloo, T., Abbasbandy, S., Sedaghgatfar, O., Darabi, P.: A new method for solving fuzzy integro-differential equation under generalized differentiability. Neural Comput. Appl. **21**(SUPPL. 1), 191–196 (2012)
16. Bede, B., Gal, S.G.: Generalizations of the differentiability of fuzzy-number-valued functions with applications to fuzzy differential equations. Fuzzy Set Syst. **151**, 581–599 (2005)
17. Bede, B., Gal, S.G.: Remark on the new solutions of fuzzy differential equations. Chaos Solitons Fractals (2006)
18. Bede, B., Stefanini, L.: Solution of fuzzy differential equations with generalized differentiability using LU-parametric representation. EUSFLAT **1**, 785–790 (2011)
19. Bede, B., Stefanini, L.: Generalized differentiability of fuzzy-valued functions. Fuzzy Sets Syst. **230**, 119–141 (2013)
20. Chang, S., Zadeh, L.: On fuzzy mapping and control. IEEE Trans. Syst. Cybern. **2**, 30–34 (1972)
21. Chalco-Cano, Y., Romoman-Flores, H.: On new solutions of fuzzy differential equations. Chaos, Solitons and Fractals **38**, 112–119 (2008)
22. Dubois, D., Prade, H.: Toward fuzzy differential calculus: part 3. Differentiation. Fuzzy Sets Syst. 225–233 (1982)
23. Epperson, J.F.: An Introduction to Numerical Methods and Analysis. Wiley (2007)
24. Goetschel, R., Voxman, W.: Elementary fuzzy calculus. Fuzzy Sets Syst. **24**, 31–43 (1987)
25. Hajighasemi, S., Allahviranloo, T., Khezerloo, M., Khorasany, M., Salahshour, S.: Existence and uniqueness of solutions of fuzzy volterra integro-differential equations. Inf. Process. Manag. Uncertain. Knowl. Based Syst. **81**, 491–500 (2010)
26. Hukuhara, M.: Integration des applications mesurables dont la valeur est un compact convex. Funkcial. Ekvac. **10**, 205–229 (1967)
27. Jafari, R., Razvarz, S.: Solution of fuzzy differential equations using fuzzy Sumudu transforms. Math. Comput. Appl. **23**, (2018). https://doi.org/10.3390/mca23010005
28. Kaleva, O.: Fuzzy differential equations. Fuzzy Sets Syst. **24**, 301–317 (1987)
29. Kandel, A., Byatt, W.J.: Fuzzy differential equations. In: Proceedings of International Conference on Cybernetics and Society, Tokyo (1978)
30. Ma, M., Friedman, M., Kandel, A.: Numerical solutions of fuzzy differential equations. Fuzzy Sets Syst. **105**, 133–138 (1999)
31. Negoita, C.V., Ralescu, D.: Applications of Fuzzy Sets to Systems Analysis. Wiley, New York (1975)
32. Puri, M.L., Ralescu, D.A.: Differentials of fuzzy functions. J. Math. Anal. Appl. **114**, 409–422 (1986)
33. Rabiei, F., Ismail, F., Ahmadian, A., Salahshour, S.: Numerical solution of second-order fuzzy differential equation using improved Runge-Kutta Nystrom method. Math. Probl. Eng. **2013**, 1–10 (2013)
34. Salahshour, S., Ahmadian, A., Abbasbandy, S., Baleanud, D.: M-fractional derivative under interval uncertainty: theory, properties and applications. Chaos, Solitons and Fractals **117**, 84–93 (2018)
35. Stefanini, L.: A generalization of Hukuhara difference for interval and fuzzy arithmetic. Advances in Soft Computing, vol. 48, Springer (2008). An extended version is available online at the RePEc service. https://econpapers.repec.org/paper/urbwpaper/08-5f01.htm
36. Seikkala, S.: On the fuzzy initial value problem. Fuzzy Sets Syst. **24**, 319–330 (1987)
37. Stefanini, L., Bede, B.: Generalized Hukuhara differentiability of interval-valued functions and interval differential equations. Nonlinear Anal. **71**, 1311–1328 (2009)
38. Tapaswini, S., Chakraverty, S.: A new approach to fuzzy initial value problem by improved Euler method. Fuzzy Inf. Eng. **3**, 293–312 (2012)

Solution Strategy for Fuzzy Fractional Order Linear Homogeneous Differential Equation by Caputo-H Differentiability and Its Application in Fuzzy EOQ Model

Mostafijur Rahaman, Sankar Prasad Mondal, A. El Allaoui, Shariful Alam, Ali Ahmadian, and Soheil Salahshour

Abstract This chapter aims to describe a stock dependent memory concerned EOQ model in fuzzy uncertain situation. Before developing the EOQ model, a brief theory of fuzzy fractional linear homogeneous differential equation inspired by Caputo H differentiability has been established in this paper. In the fuzzy fractional linear homogeneous differential equation, both the co-efficient and the initial value are assumed to be fuzzy number.

Keywords Riemann-Liouville differentiability · Caputo differentiability · Laplace transforms · EOQ model · Stock · Memory

1 Introduction

Fractional calculus (FC) is a topic of much enthusiasm for the current era of mathematical research having an origin in distant past. For the mastery of the FC to analyze the dynamical behavior of the physical problems in real world, it has gained much interest for the modeling of problems in the fields of science and technology [1–4].

M. Rahaman · S. Alam
Department of Mathematics, Indian Institute of Engineering Science and Technology, Shibpur, Howrah, India

S. P. Mondal
Department of Applied Science, Maulana Abul Kalam Azad University of Technology, West Bengal, Kolkata, India

A. El Allaoui
LMACS, Laboratoire de Mathematiques Appliquees & Calcul Scientique, Sultan Moulay Slimane University, PO Box 523, 23000 Beni Mellal, Morocco

A. Ahmadian (✉)
Institute of IR 4.0, The National University of Malaysia, UKM, 43600 Bangi, Selangor, Malaysia

S. Salahshour
Faculty of Engineering and Natural Sciences, Bahcesehir University, Istanbul, Turkey

T. Allahviranloo and S. Salahshour (eds.), *Advances in Fuzzy Integral and Differential Equations*, Studies in Fuzziness and Soft Computing 412,
https://doi.org/10.1007/978-3-030-73711-5_5

The notion of the fractional derivative seems to be a global whereas that of integer order derivative is local.

The demonstration of FDE considering Riemann-Liouville differentiability in uncertainty was first proposed by Agarwal et al. [5]. This creates a new domain of research to study of FDE in the uncertain (fuzzy and interval) environment. Fuzzy set theory and differential equation become more popular now days. Hoa et al. [6] described a fuzzy fractional initial valued problems under Caputo Differentiability. Again, an interval counterpart with generalized Hukuhara fractional differentiability was established by Lupulescu et al. [7]. Laplace transform on the fuzzy domain has been established by Allahviranloo and Ahmadi [8]. Subsequently, application of the fuzzy Laplace transformation was discussed by Salahsour and Allahviranloo [9]. Moreover, the Riemann-Liouville and Caputo fractional derivative counterpart of the Laplace transform under uncertainty was established and application on the fractional differential equation was described in the following studies [10–14].

On the inventory control problem is one of the popular research domains. Generally, the inventory models are described by the integer order differential equations. But, involvement of memory effect on the model due to human's participation cannot be described by the traditional differential equations. In this situation, fractional differential equation takes a crucial role. Very recently, some development [15–17] in this direction has been done. On the other hand, some studies [18–45] on fuzzy set theory and its application on inventory have been done in recent past. After reviewing briefly the existing literature our observations are the followings:

(a) Though Salahsour et al. [13, 14] gave the approach to solve the fuzzy fractional differential equation under Caputo H differentiability. The researcher did not consider the coefficient to be fuzzy. They only considered the initial value to be fuzzy number. Here, coefficients as well as the initial value are considered to be fuzzy numbers.
(b) Till now, the study of memory affected inventory model in uncertain environment is almost rare. This current study makes an initiation of the study of the memory effect on an EOQ model with stock dependent demand under fuzzy uncertainty.

Rest of the article is organized as following: The Sect. 2 presents the preliminary concepts of the Riemann-Liouville and Caputo differentiability in fuzzy uncertainty. The Sect. 3 contributes for the analytical solution of fuzzy fractional linear homogeneous equation with fuzzy uncertainty involved in both the initial value and the coefficient. An EOQ model with stock depended demand is described as an application of the proposed theory in the Sect. 4. Finally, the article is ended with the conclusion in the Sect. 5.

2 Preliminaries

In the discussion, the symbols $\widetilde{CF}[0, a]$ and $\widetilde{LF}[a, b]$ are used to denote the space of all fuzzy valued continuous function and the space of all fuzzy valued Labesque integrable functions on closed interval [0, a], where a > 0 is any real number.

Definition 2.1 ([12, 13]) Let, $\tilde{y} \in \widetilde{CF}[0, a] \cap \widetilde{LF}[0, a]$ and $x \in (0, a)$. The Caputo H differentiability of $\tilde{y}$ of order α (where $0 < \alpha < 1$) at x

$$^{C}D_x^{\alpha}\tilde{y}(x) = \frac{1}{\Gamma(1-\alpha)}\int_0^x (x-u)^{-\alpha}\tilde{y}'(u)du \tag{2.1}$$

exists finitely, where $\tilde{y}'$ denote the fuzzy gH differentiability of $\tilde{y}$.

(i) When $\tilde{y}$ is (i)-gH differentiable with the existence of (2.1), then $\tilde{y}$ is called $^{C}[(i)-\alpha]$ differentiable.
(ii) When $\tilde{y}$ is (ii)-gH differentiable with the existence of (2.1), then $\tilde{y}$ is called $^{C}[(ii)-\alpha]$ differentiable.

Theorem 2.1 ([13])*Let,* $\tilde{y} \in \widetilde{CF}[0, a] \cap \widetilde{LF}[0, a]$, $0 < \alpha < 1$ *be any real number and* $x \in (0, a)$.

(a) *If* $\tilde{y}$ *is* $^{C}[(i)-\alpha]$ *differentiable then*

$$^{C}D_x^{\alpha}\tilde{y}(x;r) = \left[{}^{C}D_x^{\alpha}\underline{y(x;r)}, {}^{C}D_x^{\alpha}\overline{y(x;r)}\right]$$

(b) *If* $\tilde{y}$ *is* $^{C}[(ii)-\alpha]$ *differentiable then*

$$^{C}D_x^{\alpha}\tilde{y}(x;r) = \left[{}^{C}D_x^{\alpha}\overline{y(x;r)}, {}^{C}D_x^{\alpha}\underline{y(x;r)}\right]$$

where, $^{C}D_x^{\alpha}\underline{y(x;r)} = \frac{1}{\Gamma(1-\alpha)}\int_0^x (x-u)^{-\alpha}\underline{y'(u;r)}du$ *and* $^{C}D_x^{\alpha}\overline{y(x;r)} = \frac{1}{\Gamma(1-\alpha)}\int_0^x (x-u)^{-\alpha}\overline{y'(u;r)}du$

Theorem 2.2 ([14]) $\tilde{y} \in \widetilde{CF}[0, \infty] \cap \widetilde{LF}[0, \infty]$, $0 < \alpha < 1$ *be any real number. Then,*

(a) $\mathbb{L}\left[{}^{C}D_x^{\alpha}\tilde{y}(x;r); s\right] = s^{\alpha}\mathbb{L}\left[\tilde{y}(x;r); s\right]\Theta_h\tilde{y}(0;r)$, *when* $\tilde{y}$ *is* $^{C}[(i)-\alpha]$ *differentiable.*
(b) $\mathbb{L}\left[{}^{C}D_x^{\alpha}\tilde{y}(x;r); s\right] = -\{\tilde{y}(0;r)\}\Theta_h\left\{-s^{\alpha}\mathbb{L}\left[\tilde{y}(x;r); s\right]\right\}$, *when* $\tilde{y}$ is $^{C}[(ii)-\alpha]$ *differentiable.*

3 Fractional Homogeneous Linear Differential Equations of Order α

Consider, the following fuzzy fractional differential equation

$$^{C}D_{x}^{\alpha}\tilde{y}(x) = \tilde{A}\tilde{y}(x), \text{ with the initial state, } \tilde{y}(0) = \tilde{y}_0 \tag{3.1}$$

Let, the solution of (3.1) is given in terms of r-cut by $\tilde{y}(x) = \left[\underline{y(x;r)}, \overline{y(x;r)}\right]$.

Here, $\tilde{y}_0$ and $\tilde{A}$ are fuzzy numbers given by $\tilde{y}_0 = \left[\underline{y_0(r)}, \overline{y_0(r)}\right]$ and $\tilde{A} = \left[\underline{A(r)}, \overline{A(r)}\right]$ in the r-cut representation. Now following two cases are considered.

3.1 *Case-1*

When the co-efficient $\tilde{A} > 0$, taking the fuzzy Laplace transformation of (3.1) we get,

$$\mathbb{L}\left[^{C}D_{x}^{\alpha}\tilde{y}(x)\right] = \tilde{A}\mathbb{L}\left[\tilde{y}(x)\right] \tag{3.2}$$

Now, we consider the following two sub cases under case-1.
Sub case 1.1 When $\tilde{y}(x)$ is $^{C}[(i) - \alpha]$ differentiable
Then, from (3.2) we get,

$$s^{\alpha}\mathbb{L}\left[\tilde{y}(x)\right]\Theta_h\tilde{y}_0 = \tilde{A}\mathbb{L}\left[\tilde{y}(x)\right]$$

That gives

$$\begin{cases} s^{\alpha}L\left[\underline{y(x;r)}\right] - s^{\alpha-1}\underline{y_0(r)} = \underline{A(r)}L\left[\underline{y(x;r)}\right] \\ s^{\alpha}L\left[\overline{y(x;r)}\right] - s^{\alpha-1}\overline{y_0(r)} = \overline{A(r)}L\left[\overline{y(x;r)}\right] \end{cases}$$

$$\text{i.e.,} \begin{cases} L\left[\underline{y(x;r)}\right] = \frac{s^{\alpha-1}\underline{y_0(r)}}{s^{\alpha}-\underline{A(r)}} \\ L\left[\overline{y(x;r)}\right] = \frac{s^{\alpha-1}\overline{y_0(r)}}{s^{\alpha}-\overline{A(r)}} \end{cases} \tag{3.3}$$

Taking inverse Laplace transform of (3.4), the solution is given by

$$\begin{cases} \underline{y(x;r)} = \underline{y_0(r)}E_{\alpha,\alpha}\left(\underline{A(r)}x^{\alpha}\right) \\ \overline{y(x;r)} = \overline{y_0(r)}E_{\alpha,\alpha}\left(\overline{A(r)}x^{\alpha}\right)\} \end{cases} \tag{3.5}$$

Sub case 1.2 When $\tilde{y}(x)$ is $^{C}[(ii) - \alpha]$ differentiable
Then, from (3.2) we get,

$$\{-\tilde{y}_0\}\Theta_h\{-s^\alpha \mathbb{L}[\tilde{y}(x)]\} = \tilde{A}\mathbb{L}[\tilde{y}(x)]$$

That gives

$$\begin{cases} s^\alpha L\left[\overline{y(x;r)}\right] - s^{\alpha-1}\overline{y_0(r)} = \underline{A(r)}L\left[\underline{y(x;r)}\right] \\ s^\alpha L\left[\underline{y(x;r)}\right] - s^{\alpha-1}\underline{y_0(r)} = \overline{A(r)}L\left[\overline{y(x;r)}\right] \end{cases} \tag{3.6}$$

Solving the system (3.6) we get,

$$\begin{aligned} L\left[\underline{y(x;r)}\right] &= \frac{\overline{A(r)}\,\overline{y_0(r)}s^{\alpha-1} + \underline{y_0(r)}s^{2\alpha-1}}{s^{2\alpha} - \underline{A(r)}\,\overline{A(r)}} \\ &= \frac{\overline{y_0(r)}}{2}\sqrt{\frac{\overline{A(r)}}{\underline{A(r)}}}\left\{\frac{s^{\alpha-1}}{s^\alpha - \sqrt{\underline{A(r)}\,\overline{A(r)}}} + \frac{s^{\alpha-1}}{s^\alpha + \sqrt{\underline{A(r)}\,\overline{A(r)}}}\right\} + \frac{\underline{y_0(r)}s^{2\alpha-1}}{s^{2\alpha} - \underline{A(r)}\,\overline{A(r)}} \end{aligned} \tag{3.7}$$

And

$$\begin{aligned} L\left[\overline{y(x;r)}\right] &= \frac{\underline{A(r)}\,\underline{y_0(r)}s^{\alpha-1} + \overline{y_0(r)}s^{2\alpha-1}}{s^{2\alpha} - \underline{A(r)}\,\overline{A(r)}} \\ &= \frac{\underline{y_0(r)}}{2}\sqrt{\frac{\overline{A(r)}\,\underline{A(r)}}{\overline{A(r)}}}\left\{\frac{s^{\alpha-1}}{s^\alpha - \sqrt{\underline{A(r)}\,\overline{A(r)}}} + \frac{s^{\alpha-1}}{s^\alpha + \sqrt{\underline{A(r)}\,\overline{A(r)}}}\right\} + \frac{\overline{y_0(r)}s^{2\alpha-1}}{s^{2\alpha} - \underline{A(r)}\,\overline{A(r)}} \end{aligned} \tag{3.8}$$

Taking the inverse Laplace transformation of (3.7) we get,

$$\begin{aligned} \underline{y(x;r)} = &\frac{\overline{y_0(r)}}{2}\sqrt{\frac{\overline{A(r)}}{\underline{A(r)}}}\left\{E_{\alpha,\alpha}\left(\sqrt{\underline{A(r)}\,\overline{A(r)}}x^\alpha\right) + E_{\alpha,\alpha}\left(-\sqrt{\underline{A(r)}\,\overline{A(r)}}x^\alpha\right)\right\} \\ &+ \underline{y_0(r)}E_{2\alpha,2\alpha}\left(\underline{A(r)}\,\overline{A(r)}x^{2\alpha}\right) \end{aligned} \tag{3.9}$$

And
Taking the inverse Laplace transformation of (3.8) we get,

$$\begin{aligned} \overline{y(x;r)} = &\frac{\underline{y_0(r)}}{2}\sqrt{\frac{\overline{A(r)}\,\underline{A(r)}}{\overline{A(r)}}}\left\{E_{\alpha,\alpha}\left(\sqrt{\underline{A(r)}\,\overline{A(r)}}x^\alpha\right) + E_{\alpha,\alpha}\left(-\sqrt{\underline{A(r)}\,\overline{A(r)}}x^\alpha\right)\right\} \\ &+ \overline{y_0(r)}E_{2\alpha,2\alpha}\left(\underline{A(r)}\,\overline{A(r)}x^{2\alpha}\right) \end{aligned} \tag{3.10}$$

3.2 Case-2

When the coefficient $\tilde{A} < 0$, let, $\tilde{A} = -\tilde{A}'$, where $\tilde{A}' = \left[\underline{A'(r)}, \overline{A'(r)}\right] > 0$.

Taking the fuzzy Laplace transformation of (3.1) we get,

$$\mathbb{L}\left[{}^{C}D_{x}^{\alpha}\widetilde{y(x)}\right] = \tilde{A}\mathbb{L}\left[\widetilde{y(x)}\right] \tag{3.11}$$

Now, we consider the following two sub cases under case-2.

Sub case 2.1 When $\tilde{y}(x)$ is ${}^{C}[(i)-\alpha]$ differentiable

Then, from (3.11) we get,

$$s^{\alpha}\mathbb{L}[\tilde{y}(x)]\Theta_{h}\tilde{y}_{0} = -\tilde{A}'\mathbb{L}[\tilde{y}(x)]$$

That gives

$$\begin{cases} s^{\alpha}L\left[\overline{y(x;r)}\right] - s^{\alpha-1}\overline{y_{0}(r)} = -\underline{A'(r)}L\left[\underline{y(x;r)}\right] \\ s^{\alpha}L\left[\underline{y(x;r)}\right] - s^{\alpha-1}\underline{y_{0}(r)} = -\overline{A'(r)}L\left[\overline{y(x;r)}\right] \end{cases} \tag{3.12}$$

Solving the system (3.12) we get,

$$\begin{aligned} L\left[\underline{y(x;r)}\right] &= \frac{\underline{y_{0}(r)}s^{2\alpha-1} - \overline{A'(r)}\overline{y_{0}(r)}s^{\alpha-1}}{s^{2\alpha} - \underline{A'(r)}\overline{A'(r)}} \\ &= \frac{\underline{y_{0}(r)}s^{2\alpha-1}}{s^{2\alpha} - \underline{A'(r)}\overline{A'(r)}} - \frac{\overline{y_{0}(r)}}{2}\sqrt{\frac{\overline{A'(r)}}{\underline{A'(r)}}}\left\{\frac{s^{\alpha-1}}{s^{\alpha} - \sqrt{\underline{A'(r)}\overline{A'(r)}}} + \frac{s^{\alpha-1}}{s^{\alpha} + \sqrt{\underline{A'(r)}\overline{A'(r)}}}\right\} \end{aligned} \tag{3.13}$$

And

$$\begin{aligned} L\left[\overline{y(x;r)}\right] &= \frac{\overline{y_{0}(r)}s^{2\alpha-1} - \underline{A'(r)}\underline{y_{0}(r)}s^{\alpha-1}}{s^{2\alpha} - \underline{A'(r)}\overline{A'(r)}} \\ &= \frac{\overline{y_{0}(r)}s^{2\alpha-1}}{s^{2\alpha} - \underline{A'(r)}\overline{A'(r)}} - \frac{\underline{y_{0}(r)}}{2}\sqrt{\frac{\underline{A'(r)}}{\overline{A'(r)}}}\left\{\frac{s^{\alpha-1}}{s^{\alpha} - \sqrt{\underline{A'(r)}\overline{A'(r)}}} + \frac{s^{\alpha-1}}{s^{\alpha} + \sqrt{\underline{A'(r)}\overline{A'(r)}}}\right\} \end{aligned} \tag{3.14}$$

Taking the inverse Laplace transformation of (3.13) we get,

$$\begin{aligned} \underline{y(x;r)} = {} & \underline{y_{0}(r)}E_{2\alpha,2\alpha}\left(\underline{A'(r)}\overline{A'(r)}x^{2\alpha}\right. \\ & - \frac{\overline{y_{0}(r)}}{2}\sqrt{\frac{\overline{A'(r)}}{\underline{A'(r)}}}\left\{E_{\alpha,\alpha}\left(\sqrt{\underline{A'(r)}\overline{A'(r)}}x^{\alpha}\right) + E_{\alpha,\alpha}\left(-\sqrt{\underline{A'(r)}\overline{A'(r)}}x^{\alpha}\right)\right\} \end{aligned} \tag{3.15}$$

And
Taking the inverse Laplace transformation of (3.16) we get,

$$\overline{y(x;r)} = \overline{y_0(r)} E_{2\alpha,2\alpha}\left(\underline{A'(r)}\,\overline{A'(r)} x^{2\alpha}\right. \\ - \frac{\underline{y_0(r)}}{2}\sqrt{\frac{\underline{A'(r)}}{\overline{A'(r)}}}\left\{E_{\alpha,\alpha}\left(\sqrt{\underline{A'(r)}\,\overline{A'(r)}}x^{\alpha}\right) + E_{\alpha,\alpha}\left(-\sqrt{\underline{A'(r)}\,\overline{A'(r)}}x^{\alpha}\right)\right\} \tag{3.16}$$

Sub case 2.2 When $\tilde{y}(x)$ is ${}^{C}[(ii)-\alpha]$ differentiable
Then, from (3.17) we get,

$$\{-\tilde{y}_0\}\Theta_h\{-s^{\alpha}\mathbb{L}[\tilde{y}(x)]\} = -\tilde{A}'\mathbb{L}[\tilde{y}(x)]$$

This gives

$$\begin{cases} s^{\alpha}L\left[\underline{y(x;r)}\right] - s^{\alpha-1}\underline{y_0(r)} = -\underline{A'(r)}L\left[\underline{y(x;r)}\right] \\ s^{\alpha}L\left[\overline{y(x;r)}\right] - s^{\alpha-1}\overline{y_0(r)} = -\overline{A'(r)}L\left[\overline{y(x;r)}\right] \end{cases} \tag{3.17}$$

Then following sub case 1.1 we get,

$$\begin{cases} \underline{y(x;r)} = \underline{y_0(r)}E_{\alpha,\alpha}\left(-\underline{A'(r)}x^{\alpha}\right) \\ \overline{y(x;r)} = \overline{y_0(r)}E_{\alpha,\alpha}\left(-\overline{A'(r)}x^{\alpha}\right) \end{cases} \tag{3.18}$$

4 Application: EOQ Model with Stock Depended Demand

In this section we want to develop a memory concerned EOQ model with the help of fractional differential equation under the following assumptions and notations.

Assumptions

- (i) Fuzzy demand is stock dependent i.e., $\tilde{D}(t) = \tilde{\gamma}\tilde{q}(t)$, where $\tilde{\gamma}$ is a fuzzy constant.
- (ii) Shortage is not allowed.
- (iii) Lead time is zero.
- (iv) Planning horizon is infinite.
- (v) The retailing process in memory concerned.

Notations

$\tilde{D}$ The demand (Units per cycle)
$\widetilde{s_{up}}$ Fuzzy ordering cost per cycle ($)

$\widetilde{h}_c$	Fuzzy holding cost per unit time per unit quantity ($)
T	The cycle time (Weeks)
$\tilde{Q}$	Fuzzy Order quantity (Units)
$\widetilde{TAC}$	Total average cost ($/Weeks)
α	The order of differentiation
β	The order of integration.

4.1 Mathematical Modelling

Suppose, an economic order quantity model starts with the order quantity Q at time $t = 0$. Meeting up the stock depended demand the inventory level gradually decreases and stops at zero level completing the cycle. If the whole process is considered in a memory effected situation with the fuzzy order quantity and the proportional constant γ. Then, the EOQ model can be described by the following fuzzy fractional differential equation

$$\begin{cases} {}^{C}D_t^{\alpha}\tilde{q}(t) = -\tilde{\gamma}\tilde{q}(t), \text{ for } 0 < t < T \\ \tilde{q}(0) = \tilde{Q}, \tilde{q}(T) = \tilde{0} \end{cases} \tag{4.1}$$

In the r-cut representation, the fuzzy parameters are given by $\tilde{Q} = \left[\underline{Q(r)}, \overline{Q(r)}\right]$, $\tilde{D} = \left[\underline{D(r)}, \overline{D(r)}\right]$, $\tilde{\gamma} = \left[\underline{\gamma(r)}, \overline{\gamma(r)}\right]$, $\tilde{s}_{up} = \left[\underline{s_{up}(r)}, \overline{s_{up}(r)}\right]$ and $\tilde{h}_c = \left[\underline{h_c(r)}, \overline{h_c(r)}\right]$

Let, $\tilde{q}(t) = \left[\underline{q(t;r)}, \overline{q(t;r)}\right]$ be the solution of (5.1) in the r-cut representation.

We consider that $\tilde{q}(t)$ is ${}^{C}[(i) - \alpha]$ differentiable. Then, following the Subcase 1.1 in the Sect. 3 we get,

$$\begin{cases} \underline{q(t;r)} = \underline{Q(r)}E_{\alpha,\alpha}\left(\underline{\gamma(r)}t^{\alpha}\right) \\ \overline{q(t;r)} = \overline{Q(r)}E_{\alpha,\alpha}\left(\overline{\gamma(r)}t^{\alpha}\right) \end{cases} \tag{4.2}$$

Set up cost,

$$s_{up} = \left[\underline{s_{up}(r)}, \overline{s_{up}(r)}\right] \tag{4.3}$$

Holding cost, $HC = \left[\underline{HC(T;r)}, \overline{HC(T;r)}\right]$

where

$$\underline{HC(T;r)} = \frac{\underline{h_c(r)}}{\Gamma(\beta)}\int_0^T (T-x)^{\beta-1}\underline{Q(r)}E_{\alpha,\alpha}\left(\underline{\gamma(r)}x^{\alpha}\right)dx$$

$$= \frac{\underline{h_c(r)Q(r)}}{\Gamma(\beta)} \int_0^T (T-x)^{\beta-1}\left[1+\frac{\underline{\gamma(r)}x^{\alpha}}{\Gamma(\alpha+1)}+\frac{\underline{\gamma(r)}^2x^{2\alpha}}{\Gamma(2\alpha+1)}\right]dx$$

$$= \frac{\underline{h_c(r)Q(r)}}{\Gamma(\beta)}\left[\frac{T^{\beta}}{\beta}+\frac{T^{\alpha+\beta}\underline{\gamma(r)}\mathrm{B}(\alpha+1,\beta)}{\Gamma(\alpha+1)}+\frac{\underline{\gamma(r)}^2T^{2\alpha+\beta}\mathrm{B}(2\alpha+1,\beta)}{\Gamma(2\alpha+1)}\right]$$

$$= \underline{h_c(r)Q(r)}\left[\frac{T^{\beta}}{\Gamma(\beta+1)}+\frac{T^{\alpha+\beta}\underline{\gamma(r)}}{\Gamma(\alpha+\beta+1)}+\frac{\underline{\gamma(r)}^2T^{2\alpha+\beta}}{\Gamma(2\alpha+\beta+1)}\right] \tag{4.4}$$

and

$$\overline{HC(T;r)} = \frac{\overline{h_c(r)}}{\Gamma(\beta)}\int_0^T (T-x)^{\beta-1}\overline{Q(r)}E_{\alpha,\alpha}\left(\overline{\gamma(r)}x^{\alpha}\right)dx$$

$$= \frac{\overline{h_c(r)Q(r)}}{\Gamma(\beta)}\int_0^T (T-x)^{\beta-1}\left[1+\frac{\overline{\gamma(r)}x^{\alpha}}{\Gamma(\alpha+1)}+\frac{\overline{\gamma(r)}^2x^{2\alpha}}{\Gamma(2\alpha+1)}\right]dx$$

$$= \frac{\overline{h_c(r)Q(r)}}{\Gamma(\beta)}\left[\frac{T^{\beta}}{\beta}+\frac{T^{\alpha+\beta}\overline{\gamma(r)}\mathrm{B}(\alpha+1,\beta)}{\Gamma(\alpha+1)}+\frac{\overline{\gamma(r)}^2T^{2\alpha+\beta}\mathrm{B}(2\alpha+1,\beta)}{\Gamma(2\alpha+1)}\right]$$

$$= \overline{h_c(r)Q(r)}\left[\frac{T^{\beta}}{\Gamma(\beta+1)}+\frac{T^{\alpha+\beta}\overline{\gamma(r)}}{\Gamma(\alpha+\beta+1)}+\frac{\overline{\gamma(r)}^2T^{2\alpha+\beta}}{\Gamma(2\alpha+\beta+1)}\right] \tag{4.5}$$

Therefore, total average cost during the whole period,

$$TAC = \left[\underline{TAC(T;r)}, \overline{TAC(T;r)}\right]$$

where

$$\underline{TAC(T;r)} = \frac{\underline{s_{up}(r)}}{T} + \underline{h_c(r)Q(r)}\left[\frac{T^{\beta-1}}{\Gamma(\beta+1)}+\frac{T^{\alpha+\beta-1}\underline{\gamma(r)}}{\Gamma(\alpha+\beta+1)}+\frac{\underline{\gamma(r)}^2T^{2\alpha+\beta-1}}{\Gamma(2\alpha+\beta+1)}\right]$$

and

$$\overline{TAC(T;r)} = \frac{\overline{s_{up}(r)}}{T} + \overline{h_c(r)Q(r)}\left[\frac{T^{\beta-1}}{\Gamma(\beta+1)}+\frac{T^{\alpha+\beta-1}\overline{\gamma(r)}}{\Gamma(\alpha+\beta+1)}+\frac{\overline{\gamma(r)}^2T^{2\alpha+\beta-1}}{\Gamma(2\alpha+\beta+1)}\right]$$

Therefore, the optimization problem can be given by

$$\begin{cases} Find\, r \\ Minimize\, \underline{TAC}(T;r) \\ Minimize \overline{TAC}(T;r) \\ 0 < r < 1 \\ \underline{TAC}(T;r) \leq \overline{TAC}(T;r) \\ T > 0. \end{cases}$$

4.2 Numerical Illustration

Considering the following numerical values of the fuzzy parameters as TFN: $\widetilde{s_{up}} = \langle 500, 550, 600\rangle$, $\widetilde{h_c} = \langle 1.4, 1.5, 1.6\rangle$, $\tilde{Q} = \langle 200, 225, 250\rangle$, $\tilde{\gamma} = \langle 0.4, 0.5, 0.6\rangle$. Then, $s_{up}(r) = 500 + 50r$, $\overline{s_{up}(r)} = 600 - 50r$, $\underline{h_c(r)} = 1.4 + 0.1r$, $\overline{h_c(r)} = 1.6 - 0.1r$, $\underline{Q(r)} = 200 + 25r$, $\overline{Q(r)} = 250 - 25r$, $\underline{\gamma(r)} = 0.4 + 0.1r$, $\overline{\gamma(r)} = 0.6 - 0.1r$. Then the optimal value of the objective function, cycle time and inspiration level with respect to the variation of values of the differential and integral memory index are given in the Table 1.

4.3 Graphical Analysis

The data regarding the optimal values of the objective functions and decision variables from Table 1 is presented graphically in this subsection. Figure 1 represents the relationship between the total average cost and total cycle time graphically. In the Fig. 1, the Series 1 and Series 2 represent the graphs for $\underline{TAC}(T;r)$ and $\overline{TAC}(T;r)$ respectively. The three dimensional relationship between the memory index and the total average cost is presented in the Fig. 2. Also, in the Fig. 3, the three dimensional relationship between the memory index and total cycle time is shown.

The lower values of the memory index imply stronger memory. So, with respect to memory our major observations are the followings:

(a) When, the model goes through the stronger memory with both integral and differential memory index, the total average cost increase initially and finally decreases with minimum cost for the both the memory index 0.4. In the absent of differential memory index, the lower values of the integral memory index only results in the decreasing of cost values.

(b) In the absent of integral memory index, the lower values of the differential memory index only results in the increasing of cost values initially and in the decreasing cost value finally.

In this model, the total average cost can be minimized letting the inventory for a long run. The total average cost is given in terms of r-cut for different values of r.

Table 1 Optimal values of the objective function and cycle time with respect to memory index

Differential memory index (α)	Integral memory index (β)	r	T^*	$\underline{TAC(T;r)}^*$	$\overline{TAC(T;r)}^*$
1	1	1	1.738	885.56	885.56
		0.8	1.700	851.38	921.10
		0.6	1.663	819.21	957.23
		0.4	1.629	788.85	993.95
		0.2	1.596	760.13	1031.26
0.8	0.8	1	2.068	918.39	918.39
		0.8	2.013	878.15	960.42
		0.6	1.961	840.45	1003.32
		0.4	1.912	805.05	1047.09
		0.2	1.866	771.75	1091.75
0.6	0.6	1	3.408	856.84	856.84
		0.8	3.295	812.10	903.72
		0.6	3.189	770.18	951.89
		0.4	3.090	730.84	1001.38
		0.2	2.997	693.88	1052.20
0.4	0.4	1	33.705	461.58	461.58
		0.8	32.353	430.30	494.33
		0.6	31.103	400.60	528.43
		0.4	29.946	372.42	563.88
		0.2	28.871	345.70	600.73
0.2	0.2	1	No feasible solution		
		0.8			
		0.6			
		0.4			
		0.2			
1	0.8	1	1.932	890.42	890.42
		0.8	1.887	853.41	929.02
		0.6	1.845	818.67	968.36
		0.4	1.804	786.00	1008.43
		0.2	1.766	755.20	1049.23
1	0.6	1	2.172	875.23	875.23
		0.8	2.120	836.04	916.19
		0.6	2.070	799.34	958.04
		0.4	2.023	764.89	1000.78
		0.2	1.979	732.50	1044.42
1	0.4	1	2.457	839.56	839.56
		0.8	2.398	799.21	881.79
		0.6	2.341	761.46	925.07
		0.4	2.288	726.06	969.40
		0.2	2.237	692.80	1014.79
1	0.2	1	2.789	800.89	800.89
		0.8	2.720	759.56	844.21
		0.6	2.656	720.84	888.75
		0.4	2.594	684.53	934.51
		0.2	2.537	650.42	981.51

(continued)

Table 1 (continued)

Differential memory index (α)	Integral memory index (β)	r	T^*	$\underline{TAC(T;r)}^*$	$\overline{TAC(T;r)}^*$
0.8	1	1	1.776	920.36	920.36
		0.8	1.733	883.28	958.97
		0.6	1.691	848.46	998.28
		0.4	1.652	815.67	1038.29
		0.2	1.615	784.72	1078.99
0.6	1	1	1.916	950.83	950.83
		0.8	1.863	910.29	993.13
		0.6	1.812	872.25	1036.28
		0.4	1.764	836.49	1080.28
		0.2	1.719	802.80	1125.13
0.4	1	1	2.338	962.13	962.13
		0.8	2.261 2.188	917.85	1008.36
		0.6	2.119	876.26	1055.68
		0.4	2.055	837.15	1104.09
		0.2		800.32	

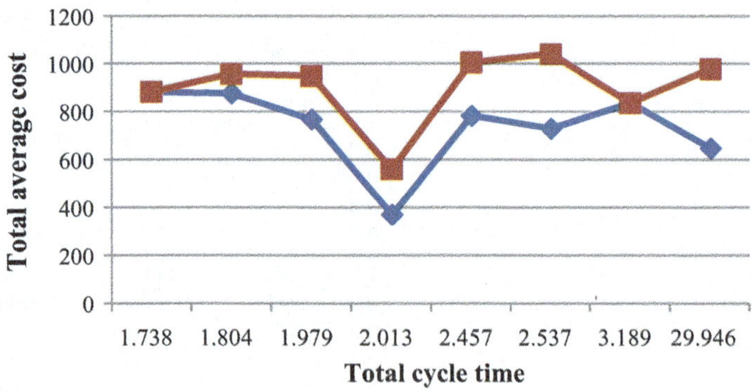

Fig. 1 Total average cost versus total time cycle

5 Conclusion

This article presents a demonstration of memory affected EOQ model through the fuzzy fractional differential equation approach. Here, the approximation of Mittag-Leffler function is considered to discuss analytically. In future, more realistic models with complications may be discussed in this direction.

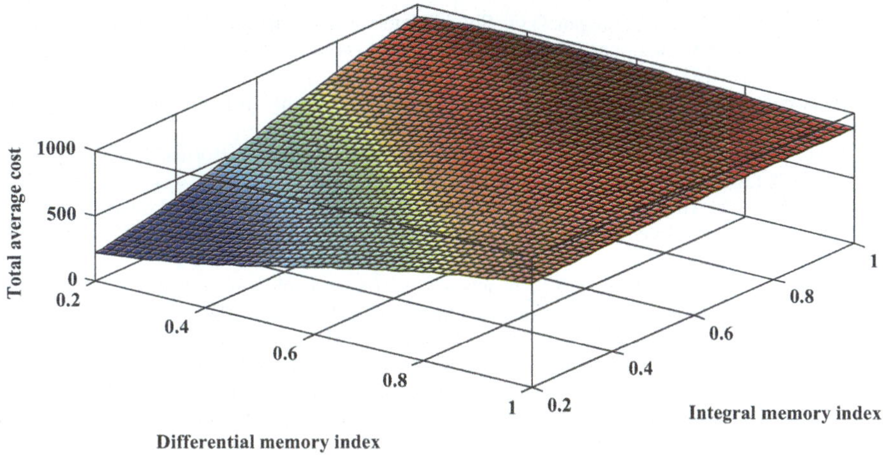

Fig. 2 Total average cost with respect to memory index

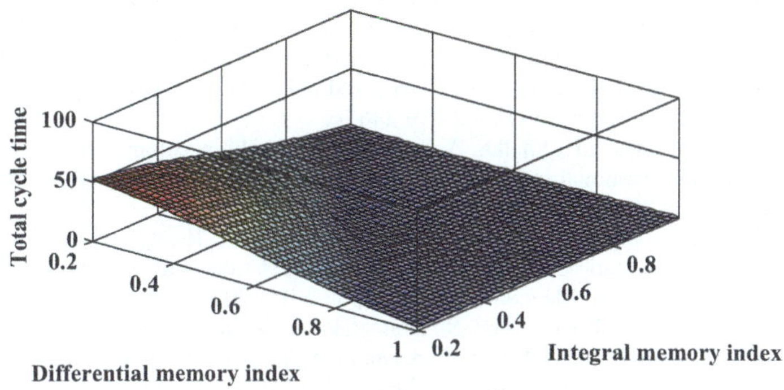

Fig. 3 Total cycle time respect to memory index

References

1. Diethelm, K., Baleanu, D., Scalas, E.: Fractional Calculus: Models and Numerical Methods. World Scientific (2012)
2. Mainardi, F.: Fractional Calculus, Fractals and Fractional Calculus in Continuum Mechanics (1997)
3. Agila, A., Baleanu, D., Eid, R., Iranfoglu, B.: Applications of the extended fractional Euler-Lagrange equations model to freely oscillating dynamical systems. Rom. J. Phys. **61**, 350–359 (2016)
4. Podlubny, I.: Fractional Differential Equations. Academic Press, San Diego, CA (1999)
5. Agarwal, R.P., Lakshmikantham, V., Nieto, JJ.: On the concept of solutions for fractional differential equations with uncertainty. Non Linear Annal. **72**, 2859–2862 (2010)
6. Hoa, N.V., Lupulescu, V., Regan, D.O.: A note on initial value problems for fractional fuzzy differential equations. Fuzzy Sets Syst. (2017). https://doi.org/10.1016/j.fss.2017.10.002

7. Lupulescu, V., Hoa, N.V., Regan, D.O.: Solving interval valued fractional initial value problems under Caputo gH- fractional differentiability. Fuzzy Sets Syst. (2016). https://doi.org/10.1016/j.fss.2016.09.015
8. Allahviranloo, T., Ahmadi, M.B.: Fuzzy Laplace transforms. Soft. Comput. **14**, 235–243 (2010). https://doi.org/10.1007/s00500-008-0397-6
9. Salahshour, S., Allahviranloo, T.: Applications of fuzzy Laplace transforms. Soft. Comput. **17**, 145–158 (2013). https://doi.org/10.1007/s00500-012-0907-4
10. Salahshour, S., Allahviranloo, T., Abbasbandy, S.: Solving fuzzy fractional differential equations by fuzzy Laplace transforms. Commun Nonlinear Sci. Numer. Simulat. **17**, 1372–1381 (2012)
11. Allahviranloo, T., Salahshour, S., Abbasbandy, S.: Explicit solutions of fractional differential equations with uncertainty. Soft Comput. https://doi.org/10.1007/s00500-011-0743-y
12. Mazandrarami, M., VahidianKamyad, A.: Modified fractional Euler method for solving fuzzy fractional initial value problem. Commun. Nonlinear Sci. Numer. Simul. **18**, 12–21 (2013)
13. Salahshour, S., Allahviranloo, T., Abbasbandy, S., Baleanu, D.: Existence and uniqueness results for fractional differential equations with uncertainty. Adv. Diff. Equ. **112** (2012). https://doi.org/10.1186/1687-1847-2012-112
14. Salahshour, S., Ahmadian, A., Senu, N., Baleanu, D., Agarwal, P.: On analytical solutions of the fractional differential equations with uncertainty: application to the Basset problem. Entropy **17**, 885–902 (2015). https://doi.org/10.3390/e17020885
15. Pakhira, R., Ghosh, U., Sarkar, S.: Application of memory effects in an inventory model with price dependent demand rate during shortage. Int. J. Educ. Manage. Eng. **3**, 51–64 (2019)
16. Rahaman, M., Mondal, S.P., Shaikh, A.A., et al.: Arbitrary-order economic production quantity model with and without deterioration: generalized point of view. Adv. Differ. Equ. **2020**, 16 (2020). https://doi.org/10.1186/s13662-019-2465-x
17. Rahaman, M., Mondal, S.P., Shaikh, A.A., et al.: Artificial bee colony optimization-inspired synergetic study of fractional-order economic production quantity model. Soft. Comput. (2020). https://doi.org/10.1007/s00500-020-04867-y
18. Chakraborty, A., Maity, S., Jain, S., Mondal, SP., Alam, S.: Hexagonal fuzzy number and its distinctive representation, ranking, defuzzification technique and application in production inventory management problem. Granular Comput. 1–15 (2020)
19. Chakraborty, A., Mondal, S.P., Alam, S., Ahmadian, A., Senu, N., De, D., Salahshour, S.: The pentagonal fuzzy number: its different representations, properties, ranking, defuzzification and application in game problems. Symmetry **11**(2), 248 (2019)
20. Mondal, S.P., Mandal, M.: Pentagonal fuzzy number, its properties and application in fuzzy equation. Future Comput. Inf. J. **2**(2), 110–117 (2017)
21. Mondal, S.P.: Differential equation with interval valued fuzzy number and its applications. Int. J. Syst. Assurance Eng. Manage. **7**(3), 370–386 (2016)
22. Allahviranloo, T., Abbasbandy, S., Sedaghgatfar, O., Darabi, P.: A new method for solving fuzzy integro-differential equation under generalized differentiability. Neural Comput. Appl. **21**, 191–196 (2012)
23. Hajighasemi, S., Allahviranloo. T., Khezerloo, M., Khorasany M., Salahshour S.: Existence and uniqueness of solutions of fuzzy volterra integro-differential equations. In: Communications in Computer and Information Science, Vol. 81, Part 2, 2010, pp. 491–500, 13th International Conference on Information Processing and Management of Uncertainty, IPMU 2010, Dortmund, Germany, 28 June 2010 through 2 July 2010, Code 98056 (Conference Paper)
24. Gouyandeh, Z., Allahviranloo, T., Abbasbandy, S.: A fuzzy solution of heat equation under generalized Hukuhara differentiability by fuzzy Fourier transform. Fuzzy Sets Syst. **309**(15), 81–97 (2017)
25. Allahviranloo, T., Abbasbandy, S., Rouhparvar, H.: The exact solutions of fuzzy wave-like equations with variable coefficients by a variational iteration method. Appl. Soft Comput. J. **11**(2), 2186–2192 (2011)
26. Chehlabi, M., Allahviranloo, T.: Concreted solutions to fuzzy linear fractional differential equations. Appl. Soft Comput. J. **44**, 108–116 (2016)

27. Allahviranloo, T., Lotfi, F.H., Kiasari, M.K., Khezerloo, M.: On the fuzzy solution of LR fuzzy linear systems. Appl. Math. Model. **37**(3), 1170–1176 (2013)
28. Allahviranloo, T., Amirteimoori, A., Khezerloo, M., Khezerloo, S.: A new method for solving fuzzy Volterra integro-differential equations. Aust. J. Basic Appl. Sci. **5**(4), 154–164 (2011)
29. Allahviranloo, T., Salahshour, S.: Fuzzy symmetric solutions of fuzzy linear systems. J. Comput. Appl. Math. **235**(16), 4545–4553 (2011)
30. Allahviranloo, T., Salahshour, S., Khezerloo, M.: Maximal- and minimal symmetric solutions of fully fuzzy linear systems. J. Comput. Appl. Math. **235**(16), 4652–4662 (2011)
31. Allahviranloo, T., Ahmady, E., Ahmady, N.: Nth-order fuzzy linear differential equations. Inf. Sci. **178**(5), 1309–1324 (2008)
32. Allahviranloo, T., Abbasbandy, S., Ahmad, N., Ahmady, E.: Improved predictor-corrector method for solving fuzzy initial value problems. Inf. Sci. **179**(7), 945–955 (2009)
33. Abbasbandy, S., Allahviranloo, T.: Method applied to the Fuzzy system of Fredholm integral equations of the second kind. Int. J. Uncertain. Fuzz. Knowl. Based Syst. **14**(1), 101–110 (2006)
34. Salahshour, S., Ahmadian, A., Ismail, F., Baleanu, D., Senu, N.: A new fractional derivative for differential equation of fractional order under interval uncertainty. Adv. Mech. Eng. **7**(12), 1–11 (2015)
35. Ahmadian, A., Chan, S., Salahshour, S., Vaitheeswaran V.: FTFBE: a numerical approximation for fuzzy time-fractional Bloch equation. In: IEEE International Conference on Fuzzy Systems, 4 September 2014, Article number 6891696, pp 418–423, 2014 IEEE International Conference on Fuzzy Systems, FUZZ-IEEE 2014, Beijing, China, 6 July 2014 through 11 July 2014, Category number CFP14FUZ-ART, Code 114802
36. Salahshour, S., Ahmadian, A., Ismail, F., Baleanu, D.: A fractional derivative with non-singular kernel for interval-valued functions under uncertainty. Optik **130**(1), 273–286 (2017)
37. Ahmadian, A., Salahshour, S., Ali-Akbari, M., Ismail, F., Baleanu, D.: A novel approach to approximate fractional derivative with uncertain conditions. Chaos, Solitons Fractals **104**, 68–76 (2017)
38. Salahshour, S., Ahmadian, A., Chan, C.S.: Successive approximation method for Caputo q-fractional IVPs. Commun. Nonlinear Sci. Numer. Simul. **24**(1–3), 153–158 (2015)
39. Rabiei, F., Ismail, F., Ahmadian, A., Salahshour S.: Numerical solution of second-order fuzzy differential equation using improved Runge-kutta nystrom method. Math. Prob. Eng., Article number 803462 (2013)
40. Shahriyar, M.R.B, Ismail, F., Aghabeigi, S., Ahmadian, A., Salahshour, S.: An eigenvalue-eigenvector method for solving a system of fractional differential equations with uncertainty. Math. Prob. Eng., Article number 579761 (2013)
41. Allahviranloo, T., Khezerloo, M., Sedaghatfar, O., Salahshour, S.: Toward the existence and uniqueness of solutions of second-order fuzzy Volterra integro-differential equations with fuzzy kernel. Neural Comput. Appl. **22**(SUPPL 1), 133–141 (2013)
42. Allahviranloo, T., Salahshour, S.: A new approach for solving first order fuzzy differential equation. In: Communications in Computer and Information Science, vol. 81, Part 2, 2010, pp. 522–531, 13th International Conference on Information Processing and Management of Uncertainty, IPMU 2010, Dortmund, Germany, 28 June 2010 through 2 July 2010, Code 98056
43. Salahshour, S., Ahmadian, A., Senu, N., Baleanu, D., Agarwal, P.: On analytical solutions of the fractional differential equation with uncertainty: application to the basset problem. Entropy **17**(2), 885–902 (2015)
44. Allahviranloo, T., Abbasbandy, S., Salahshour, S., Hakimzadeh, A.: A new method for solving fuzzy linear differential equations. Computing (Vienna/New York) **92**(2), 181–197 (2011)
45. Salahshour, S., Ahmadian, A., Pansera, B., Ferrara, M.: Uncertain inverse problem for fractional dynamical systems using perturbed collage theorem. Commun. Nonlin. Sci. Numer. Simul. **94**, Article number 105553 (2021)

Homotopy Perturbation Method for Solving Fuzzy Fractional Heat-Conduction Equation

Shweta Dubey and S. Chakraverty

Abstract General heat-conduction equation and its fractional generalization are used in various applications in science and engineering. The objective of the study of fuzzy fractional heat equation is to use of Double parametric form (DPF) which deal with the uncertainty of the fractional heat equation easily and then applying Homotopy Perturbation Method (HPM) to find out the approximate solution in the form of a convergent series. The terms can be easily computed—increasing the number of terms in the approximate solution series that may be close enough to the exact solution. The results obtained here show that the introduced HPM technique is efficient and easy to implement in fuzzy fractional heat equation.

Keywords Fuzzy number · Fuzzy fractional heat equation · Double parametric form · Homotopy perturbation method

1 Introduction

Most of the differential equations may not have analytical solutions, and because of this, approximation and numerical techniques are applied widely. Fractional differential equations with uncertain parameters have many applications in science and engineering; as such, researchers have shown interest in studying these types of problems. Many scientists and engineers have studied fuzzy fractional differential equations, such as Ghazanfari and Ebrahimi in [1] and Goetschel and Voxman in [2]. Allahviranloo et al. in [3] are also deals with fuzzy fractional differential equations (FFDEs) using Adams-Bashforth, Adams-Moulton and predictor-corrector method and proved the convergence and stability of these methods. The existence, uniqueness and approximate solutions of FFDEs under Caputo's H-differentiability and the concept of Riemann-Liouville's H-differentiability to solve FFDEs by fuzzy laplace transformation method are discussed in [4, 5]. The hybrid fuzzy differential equation is studied using Euler's numerical method in [6] under Hukuhara differentiability.

S. Dubey · S. Chakraverty (✉)
Department of Mathematics, National Institute of Technology Rourkela, Rourkela 769008, Odisha, India

T. Allahviranloo and S. Salahshour (eds.), *Advances in Fuzzy Integral and Differential Equations*, Studies in Fuzziness and Soft Computing 412,
https://doi.org/10.1007/978-3-030-73711-5_6

Our aim is here to evaluate the approximate solution of the fuzzy fractional heat equation (FFHE) using the homotopy perturbation method (HPM) through a unique approach of the double parametric form (DPF) given by Chakraverty and Behera [7, 8] which deals with the uncertainty of the fractional heat equation. HPM was first introduced by He [9, 10], and frequently used to solve different types of linear and non-linear equations.

Since many non-linear equations may not contain the small parameter; however, the traditional perturbation technique depends upon small parameter, this drawback affects the application of perturbation techniques. Special techniques are used for the determination of small parameters, which is a complicated process. To remove these difficulties, HPM is used in many studies [11, 12]. HPM has a series of auxiliary parameters that regulate the convergence of solution series conveniently [13–15]. Our focus is to analyse and develop a solution that deals with the uncertainty of the fractional heat equation using DPF. Many other methods [16–19] are also used in the fuzzy fractional heat equation.

Rest of the chapter is structured as follows. In Sect. 2, some definitions like fuzzy numbers, r-cut, DPF have been involved. In Sect. 3, DPF is used in FFHE. After that in Sect. 4 HPM is applied in FFHE and an example is illustrated; then lastly, the conclusion is observed in Sect. 5.

2 Preliminaries

Here, some notation, definition, and preliminaries [20–25] have been given, which are further used in the chapter.

2.1 *Fuzzy Number*

Fuzzy set $\tilde{U}$ is a fuzzy number if the following conditions are satisfied.

1. $\tilde{U}$ is a convex fuzzy set.
2. $\tilde{U}$ is a normalized fuzzy set.
3. Its membership function is piecewise continuous.

***r*-cut**. The r-cut of a fuzzy set $\tilde{U}$ is the crisp set denoted by $\tilde{U}_\alpha$ and is defined as follows,

$$\tilde{U}_\alpha = \left\{x \in R : \mu_{\tilde{U}}(x) \geq r\right\}$$

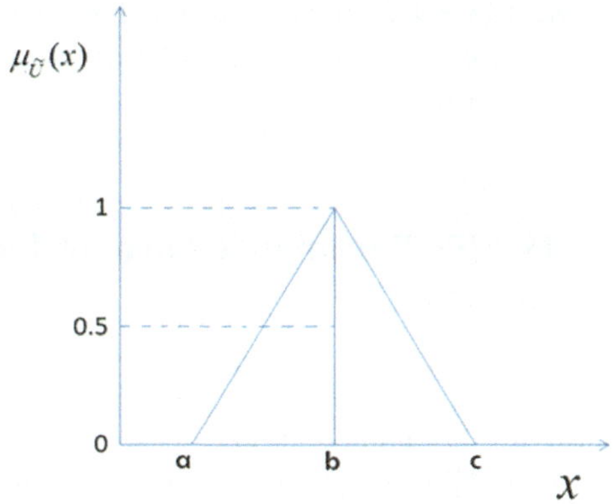

Fig. 1 Triangular Fuzzy Number $\tilde{U} = (a, b, c)$

2.2 *Triangular Fuzzy Number (TFN)*

A triangular fuzzy number $\tilde{U}$ is a convex fuzzy set $\tilde{U}$ of the real line R such that

1. There exists exactly one $x_0 \in R$ such that $\mu_{\tilde{U}}(x_0) = 1$.
2. $\mu_{\tilde{U}}(x)$ is piecewise continuous.

Consider a TFN $\tilde{U} = (a, b, c)$ as in Fig. 1. The membership function $\mu_{\tilde{U}}(x)$ of TFN $\tilde{U} = (a, b, c)$ will be defined as follows

$$\mu_{\tilde{U}}(x) = \begin{cases} 0, & x \le a, \\ \frac{x-a}{b-a}, & a \le x \le b, \\ \frac{c-x}{c-b} & b \le x \le c, \\ 0, & x \ge c \end{cases}$$

By an r-cut approach, a TFN $\tilde{U} = (a, b, c)$ can be expressed in interval form as $\left[\underline{u}(r), \overline{u}(r)\right] = [(b-a)r + a, \ -(c-b)r + c]$, where $r \in [0, 1]$.

2.3 *Double Parametric Form (DPF) of Fuzzy Number*

This form is firstly described by Chakraverty and Behera [3, 4]. Implementing the r-cut approach as given in Definition 2.1 for all the fuzzy numbers, we have $\tilde{u} = [\underline{u}(r), \overline{u}(r)]$. Now we may write the above form by using another parameter β as

$$\overline{u}(r, \beta) = \beta(\overline{u}(r) - \underline{u}(r)) + \underline{u}(r), \quad \text{where } r \text{ and } \beta \in [0, 1],$$

which is called DPF as it includes two parameters r and β.

We can substitute $\beta = 0$ and 1 for finding lower and upper bounds solutions in a single parametric form. Mathematically, symbolized as $\tilde{u}(r, 0) = \underline{u}(r)$ and $\tilde{u}(r, 1) = \overline{u}(r)$.

3 Double Parametric Form of Fuzzy Fractional Heat Equation

Using the approach of DPF (Chakraverty and Behera [3]), firstly, a single parametric form of fuzzy number has been used to convert FFHE into interval form. After that, we apply HPM to get the solution in double parametric form.

The one-dimensional fractional heat equation with the initial condition is given in crisp form in [16].

Let us consider the fuzzy form of the crisp equation as

$$\frac{\partial^{\alpha}\tilde{u}(x,t)}{\partial t^{\alpha}} - \frac{\partial^{2\alpha}\tilde{u}(x,t)}{\partial x^{2\alpha}} = \tilde{\varphi}(x,t), \qquad 0 < \alpha \leq 1, \tag{1}$$

subject to the initial conditions

$$\tilde{u}(0,t) = (0.5, \quad 1, \quad 1.5)\, f(t) \tag{2}$$

$$\frac{\partial^{\alpha}\tilde{u}(0,t)}{\partial x^{\alpha}} = (0.5, \quad 1, \quad 1.5)\, g(t) \tag{3}$$

By following Hukuhara difference [15] and single parametric form Eq. (1) can be written as

$$\left[\frac{\partial^{\alpha}\underline{u}(x,t;r)}{\partial t^{\alpha}}, \frac{\partial^{\alpha}\overline{u}(x,t;r)}{\partial t^{\alpha}}\right] - \left[\frac{\partial^{2\alpha}\underline{u}(x,t;r)}{\partial x^{2\alpha}}, \frac{\partial^{2\alpha}\overline{u}(x,t;r)}{\partial x^{2\alpha}}\right]$$
$$= \left[\underline{\varphi}(x,t;r), \overline{\varphi}(x,t;r)\right] \tag{4}$$

subject to the fuzzy initial conditions

$$[\underline{u}(0,t;r), \overline{u}(0,t;r)] = [0.5r + 0.5, 1.5 - 0.5r]\, f(t) \tag{5}$$

$$\left[\frac{\partial^{\alpha}\underline{u}(0,t;r)}{\partial x^{\alpha}}, \frac{\partial^{\alpha}\overline{u}(0,t;r)}{\partial x^{\alpha}}\right] = [0.5r + 0.5, 1.5 - 0.5r]\, g(t). \tag{6}$$

By using DPF, Eqs. (4)–(6) may be written as

$$\left[\beta\left(\frac{\partial^{\alpha}\overline{u}\ (x,t;r)}{\partial\, t^{\alpha}} - \frac{\partial^{\alpha}\underline{u}\ (x,t;r)}{\partial\, t^{\alpha}}\right) + \frac{\partial^{\alpha}\underline{u}\ (x,t;r)}{\partial\, t^{\alpha}}\right]$$

$$-\left[\beta\left(\frac{\partial^{2\alpha}\overline{u}(x,t;r)}{\partial x^{2\alpha}}-\frac{\partial^{2\alpha}\underline{u}(x,t;r)}{\partial x^{2\alpha}}\right)+\frac{\partial^{2\alpha}\underline{u}(x,t;r)}{\partial x^{2\alpha}}\right]$$
$$=\beta\left(\overline{\varphi}(x,t;r)-\underline{\varphi}(x,t;r)\right)+\underline{\varphi}(x,t;r). \tag{7}$$

with the fuzzy initial condition:

$$\beta\left(\overline{u}(0,t;r)-\underline{u}(0,t;r)\right)+\underline{u}(0,t;r)=[\beta(1-r)+0.5r+0.5]\,f(t) \tag{8}$$

$$\beta\left(\frac{\partial^{\alpha}\overline{u}(0,t;r)}{\partial x^{\alpha}}-\frac{\partial^{\alpha}\underline{u}(0,t;r)}{\partial x^{\alpha}}\right)+\frac{\partial^{\alpha}\underline{u}(0,t;r)}{\partial x^{\alpha}}$$
$$=[\beta(1-r)+0.5r+0.5]\,g(t) \tag{9}$$

where, $r, \beta \in [0, 1]$.

Here r and β control the fuzzy uncertainty of the problem.

By using DPF, Eqs. (7)–(9) have been converted as

$$\frac{\partial^{\alpha}\tilde{u}(x,t;r,\beta)}{\partial t^{\alpha}}-\frac{\partial^{2\alpha}\tilde{u}(x,t;r,\beta)}{\partial x^{2\alpha}}=\tilde{\varphi}(x,t;r,\beta) \tag{10}$$

$$\tilde{u}(0,t;r,\beta)=[\beta(1-r)+0.5r+0.5]\,f(t) \tag{11}$$

$$\frac{\partial^{\alpha}\tilde{u}(0,t;r,\beta)}{\partial x^{\alpha}}=[\beta(1-r)+0.5r+0.5]\,g(t) \tag{12}$$

On solving Eqs. (10), (11), and (12), the solution can be obtained in terms of $\tilde{u}(x,t;r,\beta)$.

To deduce the lower and upper bound solutions of (10)–(12), we substitute $\beta = 0$ *and* $\beta = 1$ respectively. Mathematically, it may be written as

$$\tilde{u}(x,t;r,0)=\underline{u}(x,t;r), \qquad \tilde{u}(x,t;r,1)=\overline{u}(x,t;r)$$

4 Homotopy Perturbation Method for the Solution of FFHE

Equations (10)–(12) may be rewritten as follows

$$\begin{aligned} D_t^{\alpha}\tilde{u}(x,t;r,\beta)-D_{xx}^{2\alpha}\tilde{u}(x,t;r,\beta)&=\tilde{\varphi}(x,t;r,\beta)\\ \tilde{u}_0(x,t;r,\beta)&=[\beta(1-r)+0.5r+0.5]\,f(t)\\ D_{x,0}^{\alpha}(x,t;r,\beta)&=[\beta(1-r)+0.5r+0.5]\,g(t) \end{aligned} \tag{13}$$

Firstly, construct homotopy which satisfies the relation,

$$D_{xx}^{2\alpha}\tilde{u}(x,t;r,\beta) = p\,(D_t^{\alpha}\tilde{u}(x,t;r,\beta)) - \tilde{\varphi}(x,t;r,\beta) \tag{14}$$

where p ∈ [0, 1] is the embedding parameter.

Let the series solution of Eq. (14) is as follows

$$\begin{aligned}\tilde{u}(x,t;r,\beta) &= \tilde{u}_0(x,t;r,\beta) + p\,\tilde{u}_1(x,t;r,\beta)\\ &\quad + p^2\,\tilde{u}_2(x,t;r,\beta) + \cdots\end{aligned} \tag{15}$$

Substituting Eq. (15) in Eq. (14), we get

$$\begin{aligned}D_{xx}^{2\alpha}(\tilde{u}_0 + p\tilde{u}_1 + p^2\tilde{u}_2 + \cdots) &= p\left[D_t^{\alpha}\left(\tilde{u}_0 + p\tilde{u}_1 + p^2\tilde{u}_{.2} + \cdots\right)\right]\\ &\quad - \tilde{\varphi}(x,t;r,\beta)\end{aligned}$$

Equating the coefficient of identical power of p from both in above, we get

$$\begin{aligned}
p^0 &: \quad D_{xx}^{2\alpha}\,\tilde{u}_0(x,t;r,\beta) = -\tilde{\varphi}(x,t;r,\beta)\\
p^1 &: \quad D_{xx}^{2\alpha}\,\tilde{u}_1(x,t;r,\beta) = D_t^{\alpha}\tilde{u}_0(x,t;r,\beta)\\
p^2 &: \quad D_{xx}^{2\alpha}\,\tilde{u}_2(x,t;r,\beta) = D_t^{\alpha}\tilde{u}_1(x,t;r,\beta)\\
p^3 &: \quad D_{xx}^{2\alpha}\,\tilde{u}_3(x,t;r,\beta) = D_t^{\alpha}\tilde{u}_2(x,t;r,\beta)\\
&.\\
&.\\
&.
\end{aligned} \tag{16}$$

and so on.

As $p = 1$, we have from Eq. (15)

$$\tilde{u}(x,t;r,\beta) = \sum_{m=0}^{\infty}\tilde{u}_m(x,t;r,\beta)$$

Using the inverse operator $D_{xx}^{-2\alpha}$ on both sides of (16) and after putting values in Eq. (15) for $p = 1$, we get the solution of FFHE as

$$\begin{aligned}\tilde{u}(x,t;r,\beta) &= [\beta(1-r) + 0.5r + 0.5]\,f(t)\\ &\quad + [\beta(1-r) + 0.5r + 0.5]\,g(t) \times \frac{x^{\alpha}}{\Gamma(\alpha+1)}\\ &\quad - D_{xx}^{-2\alpha}(\tilde{\varphi}(x,t;r,\beta)) + D_{xx}^{-2\alpha}\left[D_t^{\alpha}\,\tilde{u}(x,t;r,\beta)\right]\end{aligned} \tag{17}$$

The solution can also be expressed as

$$\tilde{u}(x,t;r,\beta) = \tilde{u}_0(x,t;r,\beta) + D_{xx}^{-2\alpha}\left[D_t^{\alpha}\,\tilde{u}(x,t;r,\beta)\right] \tag{18}$$

where,

$$\begin{aligned}\tilde{u}_0(x,t;r,\beta) &= [\beta(1-r)+0.5r+0.5]\,f(t)\\ &\quad + [\beta(1-r)+0.5r+0.5]\,g(t)\times\frac{x^{\alpha}}{\Gamma(\alpha+1)}\\ &\quad - D_{xx}^{-2\alpha}(\tilde{\varphi}(x,t;r,\beta))\end{aligned} \tag{19}$$

From Eq. (18) corresponding iterative formula can be written as

$$\tilde{u}_{n+1}(x,t;r,\beta) = D_{xx}^{-2\alpha}\left[D_t^{\alpha}\,\tilde{u}_n(x,t:r,\beta)\right],\quad n\geq 0$$

The lower and upper bounds of the solution in a single parametric form are obtained by substituting $\beta = 0\ and\ \beta = 1$ respectively in Eq. (19).

Accordingly we have

$$\begin{aligned}\underline{u}(x,t;r) &= (0.5r+0.5)\,f(t) + \frac{x^{\alpha}}{\Gamma(1+\alpha)}(0.5r+0.5)\,g(t)\\ &\quad -D_{xx}^{-2\alpha}[\tilde{\varphi}(x,t;r,0)] + D_{xx}^{-2\alpha}[D_{xx}^{2\alpha}\,\tilde{u}(x,t;r,0)]\end{aligned}$$

$$\begin{aligned}\overline{u}(x,t;r) &= (1.5-0.5r)\,f(t) + \frac{x^{\alpha}}{\Gamma(1+\alpha)}(1.5-0.5r)\,g(t)\\ &\quad -D_{xx}^{-2\alpha}[\tilde{\varphi}(x,t;r,1)] + D_{xx}^{-2\alpha}[D_{xx}^{2\alpha}\,\tilde{u}(x,t;r,1)]\end{aligned}$$

4.1 Numerical Implementation of the Proposed Method

Now, we include an example and solved it by the above-described method.

Example Let us consider a non-homogeneous local FFHE (see [16] for the crisp problem) with $\tilde{\varphi}(x,t) = 1,\ f(t) = \frac{t^{\alpha}}{\Gamma(\alpha+1)},\ g(t) = 0$, the problem takes the form

$$\frac{\partial^{\alpha}\tilde{u}(x,t)}{\partial t^{\alpha}} - \frac{\partial^{2\alpha}\tilde{u}(x,t)}{\partial x^{2\alpha}} = 1 \tag{20}$$

subject to initial conditions

$$\tilde{u}(0,t) = (0.5,\ 1,\ 1.5)\,\frac{t^{\alpha}}{\Gamma(1+\alpha)}$$

$$\frac{\partial^{\alpha}\tilde{u}(x,t)}{\partial x^{\alpha}}=0.$$

Solution Firstly, Eq. (20) can be rewritten using DPF as

$$\frac{\partial^{\alpha}\tilde{u}(x,t,r,\beta)}{\partial t^{\alpha}}-\frac{\partial^{2\alpha}\tilde{u}(x,t,r,\beta)}{\partial x^{2\alpha}}=1$$

Subject to initial conditions

$$\tilde{u}(0,t,r,\beta)=(0.5,\ 1,\ 1.5)\frac{t^{\alpha}}{\Gamma(1+\alpha)}$$
$$\frac{\partial^{\alpha}\tilde{u}(x,t,r,\beta)}{\partial x^{\alpha}}=0.$$

In view of Eq. (19), the iterative algorithm can be written as

$$\tilde{u}_0(x,t;r,\beta)=[\beta(1-r)+0.5r+0.5]\frac{t^{\alpha}}{\Gamma(\alpha+1)}-D_{xx}^{-2\alpha}[1] \tag{21}$$

$$\tilde{u}_{n+1}(x,t;r,\beta)=D_{xx}^{-2\alpha}\left[D_t^{\alpha}\,\tilde{u}_n(x,t\ :r,\beta)\right],\ \ n\geq 0 \tag{22}$$

Therefore from Eqs. (21) and (22), we get the iterative terms as below

$$\tilde{u}_0(x,t;r,\beta)=[\beta(1-r)+0.5r+0.5]\frac{t^{\alpha}}{\Gamma(\alpha+1)}-\frac{x^{2\alpha}}{\Gamma(1+2\alpha)}$$

$$\tilde{u}_1(x,t;r,\beta)=D_{xx}^{-2\alpha}[(\beta(1-r)+0.5r+0.5)]$$

$$\tilde{u}_1(x,t;r,\beta)=\frac{x^{2\alpha}}{\Gamma(1+2\alpha)}\ [(\beta(1-r)+0.5r+0.5)]$$

$$\tilde{u}_2(x,t;r,\beta)=\tilde{u}_3(x,t;r,\beta)=\cdots=\tilde{u}_n(x,t;r,\beta)=0$$

Therefore, the solution of the given Eq. (20) becomes,

$$\tilde{u}(x,t;r,\beta)=[\beta(1-r)+0.5r+0.5]\frac{t^{\alpha}}{\Gamma(\alpha+1)}+[(\beta(1-r)+0.5r-0.5)]\frac{x^{2\alpha}}{\Gamma(1+2\alpha)} \tag{23}$$

For a crisp solution, putting $r=1$ and $\beta=0$ or $\beta=1$, we get

$$u\,(x,t)=\frac{t^{\alpha}}{\Gamma(\alpha+1)} \tag{24}$$

In Eq. (24) for $\alpha = 1$,

$$u\,(x,t) = t$$

Similarly for $\alpha = \frac{1}{2}$,

$$u\,(x,t) = \frac{4t^{\frac{1}{2}}}{\sqrt{\pi}}$$

which is exactly the same as given in [16] for the crisp case.

For parameters $\beta = 1$, $r = 0$, the solution will be of the form

$$\tilde{u}(x,t;0,1) = \frac{1.5t^{\alpha}}{\Gamma(1+\alpha)} + \frac{0.5x^{2\alpha}}{\Gamma(1+2\alpha)} \tag{25}$$

Putting $\alpha = \frac{1}{2}$, the solution from Eq. (25) may be written as

$$\tilde{u}(x,t;0,1) = \frac{3\sqrt{t}}{\sqrt{\pi}} + 0.5x$$

Again by Putting $\alpha = 1$, the solution from Eq. (25) can takes the form

$$\tilde{u}(x,t;0,1) = 1.5t + 0.25x^2$$

Numerical solutions of Eq. (20) for $\alpha = 1$ have been presented in Table 1 with respect to different values of t and x. From this table, it may be observed that the value of $u(x,t)$ increases continuously as the time increases for a particular value of x. Similarly for $\alpha = \frac{1}{2}$, Table 2 presents the numerical solution of Eq. (20) concerning various time and space points. The same nature of the solution has been observed from Table 2 as we got from Table 1.

Table 1 Numerical solution of Eq. (20) for $\alpha = 1$

t	$u(x,t)$					
	$x = 0$	$x = 0.2$	$x = 0.4$	$x = 0.6$	$x = 0.8$	$x = 1$
1	1.5	1.51	1.54	1.59	1.66	1.75
1.2	1.80	1.81	1.84	1.89	1.96	2.05
1.4	2.10	2.11	2.14	2.19	2.26	2.35
1.6	2.40	2.41	2.44	2.49	2.56	2.65
1.8	2.70	2.71	2.74	2.79	2.86	2.95
2	3.0	3.01	3.04	3.09	3.16	3.25

Table 2 Numerical solution of Eq. (20) for $\alpha = \frac{1}{2}$

t	$u(x,t)$					
	$x = 0$	$x = 0.2$	$x = 0.4$	$x = 0.6$	$x = 0.8$	$x = 1$
1	1.692568750	1.792568750	1.892568750	1.992568750	2.092568750	2.192568750
1.2	1.854116170	1.954116170	2.054116170	2.154116170	2.254116170	2.354116170
1.4	2.002674354	2.102674354	2.202674354	2.302674354	2.402674354	2.502674354
1.6	2.140948939	2.240948939	2.340948939	2.440948939	2.540948939	2.640948939
1.8	2.270819269	2.370819269	2.470819269	2.570819269	2.670819269	2.770819269
2	2.393653682	2.493653682	2.593653681	2.693653681	2.793653681	2.893653682

5 Conclusion

Our purpose is to describe an approximate-analytical solution for FFHE by HPM using DPF. Taking the value of $\alpha = 1$, the approximate-analytical solution is the same as the analytical solution. HPM provides rapid convergence of power series solutions. It may be noted that the use of DPF in solving uncertainty in fractional heat equation is a very effective approach. Also, it can be observed that by the use of HPM, accuracy can be increased by taking more terms in approximate solution. This shows that the HPM is capable and efficient technique that gets high precision of many problems.

References

1. Ghazanfari, B., Ebrahimi, P.: Differential transformation method for solving fuzzy fractional heat equations. Int. J. Math. Model. Comput. **5**(1 (WINTER)), 81–89 (2015)
2. Goetschel Jr., R., Voxman, W.: Elementary fuzzy calculus. Fuzzy Sets Syst. **18**(1), 31–43 (1986)
3. Allahviranloo, T., Ahmady, N., Ahmady, E.: Numerical solution of fuzzy differential equations by predictor–corrector method. Inf. Sci. **177**(7), 1633–1647 (2007)
4. Salahshour, S., Allahviranloo, T., Abbasbandy, S.: Solving fuzzy fractional differential equations by fuzzy Laplace transforms. Commun. Nonlinear Sci. Numer. Simul. **17**(3), 1372–1381 (2012)
5. Salahshour, S., Allahviranloo, T., Abbasbandy, S., Baleanu, D.: Existence and uniqueness results for fractional differential equations with uncertainty. Adv. Differ. Eq. **2012**(1), 112 (2012)
6. Allahviranloo, T., Salahshour, S.: Euler method for solving hybrid fuzzy differential equation. Soft. Comput. **15**(7), 1247–1253 (2011)
7. Behera, D., Chakraverty, S.: New approach to solve fully fuzzy system of linear equation using single and double parametric form of fuzzy numbers. Sadhana **40**(1), 35–49 (2015)
8. Chakraverty, S., Tapaswini, S., Behera, D.: Fuzzy Differential Equations and Applications for Engineers and Scientists. CRC Press (2016)
9. He, J.H.: Homotopy perturbation technique. Comput. Methods Appl. Mech. Eng. **178**(3–4), 257–262 (1999)
10. He, J.H.: A coupling method of a homotopy technique and a perturbation technique for non-linear problems. Int. J. Non-Linear Mech. **35**(1), 37–43 (2000)

11. Saadatmandi, A., Dehghan, M., Eftekhari, A.: Application of He's homotopy perturbation method for non-linear system of second-order boundary value problems. Non-linear Anal.: Real World Appl. **10**(3), 1912–1922 (2009)
12. Yıldırım, A.: Solution of BVPs for fourth-order integro-differential equations by using homotopy perturbation method. Comput. Math Appl. **56**(12), 3175–3180 (2008)
13. Turkyilmazoglu, M.: The Airy equation and its alternative analytic solution. Phys. Scr. **86**(5), (2012)
14. Turkyilmazoglu, M.: Analytic approximate solutions of rotating disk boundary layer flow subject to a uniform suction or injection. Int. J. Mech. Sci. **52**(12), 1735–1744 (2010)
15. Jameel, A.F., Ghoreishi, M., Ismail, A.I.M.: Approximate solution of high order fuzzy initial value problems. J. Uncertain Syst. **8**(2), 149–160 (2014)
16. Ghazanfari, B., Ebrahimi, P.: Differential transformation method for solving fuzzy fractional heat equations. Int. J. Math. Model. Comput. **5**(1 (WINTER)), 81–89. Science & Business Media (2015)
17. Salah, A., Khan, M., Gondal, M.A.: A novel solution procedure for fuzzy fractional heat equations by homotopy analysis transform method. Neural Comput. Appl. **23**(2), 269–271 (2013)
18. Arfan, M., Shah, K., Abdeljawad, T., Hammouch, Z.: An efficient tool for solving two-dimensional fuzzy fractional-ordered heat equation. Numer. Methods Partial Differ. Eq. (2020)
19. Chalco-Cano, Y., Roman-Flores, H.: On new solutions of fuzzy differential equations. Chaos Solitons Fractals **38**(1), 112–119 (2008)
20. Jassim, H.K.: Local fractional Laplace decomposition method for non-homogeneous heat equations arising in fractal heat flow with local fractional derivative. Int. J. Adv. Appl. Math. Mech. **2**(4), 1–7 (2015)
21. Najafi, H.S., Edalatpanah, S.A.: An improved model for iterative algorithms in fuzzy linear systems. Comput. Math. Model. **24**(3), 443–451 (2013)
22. Hanss, M.: Applied Fuzzy Arithmetic. Springer, Berlin Heidelberg (2005)
23. Jaulin, L., Kieffer, M., Didrit, O., Walter, E.: Interval analysis. In: Applied Interval Analysis, pp. 11–43. Springer, London (2001)
24. Ross, T.J.: Fuzzy Logic with Engineering Applications, vol. 2. Wiley, New York (2004)
25. Zimmermann,H.J.: Fuzzy Set Theory—And Its Applications. Springer (2011)

Fuzzy Integral Equations

Finding Optimal Results in the Homotopy Analysis Method to Solve Fuzzy Integral Equations

Mohammad Ali Fariborzi Araghi and Samad Noeiaghdam

Abstract The goal of this work is to apply the discrete stochastic arithmetic (DSA) and the fuzzy CESTAC (Controle et Estimation Stochastique des Arrondis de Calculs) method to validate the results of solving fuzzy Fredholm or Volterra integral equations (IEs) by the homotopy analysis method (HAM). The CADNA (Control of Accuracy and Debugging for Numerical Applications) library is applied to implement the fuzzy CESTAC method. Furthermore, the optimal convergence control parameter of the HAM is obtained. A theorem is proved to show the accuracy of the HAM based on the concept of common significant digits and two algorithms are proposed by applying the fuzzy CESTAC method to compute the optimal results. The results of the sample examples illustrate the efficiency of using the DSA in place of the usual computer arithmetic.

1 Introduction

Integral equations (IEs) play an important role in mathematics and different sciences such as engineering, mathematical physics and biology. Applying the analytical, semi-analytical or numerical schemes to find the solution of an integral equation have been the research interest subjects in recent years [73]. Specially, because of important and novel applications of fuzzy arithmetic [9], recently many researchers

M. A. Fariborzi Araghi (✉)
Department of Mathematics, Central Tehran Branch, Islamic Azad University, Tehran, Iran
e-mail: m_fariborzi@iauctb.ac.ir; fariborzi.araghi@gmail.com

S. Noeiaghdam
Industrial Mathematics Laboratory, Baikal School of BRICS, Irkutsk National Research Technical University, Irkutsk, Russian Federation

South Ural State University, Lenin Prospect 76, Chelyabinsk 454080, Russian Federation
e-mail: snoei@istu.edu; noiagdams@susu.ru

T. Allahviranloo and S. Salahshour (eds.), *Advances in Fuzzy Integral and Differential Equations*, Studies in Fuzziness and Soft Computing 412,
https://doi.org/10.1007/978-3-030-73711-5_7

focused on solving the fuzzy integral equations (FIEs) [50]. Solving the fuzzy Volterra integral equations was discussed in [8], the fuzzy Fredholm integral equations were studied in [6, 19, 26, 63, 64], fuzzy Volterra-Fredholm integral equations in [10, 32], nonlinear Hammerstein fuzzy integral equations [47], fuzzy Abel's integral equations [11, 21, 41] and a kind of fuzzy nonlinear integral equations [37] were described.

The homotopy is one of the main topics in the topology which has been applied to solve many mathematical and engineering problems [34–36]. Liao applied this technique to present the homotopy analysis method [42–45]. The methods based on the concept of homotopy have been applied to solve many linear and nonlinear problems. In [13, 18], the HAM was applied to solve the nonlinear Volterra-Fredholm integro-differential equations, in [7] the nonlinear Fredholm integral equations was discussed and in [61] the HAM was applied to solve the nonlinear epidemiological model of computer viruses. The first kind integral equations were solved based on the homotopy techniques in [22, 54, 62]. Also, the homotopic methods were applied to solve the Blasius equation [1], functional integral equations [2], singularly perturbed differential-difference equation with boundary layer behavior and delay [65], fractional heat-like physical model [40], nano boundary layer flows with nonlinear Navier boundary condition [69], magneto-hydrodynamic squeezing problem [68] and for Laplace transform [3].

Because of the flexibility of the HAM such as existence of auxiliary functions and parameter to find the solution and also to control the convergence region, many researchers have been interested in applying this method for solving the FIEs. In [29, 30], Ghanbari and Allahiranloo applied the HAM for solving fuzzy linear Volterra and Fredholm IEs, Molabahrami in [51] and Effati et al. in [17] focused on the linear fuzzy Fredholm IEs and Vali et al. in [70] discussed the two-dimensional fuzzy Fredholm IE. Also, Lotfi in [46] focused on the convergence and error estimate of nonlinear Fredholm fuzzy integral equations of the second kind and Salam in [67] investigated on the fractional FIEs by the HAM. Different classes of the Fuzzy Abel IE was studied in [21, 49], the fuzzy Volterra-Fredholm IEs were discussed in [31] and the first order fuzzy Volterra-Fredholm integro-differential equations were illustrated in [33]. Also, this method was used to solve the two dimensional nonlinear Volterra-Fredholm FIEs in [28].

In these works, in order to find the numerical and approximate results, the proposed algorithms are implemented based on the floating-point arithmetic as the common computer computations. Usually the commercial and mathematical packages such as Mathematica, Maple and MATLAB are used to perform the algorithms and compute the results. In this case, in order to show the accuracy of the method, either the exact solution must be accessible to compare the results or a termination criterion which depends on a positive small value like ϵ should be applied to estimate the absolute error. If this tolerance is considered to stop the calculations, for large values of the ϵ the algorithm will not be able to produce accurate results and it will be stopped at the first steps without getting a satisfactory solution. Also, for small values of the ϵ, we will have extra iterations without improving the accuracy. Hence, it is not possible to determine the optimal results. Also, due to the round-off errors and accumulation of errors, the accuracy of the intermediate and final results may be decreased or lost. The floating-point arithmetic is not able to validate the results and detect any instability during the run of a program. For these reasons, instead of applying the floating-point arithmetic and its corresponding termination criteria, the stochastic arithmetic can be substituted and in order to implement this arithmetic in discrete case, the CESTAC method is applied. Also, the CADNA library is performed instead of applying the mentioned softwares. By replacing the floating-point arithmetic by the discrete stochastic arithmetic (DSA), it is possible to estimate the optimal number of iterations and optimal approximations. Also, the computations are checked to detect any instability and controlled dynamically and therefore the results are validated. The stochastic arithmetic and the CESTAC method were studied by Laporet and Vignes [72] and the CADNA library was designated by Chesneaux [14–16].

Also, recently this method has been applied to validate the results of the numerical integration rules [5, 20, 24, 25], solving integral equations [52, 55, 58, 59], finding the optimal convergence control parameter of the HAM [53] and solving osmosis model and load leveling problem [56, 57, 60] and other topics mentioned in [4, 5, 12, 23]. In recent years, in the research center LIP6,[1] this library has been applied to validate the results in different projects.

In the fuzzy CESTAC method the termination criterion depends on the Hausdorff distance of two successive approximations. Also, the informatical zero is applied which shows that the number of common significant digits for Hausdorff distance of two successive fuzzy approximations is zero and after that the algorithm will be stopped. Using this method not only we can avoid producing redundant iterations but also we can find the optimal iteration and optimal approximation and also validate the results of numerical algorithms to solve a fuzzy integral equation. Moreover, some of numerical instabilities can be identified. CADNA is an efficient tool to implement the CESTAC method in fuzzy case. This library is performed on the Linux operating system based on C/C++, FORTRAN or ADA codes [38, 39]. In order to illustrate the applicability of the fuzzy CESTAC method, first a theorem must be proved to show the accuracy of the method by means of the concept of the common significant digits. In this paper, in order to validate the results, the HAM is considered to solve a fuzzy integral equation in the DSA. For this purpose, we introduce an algorithm to apply the fuzzy CESTAC method with an optimal termination criterion. Then, the proposed algorithm is implemented by the CADNA library. So, the optimal convergence control in the HAM and the optimal approximate fuzzy solution at optimal iteration can be achieved.

In the sequel, in Sect. 2, based on the background of the DSA and the CESTAC method in crisp case [71, 72], the fuzzy CESTAC method is introduced. The main idea is presented in Sect. 3. In this section, the HAM is applied to solve the fuzzy integral equations and also a theorem is proved to show the accuracy of the HAM to approximate the solution of a FIE. In Sect. 4, some examples are solved and the fuzzy CESTAC method is applied to find the optimal results, optimal iteration and the optimal convergence control parameter of the HAM according to the proposed algorithms. In Sect. 5, the conclusion remarks are presented.

---

[1] https://www-pequan.lip6.fr/.

2 Fuzzy CESTAC Method

Assume that $\{\widetilde{v}_J\}_{J\geq 1}$ is a fuzzy sequence obtained from an iterative or a numerical method which can be applied to find the exact fuzzy solution $\widetilde{v}$. In order to show the accuracy of this method, in the computer arithmetic, usually the following termination criterion is applied:

$$If\ \mathcal{D}_H(\widetilde{v}_J, \widetilde{v}) \leq \epsilon\ then\ stop, \tag{1}$$

where $\mathcal{D}_H(\widetilde{v}_J, \widetilde{v})$ is the Hausdorff distance [9, 27], between $\widetilde{v}_J, \widetilde{v}$ such that Eq. (1) depends on the small positive real number ϵ and accessibility of the exact solution. But how can we apply the condition (1), when the exact solution is not accessible? Also, we do not know the optimal value of ϵ. For large values of ϵ the numerical algorithm can be stopped soon without getting to a suitable approximation and for small values ϵ the extra iterations can be produced without improving the accuracy of results.

Consider the following rounding error with $\mathcal{M}$ mantissa bits

$$\mathcal{S}^{(p)} = s^{(p)} - \zeta^{(p)} 2^{E^{(p)}-\mathcal{M}} \nu^{(p)}, \quad p = 1, 2, 3, \tag{2}$$

obtained from the binary computer computations where $\tilde{s} = (s^{(1)}, s^{(2)}, s^{(3)})$ is a fuzzy number in triangular form which is reproduced by the computer precision in the form $\tilde{\mathcal{S}} = (\mathcal{S}^{(1)}, \mathcal{S}^{(2)}, \mathcal{S}^{(3)})$, $\zeta^{(p)}$ is the sign of $s^{(p)}$, $E^{(p)}$ is the binary exponents of the results and $2^{-\mathcal{M}}\nu^{(p)}$ is the omitted bit of mantissa. We know that for $\mathcal{M} = 24, 53$ the results can be obtained in single or double precision. If we choose $\nu^{(p)}$ as chance variable of uniformly distributed on interval $[-1, 1]$, we can make perturbation on the last mantissa bit of $\widetilde{s}$. Therefore, we can produce the random variables of $\widetilde{\mathcal{S}}$ such that, the average $\mu^{(p)}$ and the standard deviation $\sigma^{(p)}$ have direct role in their precision for $p = 1, 2, 3$. If we repeat this process for l times, the exact values of $s^{(p)}$ can be approximated by $\mathcal{S}_\kappa^{(p)}$, $\kappa = 1, ..., l$ and the mathematical expectation of $\mathcal{S}_k^{(p)}$ is equal with $s^{(p)}$. But all of the mentioned processes will be done by applying the CADNA library [14, 71, 72].

In this method, the iterative procedure will be stopped when the number of common significant digits for two successive approximations is zero which is called informatical zero and denoted by @.0 [71, 72]. By the following algorithm, the fuzzy CESTAC method is described where T_υ is the value of $T-$distribution with $l-1$ degree of freedom and confidence interval $1-\upsilon$.

Algorithm 1:

Step 1- Produce l samples of $\mathcal{S}^{(p)}$ as

$$\mathcal{S}^{(p)} = \left\{\mathcal{S}_1^{(p)}, \mathcal{S}_2^{(p)}, \cdots, \mathcal{S}_l^{(p)}\right\}, \ p = 1, 2, 3,$$

by means of the perturbation of the last bit of mantissa.

Step 2- Find

$$\mathcal{S}_{ave}^{(p)} = \frac{\sum_{\kappa=1}^{l} \mathcal{S}_{\kappa}^{(p)}}{l}, \ p = 1, 2, 3.$$

Step 3- Calculate

$$\left(\sigma^{(p)}\right)^2 = \frac{\sum_{\kappa=1}^{l} \left(\mathcal{S}_{\kappa}^{(p)} - \mathcal{S}_{ave}^{(p)}\right)^2}{l-1}, \ p = 1, 2, 3.$$

Step 4- Calculate the estimation of number of common significant digits of $\mathcal{S}^{(p)}$ and $\mathcal{S}_{ave}^{(p)}$ by

$$C_{\mathcal{S}_{ave}^{(p)}, \mathcal{S}^{(p)}} = \log_{10} \frac{\sqrt{l}\left|\mathcal{S}_{ave}^{(p)}\right|}{T_{\upsilon}\sigma^{(p)}}, \ p = 1, 2, 3.$$

Step 5- If

$$C_{\mathcal{S}_{ave}^{(p)}, \mathcal{S}^{(p)}} \leq 0 \quad or \quad \mathcal{S}_{ave}^{(p)} = 0,$$

then write $\widetilde{\mathcal{S}} = @.0$.

Based on Algorithm 1, Eq. (1) can be replaced by the following criterion

$$If\ \mathcal{D}_H(\widetilde{\upsilon}_J, \widetilde{\upsilon}_{J+1}) = @.0\ then\ stop, \tag{3}$$

where @.0 shows the informatical zero and only can be justified in the DSA. It means the number of significant digits of the Hausdorff distance between two fuzzy sequential results is zero. By applying condition (3), it is possible to find the optimal approximation, optimal convergence control parameter and optimal step of the HAM based on the DSA, the fuzzy CESTAC method and the CADNA library.

3 Main Idea

In the sequel, the following fuzzy integral equation of the second kind is discussed [48, 63, 66, 74]:

$$\tilde{F}(x) = \tilde{f}(x) + \lambda \int_{g_1(x)}^{g_2(x)} K(x, t, \tilde{F}(t))dt, \ g_1(x) \leq x, t \leq g_2(x). \tag{4}$$

Let

$$(\underline{f}(x;r), \overline{f}(x;r)),\ 0 \le r \le 1,$$

and

$$(\underline{F}(x;r), \overline{F}(x;r)),\ 0 \le r \le 1,\ g_1(x) \le x \le g_2(x),$$

be the parametric forms of $\tilde{f}(x)$ and $\tilde{F}(x)$ respectively. Then, the parametric form of Eq. (4) can be written as follows:

$$\begin{aligned}\underline{F}(x;r) &= \underline{f}(x;r) + \lambda \int_{g_1(x)}^{g_2(x)} \underline{K(x,t,F(t;r))}dt,\ g_1(x) \le x, t \le g_2(x),\\ \overline{F}(x;r) &= \overline{f}(x;r) + \lambda \int_{g_1(x)}^{g_2(x)} \overline{K(x,t,F(t;r))}dt,\ g_1(x) \le x, t \le g_2(x),\end{aligned} \tag{5}$$

for $0 \le r \le 1$, where

$$\underline{K(x,t,F(t;r))} = \begin{cases} K(x,t,\underline{F}(t;r)), & \underline{K(x,t,F(t;r))} \ge 0,\\ K(x,t,\overline{F}(t;r)), & \underline{K(x,t,F(t;r))} < 0,\end{cases} \tag{6}$$

and

$$\overline{K(x,t,F(t;r))} = \begin{cases} K(x,t,\overline{F}(t;r)), & \overline{K(x,t,F(t;r))} \ge 0,\\ K(x,t,\underline{F}(t;r)), & \overline{K(x,t,F(t;r))} < 0.\end{cases} \tag{7}$$

Without loss of generality, let the kernel K be positive. In this case, the operator N can be defined as follows:

$$\begin{aligned}\underline{N}[\underline{F}] &= \underline{F}(x;r) - \underline{f}(x;r) - \lambda \int_{g_1(x)}^{g_2(x)} K(x,t,\underline{F}(t;r))dt,\ g_1(x) \le x, t \le g_2(x),\\ \overline{N}[\overline{F}] &= \overline{F}(x;r) - \overline{f}(x;r) - \lambda \int_{g_1(x)}^{g_2(x)} K(x,t,\overline{F}(t;r))dt,\ g_1(x) \le x, t \le g_2(x).\end{aligned} \tag{8}$$

The zero order deformation equation can be constructed as follows [28, 30, 51]:

$$\begin{aligned}(1-q)L[\underline{\phi}^{\hbar}(x,q;r) - \underline{w}_0^{\hbar}(x;r)] &= q\hbar\mathcal{H}(x)\underline{N}[\underline{\phi}^{\hbar}(x,q;r)],\\ (1-q)L[\overline{\phi}^{\hbar}(x,q;r) - \overline{w}_0^{\hbar}(x;r)] &= q\hbar\mathcal{H}(x)\overline{N}[\overline{\phi}^{\hbar}(x,q;r)],\end{aligned} \tag{9}$$

where $q \in [0, 1]$ is the embedding parameter, $\hbar$ is nonzero auxiliary parameter to control the convergence region, L is an auxiliary linear operator, $\mathcal{H}(x)$ is an auxiliary function and $\underline{\phi}^{\hbar}(x, q; r)$ and $\overline{\phi}^{\hbar}(x, q; r)$ are the unknown functions that should be determined. Substituting Eq. (8) into (9) we get:

$$
\begin{aligned}
&(1-q)L[\underline{\phi}^{\hbar}(x,q;r) - \underline{w}_0^{\hbar}(x;r)] = q\hbar\mathcal{H}(x)[\underline{\phi}^{\hbar}(x,q;r) - \underline{f}(x;r) - \lambda\textstyle\int_{g_1(x)}^{g_2(x)} K(x,t,\underline{\phi}^{\hbar}(t,q;r))dt],\\
&(1-q)L[\overline{\phi}^{\hbar}(x,q;r) - \overline{w}_0^{\hbar}(x;r)] = q\hbar\mathcal{H}(x)[\overline{\phi}^{\hbar}(x,q;r) - \overline{f}(x;r) - \lambda\textstyle\int_{g_1(x)}^{g_2(x)} K(x,t,\overline{\phi}^{\hbar}(t,q;r))dt],
\end{aligned}
\tag{10}
$$

where $\underline{w}_0^{\hbar}(x; r)$ and $\overline{w}_0^{\hbar}(x; r)$ are the initial guesses of $\underline{F}(x; r)$ and $\overline{F}(x; r)$ respectively. By choosing $\mathcal{L} = I$ and $\mathcal{H}(x) = 1$, Eq. (10) can be written in the following form:

$$
\begin{aligned}
&(1-q)[\underline{\phi}^{\hbar}(x,q;r) - \underline{w}_0^{\hbar}(x;r)] = q\hbar[\underline{\phi}^{\hbar}(x,q;r) - \underline{f}(x;r) - \lambda\textstyle\int_{g_1(x)}^{g_2(x)} K(x,t,\underline{\phi}^{\hbar}(t,q;r))dt],\\
&(1-q)[\overline{\phi}^{\hbar}(x,q;r) - \overline{w}_0^{\hbar}(x;r)] = q\hbar[\overline{\phi}^{\hbar}(x,q;r) - \overline{f}(x;r) - \lambda\textstyle\int_{g_1(x)}^{g_2(x)} K(x,t,\overline{\phi}^{\hbar}(t,q;r))dt].
\end{aligned}
\tag{11}
$$

According to the traditional HAM [42–45], we know that for $q = 0$ and $q = 1$ we have

$$
\underline{\phi}^{\hbar}(x, 0; r) = \underline{w}_0^{\hbar}(x; r),
$$

$$
\overline{\phi}^{\hbar}(x, 0; r) = \overline{w}_0^{\hbar}(x; r),
$$

and

$$
\underline{\phi}^{\hbar}(x, 1; r) = \underline{f}(x; r) + \lambda \textstyle\int_{g_1(x)}^{g_2(x)} K(x, t, \underline{\phi}^{\hbar}(t, 1; r))dt,
$$

$$
\overline{\phi}^{\hbar}(x, 1; r) = \overline{f}(x; r) + \lambda \textstyle\int_{g_1(x)}^{g_2(x)} K(x, t, \overline{\phi}^{\hbar}(t, 1; r))dt.
$$

Thus we will be able to produce the solutions $\underline{F}(x; r)$ and $\overline{F}(x; r)$ from $\underline{w}_0^{\hbar}(x; r)$ and $\overline{w}_0^{\hbar}(x; r)$ by increasing q from 0 to 1. Writing the Taylor series for $\underline{\phi}^{\hbar}(x, q; r)$ and $\overline{\phi}^{\hbar}(x, q; r)$ with respect to q gives

$$
\begin{aligned}
&\underline{\phi}^{\hbar}(x, q; r) = \underline{w}_0^{\hbar}(x; r) + \textstyle\sum_{m=1}^{\infty} \underline{u}_m^{\hbar}(x; r)q^m,\\
&\overline{\phi}^{\hbar}(x, q; r) = \overline{w}_0^{\hbar}(x; r) + \textstyle\sum_{m=1}^{\infty} \overline{u}_m^{\hbar}(x; r)q^m,
\end{aligned}
\tag{12}
$$

where

$$
\begin{aligned}
&\underline{u}_m^{\hbar}(x; r) = \frac{1}{(m)!} \left.\frac{\partial^m \underline{\phi}^{\hbar}(x,q;r)}{\partial q^m}\right|_{q=0}, \quad m \geq 1,\\
&\overline{u}_m^{\hbar}(x; r) = \frac{1}{(m)!} \left.\frac{\partial^m \overline{\phi}^{\hbar}(x,q;r)}{\partial q^m}\right|_{q=0}, \quad m \geq 1.
\end{aligned}
\tag{13}
$$

Differentiating the zero-order deformation equation (9) m times with respect to the embedding parameter q and dividing them by $m!$ and finally setting $q = 0$, then the m-th order deformation equation can be obtained as follows:

$$\begin{aligned} &\underline{u}_m^{\hbar}(x;r) - \chi_m \underline{u}_{m-1}^{\hbar}(x;r) = \hbar[\underline{u}_{m-1}^{\hbar}(x;r) - \lambda \int_{g_1(x)}^{g_2(x)} \underline{\Phi}_{m-1}^{\hbar}(x,t;r)dt - (1-\chi_m)\underline{f}(x;r)], \\ &\overline{u}_m^{\hbar}(x;r) - \chi_m \overline{u}_{m-1}^{\hbar}(x;r) = \hbar[\overline{u}_{m-1}^{\hbar}(x;r) - \lambda \int_{g_1(x)}^{g_2(x)} \overline{\Phi}_{m-1}^{\hbar}(x,t;r)dt - (1-\chi_m)\overline{f}(x;r)], \end{aligned} \tag{14}$$

where

$$\chi_m = \begin{cases} 0, & m \le 1, \\ 1, & m > 1, \end{cases}$$

and

$$\begin{aligned} \underline{\Phi}_{m-1}^{\hbar}(x,t;r) &= \frac{1}{(m-1)!} \left. \frac{\partial^{m-1} k(x,t,\underline{\Phi}^{\hbar}(t,q;r))}{\partial q^{m-1}} \right|_{q=0}, \quad m \ge 1, \\ \overline{\Phi}_{m-1}^{\hbar}(x,t;r) &= \frac{1}{(m-1)!} \left. \frac{\partial^{m-1} k(x,t,\overline{\Phi}^{\hbar}(t,q;r))}{\partial q^{m-1}} \right|_{q=0}, \quad m \ge 1. \end{aligned} \tag{15}$$

Thus from the m-th order deformation equation (14) the following successive relation can be obtained as follows:

$$\begin{cases} \underline{u}_0^{\hbar}(x;r) = \underline{f}(x;r), \\ \overline{u}_0^{\hbar}(x;r) = \overline{f}(x;r), \\ \underline{u}_1^{\hbar}(x;r) = \hbar\left[-\lambda \displaystyle\int_{g_1(x)}^{g_2(x)} \underline{\Phi}_0^{\hbar}(x,t;r)dt\right], \\ \overline{u}_1^{\hbar}(x;r) = \hbar\left[-\lambda \displaystyle\int_{g_1(x)}^{g_2(x)} \overline{\Phi}_0^{\hbar}(x,t;r)dt\right], \\ \underline{u}_m^{\hbar}(x;r) = \underline{u}_{m-1}^{\hbar}(x;r) + \hbar\left[\underline{u}_{m-1}^{\hbar}(x;r) - \lambda \displaystyle\int_{g_1(x)}^{g_2(x)} \underline{\Phi}_{m-1}^{\hbar}(x,t;r)dt\right], \quad m \ge 2, \\ \overline{u}_m^{\hbar}(x;r) = \overline{u}_{m-1}^{\hbar}(x;r) + \hbar\left[\overline{u}_{m-1}^{\hbar}(x;r) - \lambda \displaystyle\int_{g_1(x)}^{g_2(x)} \overline{\Phi}_{m-1}^{\hbar}(x,t;r)dt\right], \quad m \ge 2, \end{cases} \tag{16}$$

where $\underline{\Phi}_0^{\hbar}(x,t;r) = K(x,t,\underline{u}_0^{\hbar}(x;r))$ and $\overline{\Phi}_0^{\hbar}(x,t;r) = K(x,t,\overline{u}_0^{\hbar}(x;r))$. Finally, the j-th order approximate solution of Eq. (5) can be obtained using following formula:

$$\begin{aligned}\underline{F}_j^{\hbar}(x;r) &= \sum_{i=0}^{j} \underline{u}_i^{\hbar}(x;r),\\ \overline{F}_j^{\hbar}(x;r) &= \sum_{i=0}^{j} \overline{u}_i^{\hbar}(x;r).\end{aligned} \tag{17}$$

The following definition can be deduced by the corresponding definition in crisp case mentioned in [16, 23]:

Definition 1 For two fuzzy numbers $\tilde{\xi}_1 = (\underline{\xi}_1(r), \overline{\xi}_1(r))$ and $\tilde{\xi}_2 = (\underline{\xi}_2(r), \overline{\xi}_2(r))$ the common proximity of two fuzzy numbers $\tilde{\xi}_1$ and $\tilde{\xi}_2$ can be estimated by finding the number of common significant digits of the lower and upper assigned functions as follows:

$$\begin{cases} C_{\underline{\xi}_1(r),\underline{\xi}_2(r)} = \log_{10}\left|\dfrac{\underline{\xi}_1(r)+\underline{\xi}_2(r)}{2(\underline{\xi}_1(r)-\underline{\xi}_2(r))}\right| = \log_{10}\left|\dfrac{\underline{\xi}_1(r)}{\underline{\xi}_1(r)-\underline{\xi}_2(r)}-\dfrac{1}{2}\right|, & \underline{\xi}_1(r)\neq\underline{\xi}_2(r),\\ C_{\overline{\xi}_1(r),\overline{\xi}_2(r)} = \log_{10}\left|\frac{\overline{\xi}_1(r)+\overline{\xi}_2(r)}{2(\overline{\xi}_1(r)-\overline{\xi}_2(r))}\right| = \log_{10}\left|\frac{\overline{\xi}_1(r)}{\overline{\xi}_1(r)-\overline{\xi}_2(r)}-\frac{1}{2}\right|, & \overline{\xi}_1(r)\neq\overline{\xi}_2(r),\\ C_{\underline{\xi}_1(r),\underline{\xi}_1(r)} = C_{\overline{\xi}_1(r),\overline{\xi}_1(r)} = +\infty, & \\ C_{\underline{\xi}_2(r),\underline{\xi}_2(r)} = C_{\overline{\xi}_2(r),\overline{\xi}_2(r)} = +\infty, & \end{cases} \tag{18}$$

where $0 \le r \le 1$.

Theorem 1 *Assume that the approximate solution of IE (5) is obtained from Eqs. (16) and (17) which depends on the convergence control parameter* $\hbar$, *then*

$$\begin{aligned} C_{\underline{F}_m^{\hbar}(x;r),\underline{F}_{m+1}^{\hbar}(x;r)} &= C_{\underline{F}_m^{\hbar}(x;r),\underline{F}^{\hbar}(x;r)} + O(\tfrac{1}{m+1}),\\ C_{\overline{F}_m^{\hbar}(x;r),\overline{F}_{m+1}^{\hbar}(x;r)} &= C_{\overline{F}_m^{\hbar}(x;r),\overline{F}^{\hbar}}(x;r) + O(\tfrac{1}{m+1}). \end{aligned} \tag{19}$$

Proof We prove the first result of Eq. (19). The second one can be proved similarly. Applying Definition 1, we can write

$$\begin{aligned} & C_{\underline{F}_m^{\hbar}(x;r),\underline{F}_{m+1}^{\hbar}(x;r)} - C_{\underline{F}_m^{\hbar}(x;r),\underline{F}^{\hbar}(x;r)}\\ &= \log_{10}\left|\frac{\underline{F}_m^{\hbar}(x;r)+\underline{F}_{m+1}^{\hbar}(x;r)}{2(\underline{F}_m^{\hbar}(x;r)-\underline{F}_{m+1}^{\hbar}(x;r))}\right| - \log_{10}\left|\frac{\underline{F}_m^{\hbar}(x;r)+\underline{F}^{\hbar}(x;r)}{2(\underline{F}_m^{\hbar}(x;r)-\underline{F}^{\hbar}(x;r))}\right|\\ &= \log_{10}\left|\frac{\underline{F}_m^{\hbar}(x;r)+\underline{F}_{m+1}^{\hbar}(x;r)}{2(\underline{F}_{m+1}^{\hbar}(x;r))}\right| - \log_{10}\left|\frac{\underline{F}_m^{\hbar}(x;r)+\underline{F}^{\hbar}(x;r)}{2(\underline{F}_m^{\hbar}(x;r)-\underline{F}^{\hbar}(x;r))}\right|\\ &= \log_{10}\left|\frac{\underline{F}_m^{\hbar}(x;r)+\underline{F}_{m+1}^{\hbar}(x;r)}{\underline{F}_m^{\hbar}(x;r)+\underline{F}^{\hbar}(x;r)}\right| + \log_{10}\left|\frac{\underline{F}_m^{\hbar}(x;r)-\underline{F}^{\hbar}(x;r)}{\underline{F}_{m+1}^{\hbar}(x;r)}\right|. \end{aligned} \tag{20}$$

It is clear that for m large enough, the first terms of Eq. (20) can be neglected. The second term of Eq. (20) is written as:

$$\log_{10}\left|\frac{\underline{F}_m^{\hbar}(x;r)-\underline{F}^{\hbar}(x;r)}{\underline{F}_{m+1}^{\hbar}(x;r)}\right| = \log_{10}\left|\frac{\sum_{i=0}^{m}\underline{F}_i^{\hbar}(x;r)-\sum_{i=0}^{\infty}\underline{F}_i^{\hbar}(x;r)}{\underline{F}_{m+1}^{\hbar}(x;r)}\right|$$

$$= \log_{10}\left|\frac{\sum_{i=m+1}^{\infty}\underline{F}_i^{\hbar}(x;r)}{\underline{F}_{m+1}^{\hbar}(x;r)}\right|$$

$$= \log_{10}\left|1+\frac{\sum_{i=m+2}^{\infty}\hbar\mathcal{L}^{-1}\big[\mathcal{H}(x)N(\underline{F}_{i-1}^{\hbar}(x;r))\big]}{\hbar\mathcal{L}^{-1}\big[\mathcal{H}(x)N(\underline{F}_m^{\hbar}(x;r))\big]}\right|.$$

According to Eq. (13), we obtain:

$$\frac{\sum_{i=m+2}^{\infty}\hbar\mathcal{L}^{-1}\big[\mathcal{H}(x)N(\underline{F}_{i-1}^{\hbar}(x;r))\big]}{\hbar\mathcal{L}^{-1}\big[\mathcal{H}(x)N(\underline{F}_m^{\hbar}(x;r))\big]} = \frac{\sum_{i=m+2}^{\infty}\hbar\mathcal{L}^{-1}\left[\mathcal{H}(x)\frac{1}{(i-1)!}\frac{\partial^{i-1}N[\underline{\phi}^{\hbar}(x,q;r)]}{\partial q^{i-1}}\bigg|_{q=0}\right]}{\hbar\mathcal{L}^{-1}\left[\mathcal{H}(x)\frac{1}{m!}\frac{\partial^{m}N[\underline{\phi}^{\hbar}(x,q;r)]}{\partial q^{m}}\bigg|_{q=0}\right]}$$

$$= \frac{\hbar\mathcal{L}^{-1}\left[\mathcal{H}(x)\frac{1}{(m+1)!}\frac{\partial^{m+1}N[\underline{\phi}^{\hbar}(x,q;r)]}{\partial q^{m+1}}\bigg|_{q=0}\right]}{\hbar\mathcal{L}^{-1}\left[\mathcal{H}(x)\frac{1}{m!}\frac{\partial^{m}N[\underline{\phi}^{\hbar}(x,q;r)]}{\partial q^{m}}\bigg|_{q=0}\right]} + \frac{\hbar\mathcal{L}^{-1}\left[\mathcal{H}(x)\frac{1}{(m+2)!}\frac{\partial^{m+2}N[\underline{\phi}^{\hbar}(x,q;r)]}{\partial q^{m+2}}\bigg|_{q=0}\right]}{\hbar\mathcal{L}^{-1}\left[\mathcal{H}(x)\frac{1}{m!}\frac{\partial^{m}N[\underline{\phi}^{\hbar}(x,q;r)]}{\partial q^{m}}\bigg|_{q=0}\right]} + \cdots$$

$$= O(\frac{1}{m+1}).$$

Therefore, we can write

$$\log_{10}\left|\frac{\underline{F}_m^{\hbar}(x;r)-\underline{F}^{\hbar}(x;r)}{\underline{F}_{m+1}^{\hbar}(x;r)}\right| = \log_{10}\left|1+O(\frac{1}{m+1})\right|.$$

We know that $O(\frac{1}{m+1}) << 1$, therefore $\log_{10}\left|1+O(\frac{1}{m+1})\right|$ leads to zero when m increases. Thus, the proof is complete. □

Theorem 1 shows that when m increases, the common proximity between two sequential fuzzy approximations $\tilde{F}_m^{\hbar}$ and $\tilde{F}_{m+1}^{\hbar}$ are almost equal with the common proximity between the m-th approximation $\tilde{F}_m^{\hbar}$ and the exact fuzzy solution $\tilde{F}^{\hbar}$ at a given point x.

Let m be a given iteration and $\tilde{F}_m^{\hbar_k}(x;r)$ and $\tilde{F}_m^{\hbar_{k+1}}(x;r)$ be the m-th order approximate solution of Eq. (5) obtained from the HAM where $\hbar_k$ and $\hbar_{k+1}$ for $k=0,1,2,\ldots$

are two successive convergence control parameters which are chosen from the region of convergence, then

$$C_{\underline{F}_m^{\hbar_k}(x;r),\underline{F}^{\hbar_k}(x;r)} \simeq C_{\underline{F}_m^{\hbar_k}(x;r),\underline{F}_m^{\hbar_{k+1}}(x;r)},$$
$$C_{\overline{F}_m^{\hbar_k}(x;r),\overline{F}^{\hbar_k}(x;r)} \simeq C_{\overline{F}_m^{\hbar_k}(x;r),\overline{F}_m^{\hbar_{k+1}}(x;r)}. \tag{21}$$

Equation (21) shows that when k increases, the common proximity between two sequential fuzzy approximations $\tilde{F}_m^{h_k}$ and $\tilde{F}_m^{h_{k+1}}$ are almost equal with the common proximity between the m-th approximation $\tilde{F}_m^{h_k}$ and the exact fuzzy solution $\tilde{F}^{h_k}$ at a given point x.

4 Numerical Examples

In this section, two IEs are presented and the fuzzy CESTAC method is applied to validate the numerical results. For each example, the optimal step of the HAM, the optimal error and the optimal approximation are obtained based on Algorithms 2 and 3. Also, the termination criterion (3) is applied to stop the numerical process.

Algorithm 2:
1- Let $k = 0$;
2- Enter $m, \hbar_0, \Delta\hbar, x, \underline{F}_m^{\hbar_k}(x;r), \overline{F}_m^{\hbar_k}(x;r)$;
3- Do the following steps, while $Distance\left(\tilde{F}_m^{\hbar_k}(x), \tilde{F}_m^{\hbar_{k+1}}(x)\right) \neq @.0$;
{
3-1- Calculate $\underline{F}_m^{\hbar_k}(x;r), \overline{F}_m^{\hbar_k}(x;r)$ for $r = 0, 0.1, \cdots, 1$ by applying Eqs. (16) and (17);
3-2- Compute $|\underline{F}_m^{\hbar_{k+1}}(x;r) - \underline{F}_m^{\hbar_k}(x;r)|, |\overline{F}_m^{\hbar_{k+1}}(x;r) - \overline{F}_m^{\hbar_k}(x;r)|$ and the maximum values between them to estimate the Hausdorff distance which is printed by the value Distance;
3-3- Print $\hbar_k, r, \underline{F}_m^{\hbar_k}(x;r), \overline{F}_m^{\hbar_k}(x;r), |\underline{F}_m^{\hbar_{k+1}}(x;r) - \underline{F}_m^{\hbar_k}(x;r)|, |\overline{F}_m^{\hbar_{k+1}}(x;r) - \overline{F}_m^{\hbar_k}$ $(x;r)|$ and the Distance;
3-4- $\hbar_{k+1} = \hbar_k + \Delta\hbar$;
3-5- $k++$.
}

Algorithm 3:
1- Enter $\hbar_{opt}$, x;
2- Let $m = 0$;
3- Compute $\underline{F}_m^{\hbar_{opt}}(x; r), \overline{F}_m^{\hbar_{opt}}(x; r)$ using Eqs. (16) and (17);
4- Do the following steps, while $Distance\left(\tilde{F}_m^{\hbar_{opt}}(x), \tilde{F}_{m+1}^{\hbar_{opt}}(x)\right) \neq @.0$;
{
4-1- Calculate $\underline{F}_m^{\hbar_{opt}}(x; r), \overline{F}_m^{\hbar_{opt}}(x; r)$ for $r = 0, 0.1, \cdots, 1$;
4-2- Compute $|\underline{F}_{m+1}^{\hbar_{opt}}(x; r) - \underline{F}_m^{\hbar_{opt}}(x; r)|, |\overline{F}_{m+1}^{\hbar_{opt}}(x; r) - \overline{F}_m^{\hbar_{opt}}(x; r)|$ and the maximum values between them to estimate the Hausdorff distance which is printed by the value Distance;
4-3- Print $m, r, \underline{F}_m^{\hbar_{opt}}(x; r), \overline{F}_m^{\hbar_{opt}}(x; r), |\underline{F}_{m+1}^{\hbar_{opt}}(x; r) - \underline{F}_m^{\hbar_{opt}}(x; r)|, |\overline{F}_{m+1}^{\hbar_{opt}}(x; r) - \overline{F}_m^{\hbar_{opt}}(x; r)|$ and the Distance;
4-4 $m = m + 1$.
}

Example 1 Let us consider the nonlinear fuzzy Fredholm IE such that

$$\underline{f}(x; r) = (1 + x)r + \frac{(x^2r^2)}{2},$$

$$\overline{f}(x; r) = 2(1 + x)r + 2x^2r^2, \quad 0 \leq r \leq 1.$$

and

$$\underline{K(x, t, F(t; r))} = x^2(1 - 2t)\underline{F}^2(t; r), \ 0 \leq x, t \leq 1,$$

$$\overline{K(x, t, F(t; r))} = x^2(1 - 2t)\overline{F}^2(t; r), \ 0 \leq x, t \leq 1.$$

The numerical results of Algorithms 2 and 3 are presented in Tables 1 and 2. In Table 1, the optimal value of the convergence control parameter $\hbar$ can be found for $\hbar_0 = -0.9$ and $\Delta\hbar = -0.02$. To this end, for a given value of iteration $m = 5$, the optimal value $\hbar_{opt} = -1$ can be found based on Algorithm 2. In Table 2, for $\hbar_{opt} = -1$, the optimal iteration and the optimal approximation can be obtained according to Algorithm 3. The optimal iteration of the HAM for solving Example 1 is $m_{opt} = 3$ and the optimal approximation for $r = 0.5$ and $x = 0.1$ is $(\underline{F}_3(x; r), \overline{F}_3(x; r)) = (0.5499255E + 000, 0.1097057E + 001)$. In Fig. 1, the $\hbar$-curves for $\underline{F}_m^{\hbar}(0.1; 0.3)$, $\overline{F}_m^{\hbar}(0.1; 0.3)$ and $m_{opt} = 3$ are presented. It is obvious that the convergence region for $\underline{F}_m^{\hbar}(0.1; 0.3)$ is $-1.5 \leq \hbar \leq -0.5$ and for $\overline{F}_m^{\hbar}(0.1; 0.3)$ we have $-1.5 \leq \hbar \leq 0$.

Example 2 Consider the following linear fuzzy Volterra IE with $K(x, t) = x - t, 0 \leq x, t \leq 1$ and

Table 1 The optimal value of $\hbar$ for solving Example 1 with $x = 0.1$

$\hbar_k$	r	$\underline{F}_m^{\hbar_k}(x;r)$	$\lvert\underline{F}_m^{\hbar_{k+1}}(x;r) - \underline{F}_m^{\hbar_k}(x;r)\rvert$	$\overline{F}_m^{\hbar_k}(x;r)$	$\lvert\overline{F}_m^{\hbar_{k+1}}(x;r) - \overline{F}_m^{\hbar_k}(x;r)\rvert$	Max
−0.90	0.0	0.0000000E+000	0.0000000E+000	0.0000000E+000	0.0000000E+000	0.0000000E+000
	0.1	0.110003E+000	0.110003E+000	0.220011E+000	0.220011E+000	0.220011E+000
	0.2	0.220011E+000	0.220011E+000	0.440021E+000	0.440021E+000	0.440021E+000
	0.3	0.330017E+000	0.330017E+000	0.660021E+000	0.660021E+000	0.660021E+000
	0.4	0.440021E+000	0.440021E+000	0.8800368E+000	0.8800368E+000	0.8800368E+000
	0.5	0.5500215E+000	0.5500216E+000	0.110010E+001	0.110010E+001	0.110010E+001
	0.6	0.660021E+000	0.660021E+000	0.1320214E+001	0.1320214E+001	0.1320214E+001
	0.7	0.770024E+000	0.770024E+000	0.154017E+001	0.154017E+001	0.154017E+001
	0.8	0.8800364E+000	0.8800365E+000	0.175915E+001	0.175915E+001	0.175915E+001
	0.9	0.990062E+000	0.990062E+000	0.197544E+001	0.197544E+001	0.197544E+001
	1.0	0.1100104E+001	0.1100104E+001	0.2184654E+001	0.2184655E+001	0.2184655E+001
−0.90	–	–	–	–	**Distance =**	**0.2184655E+001**
−0.92	0.0	0.0000000E+000	0.0000000E+000	0.0000000E+000	0.0000000E+000	0.0000000E+000
	0.1	0.1100029E+000	−0.7E−006	0.2200088E+000	−0.2E−005	−0.7E−006
	0.2	0.220008E+000	−0.22E−005	0.440016E+000	−0.4E−005	−0.22E−005
	0.3	0.3300138E+000	−0.37E−005	0.6600115E+000	−0.9E−005	−0.37E−005
	0.4	0.440016E+000	−0.4E−005	0.879989E+000	−0.47E−004	−0.4E−005
	0.5	0.5500156E+000	−0.59E−005	0.109987E+001	−0.23E−003	−0.59E−005
	0.6	0.660011E+000	−0.9E−005	0.131931E+001	−0.902E−003	−0.9E−005
	0.7	0.770004E+000	−0.2E−004	0.153720E+001	−0.2964E−002	−0.2E−004
	0.8	0.879989E+000	−0.47E−004	0.175079E+001	−0.8360E−002	−0.47E−004
	0.9	0.989955E+000	−0.10E−003	0.195372E+001	−0.2172E−001	−0.10E−003
	1.0	0.109987E+001	−0.231E−003	0.213279E+001	−0.51858E−001	−0.231E−003
−0.92	–	–	–	–	**Distance =**	**0.0000000E+000**
−0.94	0.0	0.0000000E+000	0.0000000E+000	0.0000000E+000	0.0000000E+000	0.0000000E+000
	0.1	0.1100022E+000	−0.3E−006	0.2200065E+000	−0.1E−005	−0.3E−006
	0.2	0.220006E+000	−0.1E−005	0.440011E+000	−0.2E−005	−0.2E−005
	0.3	0.3300102E+000	−0.1E−005	0.659999E+000	−0.6E−005	−0.1E−005
	0.4	0.440011E+000	−0.2E−005	0.879918E+000	−0.39E−004	−0.2E−005
	0.5	0.5500092E+000	−0.3E−005	0.109949E+001	−0.20E−003	−0.3E−005
	0.6	0.660000E+000	−0.5E−005	0.131777E+001	−0.868E−003	−0.5E−005
	0.7	0.769976E+000	−0.1E−004	0.153202E+001	−0.2958E−002	−0.1E−004
	0.8	0.879917E+000	−0.39E−004	0.173568E+001	−0.8531E−002	−0.39E−004
	0.9	0.989785E+000	−0.93E−004	0.191420E+001	−0.2244E−001	−0.93E−004
	1.0	0.109949E+001	−0.211E−003	0.203812E+001	−0.53996E−001	−0.211E−003
−0.94	–	–	–	–	**Distance =**	**0.0000000E+000**
−0.96	0.0	0.0000000E+000	0.0000000E+000	0.0000000E+000	0.0000000E+000	0.0000000E+000
	0.1	0.1100014E+000	−0.3E−006	0.2200042E+000	−0.1E−005	−0.3E−006
	0.2	0.220004E+000	−0.1E−005	0.440006E+000	−0.2E−005	−0.2E−005
	0.3	0.3300065E+000	−0.1E−005	0.659983E+000	−0.8E−005	−0.1E−005
	0.4	0.440006E+000	−0.2E−005	0.879806E+000	−0.62E−004	−0.2E−005
	0.5	0.5500020E+000	−0.3E−005	0.109888E+001	−0.33E−003	−0.3E−005
	0.6	0.659984E+000	−0.8E−005	0.131525E+001	−0.140E−002	−0.8E−005
	0.7	0.769933E+000	−0.23E−004	0.152348E+001	−0.4793E−002	−0.23E−004
	0.8	0.879806E+000	−0.61E−004	0.171068E+001	−0.1390E−001	−0.61E−004
	0.9	0.989514E+000	−0.15E−003	0.184889E+001	−0.36436E−001	−0.15E−003
	1.0	0.109888E+001	−0.339E−003	0.188199E+001	−0.87245E−001	−0.339E−003

(continued)

Table 1 (continued)

$\hbar_k$	r	$\underline{F}_m^{\hbar_k}(x;r)$	$\lvert\underline{F}_m^{\hbar_{k+1}}(x;r) - \underline{F}_m^{\hbar_k}(x;r)\rvert$	$\overline{F}_m^{\hbar_k}(x;r)$	$\lvert\overline{F}_m^{\hbar_{k+1}}(x;r) - \overline{F}_m^{\hbar_k}(x;r)\rvert$	Max
−0.96	–	–	–	–	**Distance =**	**0.0000000E+000**
−0.98	0.0	0.0000000E+000	0.0000000E+000	0.0000000E+000	0.0000000E+000	0.0000000E+000
	0.1	0.1100007E+000	−0.3E−006	0.2200020E+000	−0.1E−005	−0.3E−006
	0.2	0.220002E+000	−0.1E−005	0.440001E+000	−0.2E−005	−0.2E−005
	0.3	0.3300030E+000	−0.1E−005	0.659961E+000	−0.1E−004	−0.1E−005
	0.4	0.440001E+000	−0.2E−005	0.879634E+000	−0.94E−004	−0.2E−005
	0.5	0.5499930E+000	−0.4E−005	0.109794E+001	−0.519E−003	−0.4E−005
	0.6	0.659961E+000	−0.1E−004	0.131134E+001	−0.215E−002	−0.1E−004
	0.7	0.769869E+000	−0.35E−004	0.151028E+001	−0.7297E−002	−0.35E−004
	0.8	0.879634E+000	−0.94E−004	0.167217E+001	−0.21142E−001	−0.94E−004
	0.9	0.989093E+000	−0.23E−003	0.174856E+001	−0.55148E−001	−0.23E−003
	1.0	0.109794E+001	−0.520E−003	0.164322E+001	−0.13135E+000	−0.520E−003
−0.98	–	–	–	–	**Distance =**	**0.0000000E+000**
−1.00	0.0	0.0000000E+000	0.0000000E+000	0.0000000E+000	0.0000000E+000	0.0000000E+000
	0.1	0.1099999E+000	−0.3E−006	0.2199999E+000	−0.1E−005	−0.3E−006
	0.2	0.219999E+000	@.0	0.439996E+000	−0.2E−005	@.0
	0.3	0.3299995E+000	@.0	0.659929E+000	@.0	@.0
	0.4	0.439996E+000	−0.2E−005	0.879379E+000	−0.13E−003	−0.2E−005
	0.5	0.5499818E+000	−0.58E−005	0.109654E+001	−0.762E−003	−0.58E−005
	0.6	0.659929E+000	@.0	0.130560E+001	−0.3131E−002	@.0
	0.7	0.769774E+000	−0.51E−004	0.149095E+001	−0.1054E−001	−0.51E−004
	0.8	0.879379E+000	−0.13E−003	0.161601E+001	−0.30517E−001	−0.13E−003
	0.9	0.988470E+000	−0.33E−003	0.160291E+001	−0.79221E−001	−0.33E−003
	1.0	0.109654E+001	−0.762E−003	0.129806E+001	−0.18764E+000	−0.762E−003
−1.00	–	–	–	–	**Distance =**	**@.0**

$$\underline{f}(x;r) = (1 - x - \frac{x^2}{2})r,$$

$$\overline{f}(x;r) = (1 - x - \frac{x^2}{2})(2 - r), \quad 0 \le r \le 1.$$

In this example, the accuracy of the results are shown based on two successive iterations. The numerical results of Algorithm 2 to validate the results are presented in Table 3 for finding the optimal convergence control parameter which is $\hbar_{opt} = -0.95$ for $\hbar_0 = -0.9$ and $\Delta\hbar = -0.01$. Furthermore, in Table 4, the optimal approximation, the optimal distance and the optimal iteration of the HAM are presented using Algorithm 3. In this example, the optimal iteration is $m_{opt} = 6$ and the optimal approximation for $r = x = 0.5$ is $(\underline{F}_3(x;r), \overline{F}_3(x;r)) = (0.2394522E + 000, 0.718356E + 000)$. The $\hbar$-curves of this example for $\underline{F}^{\hbar}_{m_{opt}}(0.5; 0.3)$, $\overline{F}^{\hbar}_{m_{opt}}(0.5; 0.3)$ and $m_{opt} = 6$ are shown in Fig. 2. The convergence region for both cases is $-1.5 \le \hbar \le -0.5$.

Table 2 The optimal iteration of the HAM for solving Example 1 with $x = 0.1$, $\hbar_{opt} = -1$

m	r	$\underline{F}_m^{\hbar_k}(x;r)$	$\lvert\underline{F}_{m+1}^{\hbar_k}(x;r) - \underline{F}_m^{\hbar_k}(x;r)\rvert$	$\overline{F}_m^{\hbar_k}(x;r)$	$\lvert\overline{F}_{m+1}^{\hbar_k}(x;r) - \overline{F}_m^{\hbar_k}(x;r)\rvert$	Max
0	0.0	0.0000000E+000	0.0000000E+000	0.0000000E+000	0.0000000E+000	0.0000000E+000
	0.1	0.1100500E+000	0.1100500E+000	0.2202000E+000	0.2202000E+000	0.2202000E+000
	0.2	0.2201999E+000	0.2202000E+000	0.4407999E+000	0.4408000E+000	0.4408000E+000
	0.3	0.330449E+000	0.330449E+000	0.6617999E+000	0.6618000E+000	0.6618000E+000
	0.4	0.4408000E+000	0.4408000E+000	0.883199E+000	0.883200E+000	0.883200E+000
	0.5	0.5512500E+000	0.5512501E+000	0.110500E+001	0.110500E+001	0.110500E+001
	0.6	0.6618000E+000	0.6618001E+000	0.1327200E+001	0.1327200E+001	0.1327200E+001
	0.7	0.7724501E+000	0.7724501E+000	0.1549800E+001	0.1549800E+001	0.1549800E+001
	0.8	0.8832001E+000	0.8832001E+000	0.1772800E+001	0.1772800E+001	0.1772800E+001
	0.9	0.9940502E+000	0.9940502E+000	0.1996200E+001	0.1996200E+001	0.1996200E+001
	1.0	0.1105000E+001	0.1105000E+001	0.2220000E+001	0.2220000E+001	0.2220000E+001
0	–	–	–	–	**Distance =**	**0.2220000E+001**
1	0.0	0.0000000E+000	0.0000000E+000	0.0000000E+000	0.0000000E+000	0.0000000E+000
	0.1	0.1099967E+000	0.1099968E+000	0.219974E+000	0.219974E+000	0.219974E+000
	0.2	0.219974E+000	0.219974E+000	0.4397888E+000	0.4397888E+000	0.4397888E+000
	0.3	0.3299117E+000	0.3299117E+000	0.6592728E+000	0.6592729E+000	0.6592729E+000
	0.4	0.4397888E+000	0.4397888E+000	0.8782421E+000	0.8782422E+000	0.8782422E+000
	0.5	0.549583E+000	0.549583E+000	0.1096500E+001	0.1096500E+001	0.1096500E+001
	0.6	0.659272E+000	0.659272E+000	0.1313836E+001	0.1313837E+001	0.1313837E+001
	0.7	0.768833E+000	0.768833E+000	0.1530030E+001	0.1530030E+001	0.1530030E+001
	0.8	0.878242E+000	0.878242E+000	0.1744845E+001	0.1744845E+001	0.1744845E+001
	0.9	0.987472E+000	0.987473E+000	0.1958033E+001	0.1958033E+001	0.1958033E+001
	1.0	0.1096500E+001	0.1096500E+001	0.216933E+001	0.216933E+001	0.216933E+001
1	–	–	–	–	**Distance =**	**0.216933E+001**
2	0.0	0.0000000E+000	0.0000000E+000	0.0000000E+000	0.0000000E+000	0.0000000E+000
	0.1	0.110000E+000	0.34E−005	0.220003E+000	0.298E−004	0.298E−004
	0.2	0.2200039E+000	0.29E−004	0.4400665E+000	0.277E−003	0.277E−003
	0.3	0.3300205E+000	0.108E−003	0.660354E+000	0.1081E−002	0.1081E−002
	0.4	0.4400665E+000	0.277E−003	0.8811772E+000	0.2935E−002	0.2935E−002
	0.5	0.550166E+000	0.5833E−003	0.110301E+001	0.6516E−002	0.6516E−002
	0.6	0.6603544E+000	0.1081E−002	0.132655E+001	0.1272E−001	0.1272E−001
	0.7	0.7706732E+000	0.1839E−002	0.1552726E+001	0.22695E−001	0.22695E−001
	0.8	0.8811772E+000	0.2935E−002	0.1782714E+001	0.37869E−001	0.37869E−001
	0.9	0.9919323E+000	0.4459E−002	0.201803E+001	0.59998E−001	0.59998E−001
	1.0	0.110301E+001	0.6516E−002	0.226053E+001	0.91200E−001	0.91200E−001
2	–	–	–	–	**Distance =**	**0.91200E−001**
3	0.0	0.0000000E+000	0.0000000E+000	0.0000000E+000	0.0000000E+000	0.0000000E+000
	0.1	0.1099999E+000	@.0	0.2199993E+000	@.0	@.0
	0.2	0.219999E+000	@.0	0.4399766E+000	@.0	@.0
	0.3	0.3299946E+000	−0.258E−004	0.6598063E+000	−0.548E−003	−0.258E−004
	0.4	0.439976E+000	−0.899E−004	0.879111E+000	−0.2065E−002	−0.899E−004
	0.5	0.5499255E+000	−0.241E−003	0.1097057E+001	−0.5959E−002	−0.241E−003
	0.6	0.6598064E+000	−0.548E−003	0.131206E+001	−0.1449E−001	−0.548E−003
	0.7	0.769563E+000	−0.1109E−002	0.1521459E+001	−0.31266E−001	−0.1109E−002
	0.8	0.8791120E+000	−0.2065E−002	0.1720993E+001	−0.617209E−001	−0.2065E−002
	0.9	0.9883320E+000	−0.3600E−002	0.1904290E+001	−0.11374E+000	−0.3600E−002
	1.0	0.109705E+001	−0.59594E−002	0.206214E+001	−0.198388E+000	−0.59594E−002
3	–	–	–	–	**Distance =**	**@.0**

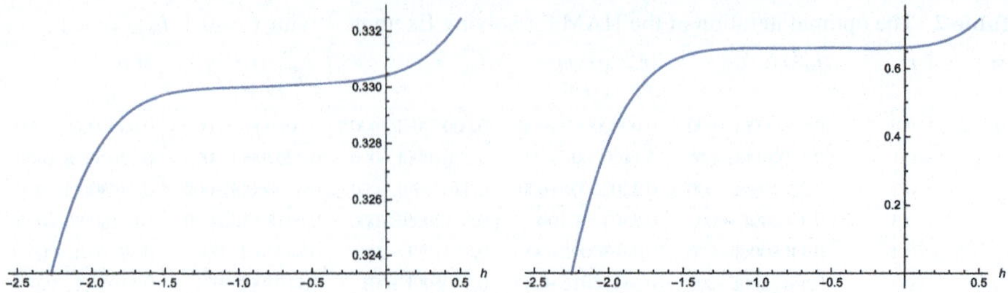

Fig. 1 The $\hbar$-curves of Example 1 for $\underline{F}^{\hbar}_{m_{opt}}(0.1; 0.3)$, $\overline{F}^{\hbar}_{m_{opt}}(0.1; 0.3)$ and $m_{opt} = 3$

Table 3 The optimal value of $\hbar$ for solving Example 2 with $x = 0.5$

$\hbar_k$	r	$\underline{F}_m^{\hbar_k}(x;r)$	$\lvert\underline{F}_m^{\hbar_{k+1}}(x;r) - \underline{F}_m^{\hbar_k}(x;r)\rvert$	$\overline{F}_m^{\hbar_k}(x;r)$	$\lvert\overline{F}_m^{\hbar_{k+1}}(x;r) - \overline{F}_m^{\hbar_k}(x;r)\rvert$	Max
−0.9	0.0	0.0000000E+000	0.0000000E+000	0.9578046E+000	0.9578046E+000	0.9578046E+000
	0.1	0.478902E−001	0.478902E−001	0.909914E+000	0.909914E+000	0.909914E+000
	0.2	0.957804E−001	0.957804E−001	0.862024E+000	0.862024E+000	0.862024E+000
	0.3	0.1436707E+000	0.1436707E+000	0.814134E+000	0.814134E+000	0.814134E+000
	0.4	0.1915609E+000	0.1915610E+000	0.766243E+000	0.766243E+000	0.766243E+000
	0.5	0.2394512E+000	0.2394512E+000	0.718353E+000	0.718353E+000	0.718353E+000
	0.6	0.2873415E+000	0.2873415E+000	0.670462E+000	0.670462E+000	0.670462E+000
	0.7	0.335231E+000	0.335231E+000	0.6225731E+000	0.6225731E+000	0.6225731E+000
	0.8	0.3831219E+000	0.3831220E+000	0.574682E+000	0.574682E+000	0.574682E+000
	0.9	0.431012E+000	0.431012E+000	0.526792E+000	0.526792E+000	0.526792E+000
	1.0	0.478902E+000	0.478902E+000	0.478902E+000	0.478902E+000	0.478902E+000
−0.9	–	–	–	–	**Distance =**	**0.9578046E+000**
−0.91	0.0	0.0000000E+000	0.0000000E+000	0.957806E+000	@.0	@.0
	0.1	0.478903E−001	@.0	0.909915E+000	@.0	@.0
	0.2	0.9578064E−001	0.1E−006	0.862025E+000	@.0	@.0
	0.3	0.143670E+000	0.2E−006	0.8141353E+000	@.0	@.0
	0.4	0.191561E+000	0.3E−006	0.766245E+000	@.0	@.0
	0.5	0.239451E+000	0.3E−006	0.7183546E+000	@.0	@.0
	0.6	0.287341E+000	0.4E−006	0.670464E+000	0.1E−005	0.1E−005
	0.7	0.3352322E+000	@.0	0.622573E+000	@.0	@.0
	0.8	0.3831225E+000	0.5E−006	0.574683E+000	@.0	@.0
	0.9	0.4310129E+000	0.7E−006	0.526793E+000	@.0	@.0
	1.0	0.4789032E+000	0.7E−006	0.478902E+000	@.0	@.0
−0.91	–	–	–	–	**Distance =**	**0.1E–005**
−0.92	0.0	0.0000000E+000	0.0000000E+000	0.957807E+000	@.0	@.0
	0.1	0.478903E−001	0.6E−007	0.909917E+000	@.0	@.0
	0.2	0.957807E−001	0.1E−006	0.862026E+000	@.0	@.0
	0.3	0.143671E+000	@.0	0.814136E+000	0.9E−006	0.9E−006
	0.4	0.191561E+000	@.0	0.766245E+000	@.0	@.0
	0.5	0.2394519E+000	0.2E−006	0.718355E+000	@.0	@.0
	0.6	0.2873422E+000	@.0	0.670465E+000	@.0	@.0
	0.7	0.335232E+000	0.3E−006	0.622574E+000	@.0	@.0
	0.8	0.3831230E+000	0.4E−006	0.574684E+000	@.0	@.0
	0.9	0.4310134E+000	0.4E−006	0.526793E+000	@.0	@.0
	1.0	0.478903E+000	@.0	0.478903E+000	@.0	@.0
−0.92	–	–	–	–	**Distance =**	**0.9E–006**

(continued)

Table 3 (continued)

$\hbar_k$	r	$\underline{F}_m^{\hbar_k}(x;r)$	$\lvert\underline{F}_m^{\hbar_{k+1}}(x;r)-\underline{F}_m^{\hbar_k}(x;r)\rvert$	$\overline{F}_m^{\hbar_k}(x;r)$	$\lvert\overline{F}_m^{\hbar_{k+1}}(x;r)-\overline{F}_m^{\hbar_k}(x;r)\rvert$	Max
−0.93	0.0	0.0000000E+000	0.0000000E+000	0.957808E+000	@.0	@.0
	0.1	0.4789040E−001	@.0	0.909917E+000	@.0	@.0
	0.2	0.9578082E−001	@.0	0.862027E+000	0.1E−005	0.1E−005
	0.3	0.1436712E+000	@.0	0.814137E+000	@.0	@.0
	0.4	0.1915616E+000	@.0	0.7662467E+000	@.0	@.0
	0.5	0.2394520E+000	0.1937150E−006	0.718356E+000	@.0	@.0
	0.6	0.287342E+000	0.2E−006	0.670465E+000	@.0	@.0
	0.7	0.335232E+000	@.0	0.622575E+000	@.0	@.0
	0.8	0.3831233E+000	@.0	0.574685E+000	@.0	@.0
	0.9	0.431013E+000	@.0	0.526794E+000	@.0	@.0
	1.0	0.4789042E+000	@.0	0.478904E+000	@.0	@.0
−0.93	–	–	–	–	**Distance =**	**0.1E–005**
−0.940	0.0	0.0000000E+000	0.0000000E+000	0.957809E+000	@.0	@.0
	0.1	0.4789043E−001	@.0	0.909918E+000	0.1E−005	0.1E−005
	0.2	0.9578087E−001	@.0	0.862028E+000	@.0	@.0
	0.3	0.1436713E+000	@.0	0.814137E+000	@.0	@.0
	0.4	0.191561E+000	@.0	0.766246E+000	@.0	@.0
	0.5	0.2394521E+000	@.0	0.718356E+000	@.0	@.0
	0.6	0.287342E+000	0.1192092E−006	0.670466E+000	@.0	@.0
	0.7	0.3352331E+000	@.0	0.622575E+000	@.0	@.0
	0.8	0.3831235E+000	@.0	0.5746849E+000	@.0	@.0
	0.9	0.4310139E+000	@.0	0.526794E+000	@.0	@.0
	1.0	0.4789044E+000	@.0	0.478904E+000	@.0	@.0
−0.94	–	–	–	–	**Distance =**	**0.1E–005**
−0.950	0.0	0.0000000E+000	0.0000000E+000	0.957808E+000	@.0	@.0
	0.1	0.4789045E−001	@.0	0.909918E+000	@.0	@.0
	0.2	0.957809E−001	@.0	0.862028E+000	@.0	@.0
	0.3	0.1436713E+000	@.0	0.814137E+000	@.0	@.0
	0.4	0.1915618E+000	@.0	0.766247E+000	@.0	@.0
	0.5	0.2394522E+000	0.7E−007	0.718356E+000	@.0	@.0
	0.6	0.287342E+000	@.0	0.670465E+000	@.0	@.0
	0.7	0.3352332E+000	@.0	0.622575E+000	@.0	@.0
	0.8	0.3831236E+000	@.0	0.574685E+000	@.0	@.0
	0.9	0.431014E+000	@.0	0.526794E+000	@.0	@.0
	1.0	0.478904E+000	@.0	0.478904E+000	@.0	@.0
−0.95	–	–	–	–	**Distance =**	**@.0**

Table 4 The optimal iteration of the HAM for solving Example 2 with $x = 0.5$, $\hbar_{opt} = -0.95$

m	r	$\underline{E}_m^{\hbar_k}(x;r)$	$\lvert\underline{E}_{m+1}^{\hbar_k}(x;r) - \underline{E}_m^{\hbar_k}(x;r)\rvert$	$\overline{F}_m^{\hbar_k}(x;r)$	$\lvert\overline{F}_{m+1}^{\hbar_k}(x;r) - \overline{F}_m^{\hbar_k}(x;r)\rvert$	Max
0	0.0	0.0000000E+000	0.0000000E+000	0.7500000E+000	0.7500000E+000	0.7500000E+000
	0.1	0.3749999E−001	0.3750000E−001	0.7125000E+000	0.7125000E+000	0.7125000E+000
	0.2	0.7499999E−001	0.7500000E−001	0.6749999E+000	0.6750000E+000	0.6750000E+000
	0.3	0.1125000E+000	0.1125000E+000	0.6375000E+000	0.6375000E+000	0.6375000E+000
	0.4	0.1500000E+000	0.1500000E+000	0.5999999E+000	0.6000000E+000	0.6000000E+000
	0.5	0.1875000E+000	0.1875000E+000	0.5625000E+000	0.5625000E+000	0.5625000E+000
	0.6	0.2250000E+000	0.2250000E+000	0.5249999E+000	0.5249999E+000	0.5249999E+000
	0.7	0.2625000E+000	0.2625000E+000	0.4874999E+000	0.4875000E+000	0.4875000E+000
	0.8	0.3000000E+000	0.3000000E+000	0.4499999E+000	0.4499999E+000	0.4499999E+000
	0.9	0.3375000E+000	0.3375000E+000	0.4124999E+000	0.4124999E+000	0.4124999E+000
	1.0	0.3750000E+000	0.3750001E+000	0.3749999E+000	0.3749999E+000	0.3750001E+000
0	–	–	–	–	**Distance =**	**0.7500000E+000**
1	0.0	0.0000000E+000	0.0000000E+000	0.9429686E+000	0.192968E+000	0.192968E+000
	0.1	0.4714843E−001	0.964844E−002	0.8958201E+000	0.183320E+000	0.183320E+000
	0.2	0.9429687E−001	0.192968E−001	0.8486717E+000	0.173671E+000	0.173671E+000
	0.3	0.1414453E+000	0.2894532E−001	0.8015233E+000	0.164023E+000	0.164023E+000
	0.4	0.1885937E+000	0.385937E−001	0.7543749E+000	0.154374E+000	0.154374E+000
	0.5	0.2357422E+000	0.482422E−001	0.7072264E+000	0.144726E+000	0.144726E+000
	0.6	0.2828906E+000	0.578906E−001	0.6600779E+000	0.135077E+000	0.135077E+000
	0.7	0.3300391E+000	0.675391E−001	0.6129295E+000	0.125429E+000	0.125429E+000
	0.8	0.3771875E+000	0.771875E−001	0.5657811E+000	0.115781E+000	0.115781E+000
	0.9	0.4243360E+000	0.868359E−001	0.518632E+000	0.106132E+000	0.106132E+000
	1.0	0.4714845E+000	0.964844E−001	0.4714841E+000	0.9648427E−001	0.9648427E−001
1	–	–	–	–	**Distance =**	**0.192968E+000**
2	0.0	0.0000000E+000	0.0000000E+000	0.9568085E+000	0.13839E−001	0.13839E−001
	0.1	0.4784041E−001	0.69198E−003	0.908968E+000	0.1314E−001	0.1314E−001
	0.2	0.9568083E−001	0.13839E−002	0.8611276E+000	0.12455E−001	0.12455E−001
	0.3	0.1435212E+000	0.20759E−002	0.8132870E+000	0.11763E−001	0.11763E−001
	0.4	0.1913617E+000	0.2767935E−002	0.765446E+000	0.1107E−001	0.1107E−001
	0.5	0.2392021E+000	0.34599E−002	0.717606E+000	0.1037E−001	0.1037E−001
	0.6	0.2870425E+000	0.41519E−002	0.669765E+000	0.96876E−002	0.96876E−002
	0.7	0.3348829E+000	0.48438E−002	0.6219254E+000	0.89959E−002	0.89959E−002
	0.8	0.3827234E+000	0.55358E−002	0.574085E+000	0.8304E−002	0.8304E−002
	0.9	0.4305638E+000	0.62278E−002	0.526244E+000	0.7611E−002	0.7611E−002
	1.0	0.4784043E+000	0.69198E−002	0.478404E+000	0.6919E−002	0.6919E−002
2	–	–	–	–	**Distance =**	**0.13839E−001**
3	0.0	0.0000000E+000	0.0000000E+000	0.9577443E+000	0.935E−003	0.935E−003
	0.1	0.478872E−001	0.467E−004	0.909857E+000	0.889E−003	0.889E−003
	0.2	0.957744E−001	0.9360E−004	0.861969E+000	0.842E−003	0.842E−003
	0.3	0.1436616E+000	0.140E−003	0.8140829E+000	0.795E−003	0.795E−003
	0.4	0.1915488E+000	0.1871E−003	0.766195E+000	0.748E−003	0.748E−003
	0.5	0.2394361E+000	0.234E−003	0.718308E+000	0.701E−003	0.701E−003
	0.6	0.2873233E+000	0.2807E−003	0.670421E+000	0.655E−003	0.655E−003
	0.7	0.335210E+000	0.327E−003	0.622533E+000	0.6082E−003	0.6082E−003
	0.8	0.3830978E+000	0.374E−003	0.574646E+000	0.5614E−003	0.5614E−003
	0.9	0.4309850E+000	0.4211E−003	0.526759E+000	0.514E−003	0.514E−003
	1.0	0.4788722E+000	0.467E−003	0.478872E+000	0.468E−003	0.468E−003
3	–	–	–	–	**Distance =**	**0.935E−003**

(continued)

Table 4 (continued)

m	r	$\underline{F}_m^{\hbar_k}(x;r)$	$\lvert\underline{F}_{m+1}^{\hbar_k}(x;r) - \underline{F}_m^{\hbar_k}(x;r)\rvert$	$\overline{F}_m^{\hbar_k}(x;r)$	$\lvert\overline{F}_{m+1}^{\hbar_k}(x;r) - \overline{F}_m^{\hbar_k}(x;r)\rvert$	Max
4	0.0	0.0000000E+000	0.0000000E+000	0.9578066E+000	0.62E−004	0.62E−004
	0.1	0.4789032E−001	0.31E−005	0.909914E+000	0.57E−004	0.57E−004
	0.2	0.957806E−001	0.62E−005	0.862025E+000	0.56E−004	0.56E−004
	0.3	0.1436709E+000	0.93E−005	0.814135E+000	0.52E−004	0.52E−004
	0.4	0.191561E+000	0.12E−004	0.7662451E+000	0.49E−004	0.49E−004
	0.5	0.239451E+000	0.15E−004	0.718354E+000	0.4E−004	0.4E−004
	0.6	0.287341E+000	0.18E−004	0.670463E+000	0.42E−004	0.42E−004
	0.7	0.3352322E+000	0.21E−004	0.622573E+000	0.39E−004	0.39E−004
	0.8	0.3831226E+000	0.24E−004	0.574683E+000	0.37E−004	0.37E−004
	0.9	0.4310130E+000	0.279E−004	0.526792E+000	0.33E−004	0.33E−004
	1.0	0.4789033E+000	0.310E−004	0.478903E+000	0.30E−004	0.30E−004
4	–	–	–	–	**Distance =**	**0.62E–004**
5	0.0	0.0000000E+000	0.0000000E+000	0.957809E+000	0.2E−005	0.2E−005
	0.1	0.4789045E−001	0.1E−006	0.9099184E+000	0.3E−005	0.3E−005
	0.2	0.9578090E−001	0.2E−006	0.8620279E+000	@.0	@.0
	0.3	0.1436713E+000	0.3E−006	0.814137E+000	0.1E−005	0.1E−005
	0.4	0.1915618E+000	0.5E−006	0.766247E+000	0.1E−005	0.1E−005
	0.5	0.2394522E+000	0.6E−006	0.7183566E+000	@.0	@.0
	0.6	0.2873427E+000	0.7E−006	0.670466E+000	0.2E−005	0.2E−005
	0.7	0.335233E+000	@.0	0.622575E+000	0.2E−005	0.2E−005
	0.8	0.3831237E+000	0.1E−005	0.574685E+000	@.0	@.0
	0.9	0.4310141E+000	0.1E−005	0.526794E+000	0.2E−005	0.2E−005
	1.0	0.4789046E+000	0.1E−005	0.478904E+000	@.0	@.0
5	–	–	–	–	**Distance =**	**0.2E–005**
6	0.0	0.0000000E+000	0.0000000E+000	0.957808E+000	@.0	@.0
	0.1	0.4789044E−001	@.0	0.909918E+000	@.0	@.0
	0.2	0.957808E−001	@.0	0.862027E+000	@.0	@.0
	0.3	0.143671E+000	@.0	0.814137E+000	@.0	@.0
	0.4	0.1915618E+000	@.0	0.766246E+000	@.0	@.0
	0.5	0.2394522E+000	@.0	0.718356E+000	@.0	@.0
	0.6	0.287342E+000	@.0	0.670465E+000	@.0	@.0
	0.7	0.3352331E+000	@.0	0.622575E+000	@.0	@.0
	0.8	0.3831236E+000	@.0	0.5746850E+000	@.0	@.0
	0.9	0.431014E+000	@.0	0.526794E+000	@.0	@.0
	1.0	0.4789045E+000	@.0	0.478904E+000	@.0	@.0
6	–	–	–	–	**Distance =**	**@.0**

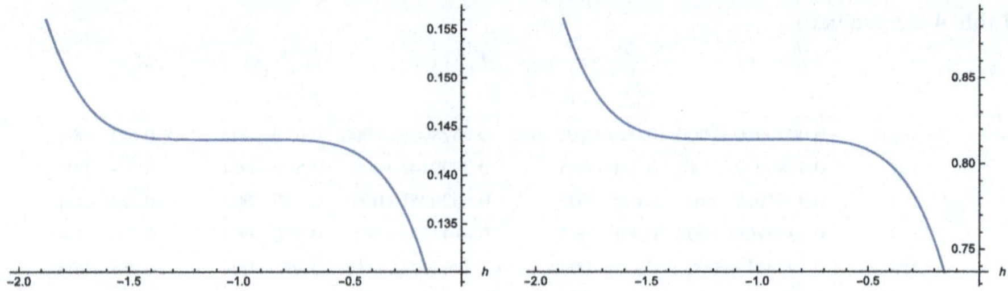

Fig. 2 The $\hbar$-curves of Example 2 for $\underline{F}^{\hbar}_{m_{opt}}(0.5; 0.3)$, $\overline{F}^{\hbar}_{m_{opt}}(0.5; 0.3)$ and $m_{opt} = 6$

5 Conclusion

Solving the fuzzy IEs is one of interesting topics in fuzzy mathematics. Despite of disadvantages of the FPA, there are a large number of researches based on this arithmetic to solve IEs. In this study, the DSA was applied to validate the numerical results of the HAM for solving second kind fuzzy IEs. Based on the CESTAC method in fuzzy case and the CADNA library, not only the disadvantages of the FPA are removed but also the optimal step of iteration, optimal approximation and optimal distance can be found. Instead of applying the traditional absolute error to show the accuracy of method, the novel condition was presented based on Hausdorff distance. Also, the proposed theorem illustrates the accuracy of the HAM to be applied for solving fuzzy IEs. Two sample IEs were considered and solved by applying the proposed algorithms to show the importance of the fuzzy CESTAC method. Consequently, it is suggested to apply the DSA for solving different kinds of fuzzy IEs with a reliable approach and detecting any instability during the performance of program.

References

1. Abbasbandy, S.: A numerical solution of Blasius equation by Adomian's decomposition method and comparison with homotopy perturbation method. Chaos Solitons Fractals **31**, 257–260 (2007)
2. Abbasbandy, S.: Application of He's homotopy perturbation method to functional integral equations. Chaos Solitons Fractals **31**, 1243–1247 (2007)
3. Abbasbandy, S.: Application of He's homotopy perturbation method for Laplace transform. Chaos Solitons Fractals **30**, 1206–1212 (2006)
4. Abbasbandy, S., Fariborzi Araghi, M.A.: The use of the stochastic arithmetic to estimate the value of interpolation polynomial with optimal degree. Appl. Numer. Math. **50**, 279–290 (2004)
5. Abbasbandy, S., Fariborzi Araghi, M.A.: A stochastic scheme for solving definite integrals. Appl. Numer. Math. **55**, 125–136 (2005)
6. Akhavan Zakeri, K., Ziari, Sh., Fariborzi Araghi, M.A., Perfilieva, I.: Efficient numerical solution to a bivariate nonlinear fuzzy Fredholm integral equation. IEEE Trans. Fuzzy Syst. (2019). https://doi.org/10.1109/TFUZZ.2019.2957100
7. Allahviranloo, T., Ghanbari, M.: Discrete homotopy analysis method for the nonlinear Fredholm integral equations. Ain Shams Eng. J. **2**(2), 133–140 (2011)

8. Allahviranloo, T., Khezerloo, M., Ghanbari, M., Khezerloo, S.: The homotopy perturbation method for fuzzy Volterra integral equations. Int. J. Comput. Cognition **8**(2), 31–37 (2010)
9. Anastassiou, G.A.: Fuzzy Mathematics: Approximation Theory. Springer, Heidelberg (2010)
10. Attari, H., Yazdani, A.: A computational method for fuzzy Volterra-Fredholm integral equations. Fuzzy Inf. Eng. **3**, 147–156 (2011)
11. Balasubramanian, K., Raja Balachandar, S., Venkatesh, S.G.: Analytical method for solving fuzzy Abel integral equation. Int. J. Pure Appl. Math. **105**(3), 393–398 (2015)
12. Barzegar Kelishami, H., Fariborzi Araghi, M.A., Allahviranloo, T.: Dynamical control of computations using the finite differences method to solve fuzzy boundary value problem. J. Intell. Fuzzy Syst. **36**(2), 1785–1796 (2019)
13. Behzadi, Sh.S., Abbasbandy, S., Allahviranloo, T., Yildirim, A.: Application of homotopy analysis method for solving a class of nonlinear Volterra-Fredholm integro-differential equations. J. Appl. Anal. Comput. **2**(2), 127–136 (2012)
14. Chesneaux, J.M.: CADNA, an ADA tool for round-off error analysis and for numerical debugging. In: Proceedings of the Congress on ADA in Aerospace, Barcelona (1990)
15. Chesneaux, J.M.: Stochastic arithmetic properties. IMACS Comput. Appl. Math. 81–91 (1992)
16. Chesneaux, J.M., Jézéquel, F.: Dynamical control of computations using the Trapezoidal and Simpson's rules. J. Univ. Comput. Sci. **4**, 2–10 (1998)
17. Effati, S., Saberinik , H., Buzhabadi, R., Taghiabadi, J.: Solving linear fuzzy Fredholm integral equations of the second kind by the homotopy analysis method. In: 4th Iranian Applied Mathematics Conference (2010)
18. Fariborzi Araghi, M.A., Behzadi, Sh.: Numerical solution of nonlinear Volterra-Fredholm integro-differential equations using homotopy analysis method. J. Appl. Math. Comput. **37**, 1–12 (2011)
19. Fariborzi Araghi, M.A., Kazemi Gelian, Gh.: Solving fuzzy Fredholm linear integral equations using Sinc method and double exponential transformation. Soft Comput. **19**(4), 1063–1070 (2015)
20. Fariborzi Araghi, M.A., Noeiaghdam, S.: A valid scheme to evaluate fuzzy definite integrals by applying the CADNA library. Int. J. Fuzzy Syst. Appl. **6**(4), 1–20 (2017)
21. Fariborzi Araghi, M.A., Noeiaghdam, S.: Homotopy analysis transform method for solving generalized Abel's fuzzy integral equations of the first kind. In: Proceeding of the 4th Iranian Joint Congress on Fuzzy and Intelligent Systems (CFIS). IEEE (2014). https://doi.org/10.1109/CFIS.2015.7391645
22. Fariborzi Araghi, M.A., Noeiaghdam, S.: A novel technique based on the homotopy analysis method to solve the first kind Cauchy integral equations arising in the theory of airfoils. J. Interpolat. Approx. Sci. Comput. **2016**(1), 1–13 (2016)
23. Fariborzi Araghi, M.A., Noeiaghdam, S.: Validation of Numerical Algorithms: Stochastic Arithmetic. Entekhab Bartar Publisher, Iran (2021) (in press)
24. Fariborzi Araghi, M.A., Noeiaghdam, S.: Finding the optimal step of fuzzy Newton-Cotes integration rules by using CESTAC method. J. Fuzzy Set Valued Anal. **2**, 62-85 (2017)
25. Fariborzi Araghi, M.A., Noeiaghdam, S.: Dynamical control of computations using the Gauss-Laguerre integration rule by applying the CADNA library. Adv. Appl. Math. Sci. **16**, 1–18 (2016)
26. Fariborzi Araghi, M.A., Parandin, N.: Numerical solution of fuzzy Fredholm integral equations by the Lagrange interpolation based on the extension principle. Soft Comput. **15**, 2449–2456 (2011)
27. Fullér, R.: Neural Fuzzy Systems, Abo Akademi University (1995)
28. Georgieva, A., Hristova, S.: Homotopy analysis method to solve two-dimensional nonlinear Volterra-Fredholm Fuzzy integral equations. Fractal Fract. **4**(9), 1–14 (2020)
29. Ghanbari, M.: Numerical solution of fuzzy linear Volterra integral equations of the second kind by homotopy analysis method. Int. J. Ind. Math. **2**(2), 73–87 (2010)
30. Ghanbari, M., Allahviranloo, T.: A new application of homotopy analysis method. Thai J. Math. **10**(1), 43–57 (2012)

31. Hamoud, A.A., Dhurgham Azeez, A., Ghadle, K.P.: A study of some iterative methods for solving fuzzy Volterra-Fredholm integral equations. Indones. J. Electr. Eng. Comput. Sci. **11**(3), 1228–1235 (2018)
32. Hamoud, A.A., Ghadle, K.P.: A study of some reliable methods for solving fuzzy Volterra-Fredholm integral equation. Acta Universitatis Apulensis **53**, 65–92 (2018)
33. Hamoud, A.A., Ghadle, K.P.: Homotopy analysis method for the first order fuzzy Volterra-Fredholm integro-differential equations. Indones. J. Electr. Eng. Comput. Sci. **11**(3), 857–867 (2018)
34. He, J.H.: Homotopy perturbation technique. Comput. Meth. Appl. Mech. Eng. **178**, 257–262 (1999)
35. He, J.H.: A coupling method of a homotopy technique and a perturbation technique for non-linear problems. Internat. J. Non- Mech. **35**, 37–43 (2000)
36. He, J.H.: Homotopy perturbation method: a new non-linear analytical technique. Appl. Math. Comput. **135**, 73–79 (2003)
37. Jalal Abdulqader, A.: Homotopy method for solving finite level fuzzy nonlinear integral equation. Appl. Comput. Math. **4**(4), 245–257 (2015)
38. Jézéquel, F., Chesneaux, J.-M., Lamotte, J.-L.: A new version of the CADNA library for estimating round-off error propagation in Fortran programs. Comput. Phys. Commun. **181**, 1927–1928 (2010)
39. Jézéquel, F., Chesneaux, J.-M.: CADNA: a library for estimating round-off error propagation. Comput. Phys. Commun. **178**, 933–955 (2008)
40. Khader, M., Kumar, S., Abbasbandy, S.: Fractional homotopy analysis transforms method for solving a fractional heat-like physical model. Walailak J. Sci. Tech. **13**(5), 337–353 (2016)
41. Kumar, H., Parida, P.K.: Solving Abel's general fuzzy linear integral equations by homotopy analysis method. Int. J. Fuzzy Comput. Model. **1**(4), 382–396 (2015)
42. Liao, S.J.: The Proposed Homotopy Analysis Techniques for the Solution of Nonlinear Problems. Shanghai Jiao Tong University, Shanghai (1992) (in English)
43. Liao, S.J.: Beyond Perturbation: Introduction to Homotopy Analysis Method. Chapman & Hall/CRC Press, Boca Raton (2003)
44. Liao, S.J.: On the homotopy analysis method for nonlinear problems. Appl. Math. Comput. **147**, 499–513 (2004)
45. Liao, S.J.: Homotopy analysis method in nonlinear differential equations. Higher Education Press, Beijing and Springer, Berlin Heidelberg (2012)
46. Lotfi, M.: Convergency and error estimate of nonlinear Fredholm fuzzy integral equations of the second kind by homotopy analysis method. In: Second International Conference on Computer Engineering and Applications. IEEE (2010). https://doi.org/10.1109/ICCEA.2010.190
47. Mihai Bica, A., Popescu, C.: Approximating the solution of nonlinear Hammerstein fuzzy integral equations. Fuzzy Sets Syst. **245**, 1–17 (2014)
48. Mirzaee, F.: Numerical solution of Fredholm fuzzy integral equations of the second kind using hybrid of block-pulse functions and Taylor series. Ain Shams Eng. J. **5**(2), 631–636 (2014)
49. Mirzaee, F., Komak Yari, M., Paripour, M.: Solving linear and nonlinear Abel fuzzy integral equations by homotopy analysis method. J. Taibah Univ. Sci. **9**(1), 104–115 (2015)
50. Mohammed, Al-S.: Reliable numerical algorithm for handling fuzzy integral equations of second kind in Hilbert spaces. Filomat **33**(2), 583–597 (2019)
51. Molabahrami, A., Shidfar, A., Ghyasi, A.: An analytical method for solving linear Fredholm fuzzy integral equations of the second kind. Comput. Math. Appl. **61**, 2754–2761 (2011)
52. Noeiaghdam, S., Fariborzi Araghi, M.A., Abbasbandy, S.: Valid implementation of Sinc-collocation method to solve the fuzzy Fredholm integral equation. J. Comput. Appl. Math. **307**, 112632 (2020). https://doi.org/10.1016/j.cam.2019.112632
53. Noeiaghdam, S., Fariborzi Araghi, M.A., Abbasbandy, S.: Finding optimal convergence control parameter in the homotopy analysis method to solve integral equations based on the stochastic arithmetic. Numer Algorithm **81**(1), 237–267 (2019)
54. Noeiaghdam, S., Fariborzi Araghi, M.A.: Homotopy regularization method to solve the singular Volterra integral equations of the first kind. Jordan J. Math. Statist. (JJMS) **11**(1), 1–12 (2018)

55. Noeiaghdam, S., Fariborzi Araghi, M.A.: A novel approach to find optimal parameter in the homotopy-regularization method for solving integral equations. Appl. Math. Inf. Sci. **14**(1), 1–8 (2020)
56. Noeiaghdam, S., Sidorov, D., Zamyshlyaeva, A., Tynda, A., Dreglea, A.: A valid dynamical control on the reverse osmosis system using the CESTAC method. Mathematics **9**, 48 (2021). https://doi.org/10.3390/math9010048
57. Noeiaghdam, S., Fariborzi Araghi, M.A.: A novel algorithm to evaluate definite integrals by the Gauss-Legendre integration rule based on the stochastic arithmetic: application in the model of osmosis system. Math. Model. Eng. Probl. **7**(4), 577–586 (2020)
58. Noeiaghdam, S., Dreglea, A., He, J.H., Avazzadeh, Z., Suleman, M., Fariborzi Araghi, M.A., Sidorov, D., Sidorov, N.: Error estimation of the homotopy perturbation method to solve second kind Volterra integral equations with piecewise smooth kernels: application of the CADNA library. Symmetry **12**, 1730 (2020). https://doi.org/10.3390/sym12101730
59. Noeiaghdam, S., Sidorov, D., Sizikov, V., Sidorov, N.: Control of accuracy on Taylor-collocation method to solve the weakly regular Volterra integral equations of the first kind by using the CESTAC method. Appl. Comput. Math. Int. J. **19**(1), 81–105 (2020)
60. Noeiaghdam, S., Sidorov, D., Muftahov, I., Zhukov, A.V.: Control of accuracy on Taylor-collocation method for load leveling problem. Bull. Irkutsk State Univ. Ser. Math. **30**, 59–72 (2019)
61. Noeiaghdam, S., Suleman, M., Budak, H.: Solving a modified non-linear epidemiological model of computer viruses by homotopy analysis method. Math. Sci. **12**, 211–222 (2018)
62. Noeiaghdam, S., Zarei, E., Barzegar Kelishami, H.: Homotopy analysis transform method for solving Abel's integral equations of the first kind. Ain Shams Eng. J. **7**, 483–495 (2016)
63. Otadi, M., Mosleh, M.: Simulation and evaluation of interval-valued fuzzy linear Fredholm integral equations with interval-valued fuzzy neural network. Neurocomputing **20512**, 519–528 (2016)
64. Parandin, N., Fariborzi Araghi, M.A.: The numerical solution of linear Fuzzy Fredholm integral equations of the second kind by using the finite and divided differences methods. Soft Comput. **15**, 729–741 (2010)
65. Sahihi, H., Abbasbandy, S., Allahviranloo, T.: Homotopy analysis method and its application for solving singularly perturbed differential-difference equation with boundary layer behavior and delay. J. New Res. Math. **6**(23), 29–38 (2020)
66. Salahshour, S., Allahviranloo, T.: Application of fuzzy differential transform method for solving fuzzy Volterra integral equations. Appl. Math. Model. **37**(31), 1016–1027 (2013)
67. Salam, A.A.: A reliable algorithm of homotopy analysis method for solving fuzzy integral equations of fractional order. Basrah J. Science **34**(3), 104–119 (2016)
68. Shaban, M., Shivanian, E., Abbasbandy, S.: Analyzing magneto-hydrodynamic (MHD) squeezing flow between two parallel disks with suction or injection by a new hybrid method based on Tau method and Homotopy analysis method (HAM). Eur. Phys. J. Plus. **128** (2013)
69. Shivanian, E., Alsulami, H.H., Alhuthali, M.S., Abbasbandy, S.: Predictor homotopy analysis method (PHAM) for nano boundary layer flows with nonlinear Navier boundary condition: existence of four solutions. Filomat **28**(8), 1687–1697 (2014)
70. Vali, M.A., Agheli, M.J., Gohari Nezhad, S.: Homotopy analysis method to solve two-dimensional fuzzy Fredholm integral equation. Gen. Math. Notes **22**(1), 31–43 (2014)
71. Vignes, J.: discrete stochastic arithmetic for validating results of numerical software. Spec. Iss. Numer. Algorithms **37**, 377–390 (2004)
72. Vignes, J.: A stochastic arithmetic for reliable scientific computation. Math. Comput. Simul. **35**, 233–261 (1993)
73. Wazwaz, A.-M.: Linear and Nonlinear Integral Equations, Methods and Applications. Higher Education Press, Beijing and Springer, Berlin Heidelberg(2011)
74. Ziari, S.: Towards the accuracy of iterative numerical methods for fuzzy Hammerstein-Fredholm integral equations. Fuzzy Sets Syst. **375**, 161–178 (2019)

Numerical Solution of Some Singular Volterra Fuzzy Integral Equations of the First Kind by Fuzzy Generalized Quadrature Formula

V. Samadpour Khalifeh Mahaleh, R. Ezzati, and S. Ziari

Abstract In this paper, we propose the fuzzy generalized quadrature formula (FGQF) for numerical solution of some singular fuzzy integral equation of Abel type. The problem of finding the solution of fuzzy singular integral equation is reduced to nonsingular system of fuzzy linear algebraic equations with positive crisp coefficients.

Keywords Fuzzy integral equations · Singular kernels · Generalized quadrature

1 Introduction

The fuzzy integral equations (FIE) are very useful for solving many problems in several applied fields such as mathematical economics; electrical engineering; medicine ; biology and optimal control theory. Since these equations usually can not be solved explicitly, so it is required to obtain approximate solutions. The Banach fixed point theorem is the main tool in studying the existence and uniqueness of the solution for fuzzy integral equations which can be found in [1, 2]. Some quadrature rules for evaluating the fuzzy integral has been investigated in [3]. Bica and Popescu [4, 5] applied the method of successive approximations for the fuzzy Hammerstein integral equation. Ezzati and Ziari [6] proved the convergence of the method of successive approximations for solving nonlinear fuzzy Fredholm integral equations of the second kind, and they proposed an iterative procedure based on the Trapezoidal quadrature.

V. Samadpour Khalifeh Mahaleh
Department of Traffic Engineering, Amin Police Science University, Tehran, Iran

R. Ezzati (✉)
Department of Mathematics, Karaj Branch, Islamic Azad University, Karaj, Iran
e-mail: reza_ezati@yahoo.com; ezati@kiau.ac.ir

S. Ziari
Department of Mathematics, Firoozkooh Branh, Islamic Azad University, Firoozkooh, Iran

T. Allahviranloo and S. Salahshour (eds.), *Advances in Fuzzy Integral and Differential Equations*, Studies in Fuzziness and Soft Computing 412,
https://doi.org/10.1007/978-3-030-73711-5_8

The method of successive approximations based on quadrature formulas and their investigating error estimation are presented in [3–13]. Other methods in the construction of the numerical approaches for fuzzy integral equations are: quadrature rules and Nyström techniques (see [14]), Adomian decomposition (see [15]), Lagrange interpolation (see [16]), Bernstein polynomials (see [17, 18]), fuzzy Haar wavelets (see [19, 20]), hybrid of block-pulse functions and Taylor series (see [21]), triangular basis functions (see [22]), Chebyshev polynomials (see [23]), fast Galerkin method (see [24]), differential transform method (see [25]) and block-pulse basis functions (see [26, 27]). Also, a full fuzzy method for solving differential equation based on Taylor expansion is presented in [28]. Due to the the error bound of midpoint rule in comparing Trapezoidal and Simpson rules, and also according to the fact that Trapezoidal and Simpson rules can not be used for approximate integrals which they are not defined in first and end points of integration, in these article we used the approximate solution of the fuzzy integral equation by using midpoint rule and compare it with these methods.

In this paper, first, we approximate the Abel fuzzy integral equation using FGQF and estimate the error approximation. Then, we use the proposed method b for solving fuzzy singular integral equation (FSIE)

$$f(x) = (FR)\int_a^x \frac{u(t)}{\sqrt{x-t}}dt, \quad x \in [a, b], \tag{1}$$

where u(t) is unknown fuzzy function and f(x) is the source fuzzy function. Equation (1) describes various problems in mechanics (such as tautohron problem), scattering and other problems.

The rest of the paper is organized as follows: In Sect. 2, we review some concept of fuzzy calculus. In Sect. 3, the FGQF for solving FSIE is introduced. The error estimation of the proposed method is presented in Sect. 4. We present one example in Sect. 5 to show the efficiency of the proposed method.

2 Preliminaries

Definition 1 (*See* [29]). A fuzzy number is a function $u : R \to [0, 1]$ having the properties:

(1) u is normal , that is $\exists\, x_0 \in R$ such that $u(x_0) = 1$,
(2) u is fuzzy convex set

$$(i.e.\ u(\lambda x + (1-\lambda)y) \geq \min\{u(x), u(y)\} \ \ \forall x, y \in R, \lambda \in [0, 1]),$$

(3) u is upper semi-continuous on R,
(4) the $\overline{\{x \in R : u(x) > 0\}}$ is compact set.

The set of all fuzzy numbers is denoted by R_F. An alternative definition which yields the same R_F is given by [30].

Definition 2 (*See* [31, 32]). An arbitrary fuzzy number is represented, in parametric form, by an ordered pair of functions $(\underline{u}(r), \overline{u}(r)), 0 \leq r \leq 1$, which satisfy the following requirements:

(1) $\underline{u}(r)$ is a bounded left continuous non-decreasing function over [0,1],
(2) $\overline{u}(r)$ is a bounded left continuous non-increasing function over [0,1],
(3) $\underline{u}(r) \leq \overline{u}(r), \quad 0 \leq r \leq 1$.

The addition and scaler multiplication of fuzzy numbers in R_F are defined as follows:

(1) $(u \oplus v)(r) = (\underline{u}(r) + \underline{v}(r), \overline{u}(r) + \overline{v}(r))$,

(2) $(\lambda \odot u)(r) = \begin{cases} (\lambda \underline{u}(r), \lambda \overline{u}(r)) & \lambda \geq 0, \\ \\ (\lambda \overline{u}(r), \lambda \underline{u}(r)) & \lambda < 0. \end{cases}$

Definition 3 (*See* [33]). For arbitrary fuzzy numbers $u = (\underline{u}(r), \overline{u}(r))$, $v = (\underline{v}(r), \overline{v}(r))$ the quantity $D(u, v) = \sup_{r \in [0,1]} \max\{|\underline{u}(r) - \underline{v}(r)| , |\overline{u}(r) - \overline{v}(r)| \}$ is the distance between u and v.
The following properties are hold (See [33]):

(1) (R_F, D) is a complete metric space,
(2) $D(u \oplus w, v \oplus w) = D(u, v) \quad \forall\, u, v, w \in R_F$,
(3) $D(k \odot u, k \odot v) = |k|\, D(u, v) \quad \forall\, u, v \in R_F \quad \forall\, k \in R$,
(4) $D(u \oplus v, w \oplus e) \leq D(u, w) + D(v, e) \quad \forall\, u, v, w, e \in R_F$.
(5) $D(a \odot u, b \odot u) \leq |a - b| D(u, 0) \quad \forall\, u \in R_F; \ \forall a, b \in R, \ ab > 0$,

Theorem 1 (See [7, 29]).

(1) The pair $(R_F, \oplus)$ is a commutative semigroup with $\tilde{0} = \chi_0$ zero element.
(2) For fuzzy numbers which are not crisp, there is no opposite element (that is, $(R_F, \oplus)$ cannot be a group).
(3) For any $a, b \in R$ with $a, b \geq 0$ or $a, b \leq 0$ and for any $u \in R_F$, we have $(a + b) \odot u = a \odot u \oplus b \odot u$.
For arbitrary $a, b \in R$, this property is not fulfilled.
(4) For any $\lambda, \mu \in R$ and $u \in R_F$, we have $\lambda \odot (u \oplus v) = \lambda \odot u \oplus \lambda \odot u$.
(5) For any $\lambda \in R$ and $u, v \in R_F$, we have $\lambda \odot (\mu \odot u) = (\lambda.\mu) \odot u$.
(6) The function of $\|.\|_F : R_F \to R$ by $\|u\|_F = D(u, \tilde{0})$ has the usual properties of the norm, that is, $\|u\|_F = 0$ if and only if $u = \tilde{o}$, $\|\lambda \odot u\|_F = |\lambda| \|u\|_F$ and $\|u \oplus v\|_F \leq \|u\|_F + \|v\|_F$.
(7) $|\|u\|_F - \|v\|_F| \leq D(u, v)$ and $D(u, v) \leq |u\|_F + \|v\|_F$ for any $u, v \in R_F$.

Definition 4 (*See* [30]). A fuzzy real number valued function $f : [a, b] \to R_F$ is said to be continuous in $x_0 \in [a, b]$, if for each $\varepsilon > 0$ there exist $\delta > 0$ such that $D(f(x), f(x_0)) < \varepsilon$, whenever $x \in [a, b]$ and $|x - x_0| < \delta$. We say that f is fuzzy continuous on $[a, b]$ if f is continuous at each $x_0 \in [a, b]$, and denote the space of all such functions by $C_F[a, b]$.

Definition 5 (*See* [7]). If $X = \{f : [a, b] \to R_F \mid \text{f is continuous}\}$, then X together with the metric

$$D^*(f, g) = \sup_{a \le s \le b} D(f(s), g(s))$$

is complete metric space.

Definition 6 (*See* [33]). Let $f : [a, b] \to R_F$, be a bounded mapping, then function $\omega_{[a,b]}(f, .) : R_+ \cup \{0\} \to R_+$ defined by

$$\omega_{[a,b]}(f, \delta) = \sup \{D(f(x), f(y)) | \quad x, y \in [a, b], \quad |x - y| \le \delta\}, \tag{2}$$

is called the modulus of oscillation of f on $[a, b]$. In addition, if $f \in C_F[a, b]$ (i.e. $f : [a, b] \to R_F$ is continuous on $[a, b]$), then $\omega_{[a,b]}(f, \delta)$ is called the modulus of continuity of f on $[a, b]$.

Some properties of the modulus of continuity are given in below.

Theorem 2 (See [6]). *The following properties hold:*

(1) $D(f(x), f(y)) \le \omega_{[a,b]}(f, |x - y|)$, *for any* $x, y \in [a, b]$,
(2) $\omega_{[a,b]}(f, \delta)$ *is increasing function of* δ,
(3) $\omega_{[a,b]}(f, 0) = 0$,
(4) $\omega_{[a,b]}(f, \delta_1 + \delta_2) \le \omega_{[a,b]}(f, \delta_1) + \omega_{[a,b]}(f, \delta_2)$ *for any* $\delta_1, \delta_2 \ge 0$,
(5) $\omega_{[a,b]}(f, n\delta) \le n\omega_{[a,b]}(f, \delta)$ *for any* $\delta \ge 0$ $n \in N$,
(6) $\omega_{[a,b]}(f, \lambda\delta) \le (\lambda + 1)\omega_{[a,b]}(f, \delta)$ *for any* $\delta, \lambda \ge 0$,
(7) if $[a, b] \subseteq [c, d]$ *then* $\omega_{[a,b]}(f, \delta) \le \omega_{[c,d]}(f, \delta)$.

Definition 7 (*See* [34]). Let $f : [a, b] \to R_F$, for $\Delta x : a = x_0 < x_1 < \cdots < x_n = b$, partition of the intervals $[a, b]$. Let us consider the intermediates points $\xi_i \in [x_{i-1}, x_i]$, $i = 1, \dots, n$, and $\delta : [a, b] \to R_+$. The division $P_x = ([x_{i-1}, x_i]; \xi_i); i = 1, \dots, n$, denoted shortly by $P_x = (\Delta^n, \xi)$ is said to be δ-fine if $[x_{i-1}, x_i] \subseteq (\xi_i - \delta(\xi_i), \xi_i + \delta(\xi_i))$.

The function f is called Henstock integrable to $I \in R_F$, if for any $\varepsilon > 0$, there is function $\delta : [a, b] \to R_+$ such that for any δ-fine division we have $D(\sum_{i=0}^{n}(x_i - x_{i-1}) \odot f(\xi_i, \eta_j), I)$, where Σ denotes the fuzzy summation. Then I is called the Henstock integral of f and denoted by $I(f) = (FH) \int_a^b f(s)ds$. If the above δ is constant functions, then one recaptures the concept of Riemann integral. In this case $I \in R_F$ will be called integral of f on $[a, b]$ and will be denoted by $(FR) \int_a^b f(s)ds$.

In [6], the authors proved that if $f \in C_F[a, b]$, its definite integral exists, and also,

$$\underline{(FR)\int_a^b f(s;r)ds} = \int_a^b \underline{f}(s;r)ds$$

$$\overline{(FR)\int_a^b f(s;r)ds} = \int_a^b \overline{f}(s;r)dsdt.$$

Lemma 1 (See [33]). *Let $f : [a, b]' \to R_F$ be fuzzy continuous and $g : [a, b]' \to R_+$ be continuous. Then $f(x) \odot g(x)$ is fuzzy continuous function $\forall x \in [a, b]$.*

Lemma 2 (See [35]). *If $f : [a, b]' \to R_F$ is a fuzzy continuous function, then*

$$(FR)\int_a^b f(s)ds = (FR)\int_a^c f(s)ds \oplus (FR)\int_c^b f(s)ds, \quad \textit{for all} \quad c \in [a, b] \tag{3}$$

Lemma 3 (See [33]). *If $f, g : [a, b]' \to R_F$ are (FR)-integrable fuzzy functions, and α, β are real numbers, then*

$$(FR)\int_a^b (\alpha f(s) \oplus \beta g(s))\, ds = \alpha(FR)\int_a^b f(s)ds \oplus \beta(FR)\int_a^b g(s)ds.$$

3 Introduce the FGQF

Let us introduce nonuniform (but coinciding) meshes on x and t as follows

$$a = x_0 = t_0 < x_1 = t_1 < \cdots < x_i = t_i < \cdots < x_{n-1} = t_{n-1}.$$

Here, $a = t_{min}$ is boundary value such as $u(a+0) = 0$. On each interval $(t_{j-1}; t_j], j = 1, 2, \ldots, n-1$ we suppose approximately

$$u(t) = u(t_j) = u_j = \text{const.} \tag{4}$$

We prove the following theorem:

Theorem 3 *Under condition (3),*

$$(FR)\int_{t_{j-1}}^{t_j} \frac{u(t)}{\sqrt{x-t}}dt = 2\left(\sqrt{x-t_{j-1}} - \sqrt{x-t_j}\right) \odot u_j, \quad a \le t_{j-1} \le t_j \le x \tag{5}$$

Proof Since

$$\int_{t_{j-1}}^{t_j} \frac{1}{\sqrt{x-t}} dt = 2\left(\sqrt{x-t_{j-1}} - \sqrt{x-t_j}\right)$$

taking into account (3), we obtain

$$(FR)\int_{t_{j-1}}^{t_j} \frac{u(t)}{\sqrt{x-t}} dt = 2\left(\sqrt{x-t_{j-1}} - \sqrt{x-t_j}\right) \odot u_j,$$

the formula (4) is called the FGQF.

Theorem 4 *Numerical solution of integral equation (3) according to the FGQF is defined as following recursion*

$$\begin{cases} u_1 = \frac{1}{2r_{1,1}} \odot f_1 \\ \\ u_i = \frac{1}{r_{i,i}} \odot \left(\frac{1}{2}f_i - \sum_{j=1}^{i-1} r_{i,j} \odot u_j\right) \quad , i = 2, 3, \ldots, n-1 \end{cases} \tag{6}$$

where $f_i = f(x_i)$, $u_j = u(t_j)$ *and*

$$r_{ij} = \sqrt{x_i - t_{j-1}} - \sqrt{x_i - t_j} \tag{7}$$

Proof by Lemma 2 we can wright

$$(FR)\int_a^{x_i} \frac{u(t)}{\sqrt{x_i-t}} dt = \sum_{j=1}^{i} \int_{t_{j-1}}^{t_j} \frac{u(t)}{\sqrt{x_i-t}} dt = 2\sum_{j=1}^{i}\left(\sqrt{x_i-t_{j-1}} - \sqrt{x_i-t_j}\right) \odot u_j = f_i,$$

thus

$$\sum_{j=1}^{i} r_{ij} \odot u_j = \frac{1}{2} f_i, \quad i = 1, 2, \ldots, n-1. \tag{8}$$

relation (7) is a fuzzy linear equations system (FLES) with crisp positive coefficient r_{ij}. FLES (7) is lower triangular and its solution can be recursively constructed. From (7) for $i = 1, 2, \ldots, n-1$ we eventually obtain $u_1, u_2, \ldots, u_{n-1}$ according to (7). As to the value of $u_0 = u(a)$, it can not be found by this scheme, but can be additionally determined as $u_0 = 0$ from physical concepts or $u_0 = u_1$ or can be derived using linear extrapolation [36], as was done in (6). □

4 Error Estimation

In this Section we derive error estimate for solution of FSIE (1) using fuzzy Taylor formula.

4.1 Fuzzy Taylor Formula

Theorem 5 (See [33]). *Let $T = [x_0, x_0 + \beta] \subset R$, with $\beta \geq 0$. We assume that $f^{(i)} : T \to R_F$ are H-differentiable for all $i = 0, 1, \ldots, n-1$, for any $x \in T$. (I.e., there exist in R_F the H-differences $f^{(i)}(x+h) - f^{(i)}(x)$ and $f^{(i)}(x) - f^{(i)}(x-h)$, $i = 0, 1, .., n-1$ for all small $h : 0 < h < \beta$. Furthermore there exist $f^{i+1}(x) \in R_F$ such that the limits in D—distance exist and*

$$f^{(i+1)}(x) = \lim_{h\to 0} \frac{f^{(i)}(x+h) - f^{(i)}(x)}{h} = \lim_{h\to 0} \frac{f^{(i)}(x) - f^{(i)}(x-h)}{h},$$

for all $i = 0, 1, .., n-1$.) Also we assume that $f^{(n)}$, is fuzzy continuous on T. Then for $s \geq a$; $s, a \in T$ we obtain

$$f(s) = f(a) \oplus f'(a) \odot (s-a) \oplus f''(a) \odot \frac{(s-a)^2}{2!} \oplus \cdots \oplus f^{(n-1)}(a) \odot \frac{(s-a)^{n-1}}{(n-1)!} \oplus R_n(s) \tag{9}$$

where

$$R_n(s) = (FR)\left(\int_a^s \left(\int_a^{s_1} \cdots \left(\int_a^{s_{n-1}} f^{(n)}(s_n)ds_n\right) ds_{n-1}\right) \cdots \right) ds_1.$$

4.2 Error Bound

Theorem 6 *Let us suppose that the unique solution of problem (1), $u(t)$ and its derivatives up to second order are H-differentiable, also we assume that $u^{(2)}$, is fuzzy continuous on $[a, x]$ and there exist $M > 0$ such that $\| u'(t_j) \| \leq M$, $j = 1, 2, \ldots, i$. Then*

$$D\left((FR)\int_a^{x_i} \frac{u(t)}{\sqrt{x_i - t}}dt,\ 2\sum_{j=1}^{i} r_{ij} \odot u_j\right) \leq \epsilon_{ij} \tag{10}$$

where

$$\epsilon_{ij} = \frac{1}{3} \| u'(t_{j-1}) \| r_{ij}\left(4x_i - 5t_{j-1} + t_j - r_{ij}^2\right)$$

Proof Since u is H-differentiable and $u^{(2)}$ is fuzzy continuous, by theorem (4) we have

$$u(t) = u(t_{j-1}) \oplus (t - t_{j-1}) \odot u'(t_{j-1}) + R_2(t)$$

where

$$R_2(t) = (FR) \int_a^t \left(\int_a^{t_1} u^{(2)}(t_2) dt_2 \right) dt_1$$

if $(t - t_{j-1}) \to 0$ we can wright

$$u(t) \simeq u(t_{j-1}) \oplus (t - t_{j-1}) \odot u'(t_{j-1})$$

thus

$$D\left((FR)\int_a^{x_i} \frac{u(t)}{\sqrt{x_i - t}} dt, 2\sum_{j=1}^{i} r_{ij} \odot u_j\right) = D\left(\sum_{j=1}^{i}(FR)\int_{t_{j-1}}^{t_j} \frac{u(t_{j-1}) \oplus (t - t_{j-1})u'(t_{j-1})}{\sqrt{x_i - t}} dt, \frac{1}{2}\sum_{j=1}^{i} r_{ij} \odot u_j\right)$$

$$\leq D\left(\sum_{j=1}^{i}(FR)\int_{t_{j-1}}^{t_j} \frac{u(t_{j-1})}{\sqrt{x_i - t}} dt, 2\sum_{j=1}^{i} r_{ij} \odot u_j\right) + D\left(\sum_{j=1}^{i}(FR)\int_{t_{j-1}}^{t_j} \frac{(t - t_{j-1})u'(t_{j-1})}{\sqrt{x_i - t}} dt, \tilde{0}\right)$$

$$= D\left(\sum_{j=1}^{i}(FR)\int_{t_{j-1}}^{t_j} \frac{(t - t_{j-1})u'(t_{j-1})}{\sqrt{x_i - t}} dt, \tilde{0}\right) = D\left(\sum_{j=1}^{i}(FR)\int_{t_{j-1}}^{t_j} \frac{(t - t_{j-1})u'(t_{j-1})}{\sqrt{x_i - t}} dt, \sum_{j=1}^{i}(FR)\int_{t_{j-1}}^{t_j} \tilde{0} dt\right)$$

$$\leq \sum_{j=1}^{i} D\left((FR)\int_{t_{j-1}}^{t_j} \frac{(t - t_{j-1})u'(t_{j-1})}{\sqrt{x_i - t}} dt, \int_{t_{j-1}}^{t_j} \tilde{0} dt\right) = \sum_{j=1}^{i}(FR)\int_{t_{j-1}}^{t_j} D\left(u'(t_{j-1})\frac{(t - t_{j-1})}{\sqrt{x_i - t}}, \tilde{0}\right)$$

$$\leq \sum_{j=1}^{i}(FR)\int_{t_{j-1}}^{t_j} |\frac{t - t_{j-1}}{\sqrt{x_i - t}}| D\left(u'(t_{j-1}), \tilde{0}\right) dt$$

$$= \| u'(t_{j-1}) \| \sum_{j=1}^{i}(FR)\int_{t_{j-1}}^{t_j} |\frac{t - t_{j-1}}{\sqrt{x_i - t}}| dt$$

$$= \frac{1}{3} \| u'(t_{j-1}) \| r_{ij}\left(4x_i - 5t_{j-1} + t_j - r_{ij}^2\right) = \varepsilon_{ij}$$ □

5 Numerical Examples

To illustrate the efficiency of the presented method in the previous Section, we give two examples. Also, we compare the numerical solution obtained by using the proposed method with the exact solution.

Table 1 The accuracy on the level sets for Example 1 in $x = 0.5$.

	n = 50,	m = 3
r-level	$\lvert \underline{u} - \underline{y}_m \rvert$	$\lvert \overline{u} - \overline{y}_m \rvert$
0.0	4.23672e-6	1.70810e-3
0.2	8.40172e-5	1.30622e-3
0.4	3.75030e-4	9.70821e-3
0.6	9.31261e-4	6.96268e-3
0.8	1.81064e-4	4.76915e-4
1.0	3.07082e-4	3.07082e-4

Example 1 Consider the following Abel fuzzy integral equation:

$$f(x) = (FR)\int_0^x \frac{u(t)}{\sqrt{x-t}}dt,$$

where

$$f(x) = (\underline{f}(x), \overline{f}(x)) = \left(\frac{4}{3}rx^{\frac{3}{2}}, \frac{4}{3}(2-r)x^{\frac{3}{2}}\right).$$

The exact solution in this case is given by

$$u(x) = (\underline{u}(x), \overline{u}(x)) = (rx, (2-r)x) \quad \text{and} \quad 0 \leq r \leq 1.$$

To obtain numerical solution, we use the proposed method in Sect. 4. So, to compare the numerical and exact solutions, one can see Table 1.

References

1. Balachandran, K., Kanagarajan, K.: Existence of solutions of general nonlinear fuzzy Volterra-Fredholm integral equations. J. Appl. Math. Stoch. Anal. **3**, 333–343 (2005)
2. Subrahmanyam, P.V., Sudarsanam, S.K.: A note on fuzzy Volterra integral equations. Fuzzy Sets Syst. **95**, 237–240 (1996)
3. Bede, B., Gal, S.G.: Quadrature rules for integrals of fuzzy-number-valued functions. Fuzzy Sets Syst. **145**, 359–380 (2004)
4. Bica, A.M., Popescu, C.: Approximating the solution of nonlinear Hammerstein fuzzy integral equations. Fuzzy Sets Syst. **95**, 1–17 (2014)
5. Bica, A.M., Popescu, C.: Numerical solutions of the nonlinear fuzzy Hammerstein-Volterra delay integral equations. Inf. Sci. **95**, 236–255 (2013)
6. Ezzati, R., Ziari, S.: Numerical solution of nonlinear fuzzy Fredholm integral equations using iterative method. Appl. Math. Comput **225**, 33–42 (2013)
7. Bica, A.M.: Error estimation in the approximation of the solution of nonlinear fuzzy Fredholm integral equations. Inf. Sci. **178**, 1279–1292 (2008)

8. Bica, A.M., Popescu, C.: Iterative numerical method for nonlinear fuzzy Volterra integral equations. J. Intell. Fuzzy Syst. **32**(3), 1639–1648 (2017)
9. Park, J.Y., Lee, S.Y., Jeong, J.U.: The approximate solutions of fuzzy functional integral equations. Fuzzy Sets Syst. **110**(2000), 79–90 (2000)
10. Samadpour Khalifeh Mahaleh, V., Ezzati, R.: Numerical solution of linear fuzzy Fredholm integral equations of second kind using iterative method and midpoint quadrature formula. J. Intell. Fuzzy Syst. **33**, 1293–1302 (2017)
11. Ziari, S., Bica, A.M.: New error estimate in the iterative numerical method for nonlinear fuzzy Hammerstein-Fredholm integral equations. Fuzzy Sets Syst. **295**, 136–152 (2016)
12. Ziari, S.: Towards the accuracy of iterative numerical methods for fuzzy Hammerstein-Fredholm integral equations. Fuzzy Sets Syst. **375**, 161–178 (2019)
13. Ziari, S., Bica, A.M.: An iterative numerical method to solve nonlinear fuzzy Volterra—Hammerstein integral equations. J. Intell. Fuzzy Syst. **37**(5), 6717–6729 (2019)
14. Abbasbandy, S., Babolian, E., Alavi, M.: Numerical method for solving linear Fredholm fuzzy integral equations of the second kind. Chaos Solitons Fract. **31**(1), 138–146 (2007)
15. Abbasbandy, S., Allahviranloo, T.: The Adomian decomposition method applied to the fuzzy system of Fredholm integral equations of the second kind. Int. J. Uncertain. Fuzziness Knowl. Based Syst. **14**(2006), 101–110 (2006)
16. Fariborzi Araghi, M.A., Parandin, N.: Numerical solution of fuzzy Fredholm integral equations by the Lagrange interpolation based on the extension principle. Soft Comput. **15**, 2449–2456 (2011)
17. Ezzati, R., Ziari, S.: Numerical solution and error estimation of fuzzy Fredholm integral equation using fuzzy Bernstein polynomials. Aust. J. Basic Appl. Sci. **5**(9), 2072–2082 (2011)
18. Ziari, S., Bica, A. M., Ezzati, R.: Iterative fuzzy Bernstein polynomials method for nonlinear fuzzy Volterra integral equations. Comp. Appl. Math. **39**(316) (2020). https://doi.org/10.1007/s40314-020-01361-x
19. Mokhtarnejad, F., Ezzati, R.: The numerical solution of nonlinear Hammerstein fuzzy integral equations by using fuzzy wavelet like operator. J. Intell. Fuzzy Syst. **28**(4), 1617–1626 (2015)
20. Ziari, S., Ezzati, R., Abbasbandy, S.: Numerical solution of linear fuzzy Fredholm integral equations of the second kind using fuzzy Haar wavelet. In: Greco, S., Bouchon-Meunier, B., Coletti, G., Fedrizzi, M., Matarazzo, B., Yager, R.R. (eds.) Advances in Computational Intelligence. IPMU 2012. Communications in Computer and Information Science, vol 299. Springer, Berlin, Heidelberg. https://doi.org/10.1007/978-3-642-31718-7_9
21. Baghmisheh, M., Ezzati, R.: Numerical solution of nonlinear fuzzy Fredholm integral equations of the second kind using hybrid of block-pulse functions and Taylor series. Adv. Diff. Equ. (2015). https://doi.org/10.1186/s13662-015-0389-7
22. Baghmisheh, M., Ezzati, R.: Error estimation and numerical solution of nonlinear fuzzy Fredholm integral equations of the second kind using triangular functions. J. Intell. & Fuzzy Syst. **30**(2), 639–649 (2016)
23. Allahviranloo, T., Behzadi, S.S.: The use of airfoil and Chebyshev polynomials methods for solving fuzzy Fredholm integro-differential equations with Cauchy kernel. Soft Comput. **18**, 1885–1897 (2014)
24. Behzadi, S.S., Allahviranloo, T., Abbasbandy, S.: The use of fuzzy expansion method for solving fuzzy linear Volterra-Fredholm integral equations. J. Intell. & Fuzzy Syst. **26**(4), 1817–1822 (2014)
25. Salahshour, S., Allahviranloo, T.: Application of fuzzy differential transform method for solving fuzzy Volterra integral equations. Appl. Math. Model. **37**, 1016–1027 (2013)
26. Ziari, S., Ezzati, R.: Fuzzy block-pulse functions and its application to solve linear fuzzy Fredholm integral equations of the second kind. In: Carvalho, J., Lesot, M.J., Kaymak, U., Vieira, S., Bouchon-Meunier, B., Yager, R. (eds.) Information Processing and Management of Uncertainty in Knowledge-Based Systems. IPMU 2016. Communications in Computer and Information Science, vol. 611. Springer, Cham. https://doi.org/10.1007/978-3-319-40581-0_67

27. Ziari, S., Perfilieva, I., Abbasbandy, S.: Block-pulse functions in the method of successive approximations for nonlinear fuzzy Fredholm integral equations. Differ. Equ. Dyn. Syst. (2019). https://doi.org/10.1007/s12591-019-00482-y
28. Allahviranloo, T., Gouyandehv, Z., Armand, A.: A full fuzzy method for solving differential equation based on Taylor expansion. J. Intell. Fuzzy Syst. **29**, 1039–1055 (2015)
29. Anastassiou, G.A., Gal, S.G.: On a fuzzy trigonometric approximation theorem of Weirstrass-type. J. Fuzzy Math. Los Angeles **9**, 701–708 (2001)
30. Kaleva, O.: Fuzzy differential equations. Fuzzy Sets Syst. **24**, 301–317 (1987)
31. Goetschel, R., Voxman, W.: Elementary fuzzy calculus. Fuzzy Sets Syst. **18**, 31–43 (1986)
32. Friedman, M., Ma, M., Kandel, A.: Numerical solutions of fuzzy differential and integral equations. Fuzzy Sets Syst. **106**, 35–48 (1999)
33. Anastassiou, G.A.: Fuzzy Mathematics: Approximation Theory. Springer, Berlin Heidelberg (2010)
34. Wu, C., Gong, Z.: On Henstock integral of fuzzy-number-valued functions. Fuzzy Sets Syst. **120**, 523–532 (2001)
35. Gal, S.G.: Approximation theory in fuzzy setting. In: Handbook of Analytic-Computational Methods in Applied Mathematics. Chapman & Hall/CRC Press, Boca Raton, pp. 617–666 (2000)
36. Sizikov, V.S.: Mathematical Methods for Processing the Results of Measurements, Politekhnika (2001)

Successive Approximations Method for Fuzzy Fredholm-Volterra Integral equations of the Second Kind

S. Ziari, A. M. Bica, and R. Ezzati

Abstract In this paper, the successive approximations technique based on the trapezoidal quadrature rule is used for solving the fuzzy Fredholm-Volterra integral equations in two dimensions. We first present the way to approximate the value of the integral of any fuzzy-valued function based on the quadrature rule, that can be sequentially applied to evaluate the multiple integral. The convergence of the method will be investigated by giving error bounds for the approximate solution. Finally, a numerical experiment is included to demonstrate that the numerical results are consistent with the theoretical results.

Keywords Fuzzy Fredholm-Volterra integral equations · Fixed-point theory · Successive approximations method · Trapezoidal quadrature rule · Numerical approximation.

1 Introduction

This paper is concerned with successive approximations method based on the fuzzy transform basis functions and trapezoidal quadrature rule for fuzzy Fredholm-Volterra integral equations as follows

$$z(x, y) = f(x, y) \oplus (FR) \int_c^d (FR) \int_a^x H(x, y, s, t) \odot \psi(z(s, t))\, dsdt, \qquad (1)$$

S. Ziari (✉)
Department of Mathematics, Firoozkooh Branch, Islamic Azad University, Firoozkooh, Iran
e-mail: shok_ziari@yahoo.com

A. M. Bica
Department of Mathematics and Informatics, University of Oradea, Universitatii 1, 410087 Oradea, Romania

R. Ezzati
Department of Mathematics, Karaj Branch, Islamic Azad University, Karaj, Iran

T. Allahviranloo and S. Salahshour (eds.), *Advances in Fuzzy Integral and Differential Equations*, Studies in Fuzziness and Soft Computing 412,
https://doi.org/10.1007/978-3-030-73711-5_9

where $(x, y) \in [a, b] \times [c, d]$ and $f(x, y)$ is bounded and continuous on $[a, b] \times [c, d]$. The existence and uniqueness results for fuzzy Fredholm-Volterra integral equations have been studied in [28]. Numerical methods for solving fuzzy integral equations have been studied by many authors. The study of fuzzy integral equations have been considered in the works of Kaleva [27] and Siekkala [46], which they converted the initial value problem for first-order fuzzy differential equations into the equivalent fuzzy Volterra integral equation. The existence and uniqueness of results for the solution of fuzzy integral equation have been studied in [8, 9, 25, 36–38] by using the Banach's fixed point principle. Mordeson [31] have been started the approach of the existence and uniqueness of solutions for fuzzy Fredholm integral equations in a distinct study. Nieto and Rodriguez-Lopez [32] have obtained the boundedness of the solution of fuzzy integral equations. Numerical methods for approximating the solution of fuzzy integral equations have been studied by many authors based on quadrature rules and successive approximations providing a direct method based on solving system of equations. The method of quadrature rules and successive approximations with the investigation of their error estimation are introduced in [11, 13, 15, 20, 39, 44, 50, 51, 54–56]. In recent years several numerical iterative approaches have been proposed based on the successive approximations method and other techniques such as the Lagrange interpolation [24], divided and finite differences [35], fuzzy Haar wavelets [29], hybrid block-pulse functions and Taylor series [6], triangular basis functions [7], hybrid of Block-Pulse functions and Bernstein polynomials [45] and block-pulse basis functions [52, 55]. Some other existing numerical and analytical methods in literature are: Adomian decomposition [1], Nyström methods [2], Chebyshev interpolation [10], Galerkin type techniques [12], Bernstein polynomials [21], fuzzy Haar wavelets [29, 52] and block pulse functions [40], homotopy analysis [30] and homotopy perturbation [5]. Recently, Seifi et al. [47] proposed a new numerical approach based on Fibonacci polynomials to solve first-order fuzzy Fredholm–Volterra integro-differential equations of the second kind, by converting into a equation involving two-dimensional integrals. The study of two-dimensional fuzzy integral equations by considering the iterative method was developed by Sadatrasoul and Ezzati [42]. Indeed, they developed the iterative method in [21] for two dimensional integral equations. Subsequently, Bica and Popescu [16] applied the successive approximations method based on fuzzy trapezoidal cubature formula with the error estimates in terms of Lipschitz constants. Also, Bica and Ziari [15] used the iterative method for solving fuzzy linear Volterra integral equations. Recently, Akhavan et al. [3] investigated the successive approximations and mixed trapezoidal and midpoint rules for the Fuzzy Fredholm Integral Equation in two dimensions. Many other iterative and direct numerical methods for solving these equations are: successive approximations based on optimal quadrature formula [43], bivariate Bernstein polynomials [22, 23] , bivariate block-pulse functions[52], successive approximations based on open fuzzy cubature rule [18], homotopy perturbation [40], homotopy analysis [48], bivariate Lagrange interpolation [33] and bicubic Splines interpolation [34].

In this work, we apply successive approximations method based on quadrature formula to fuzzy Fredholm-Volterra integral equations for obtaining the approximate solution. The structure of this paper is organized as follows: In Sect. 2, definitions and results related to the fuzzy set numbers and fuzzy-number-valued functions are presented. Section 3 deals with the study of quadrature formulas. Section 4 is concerned with the study of the existence and uniqueness of the solution of (1) and for the uniform boundedness and uniform Lipschitz properties of the sequence of successive approximations. In Sect. 5, we describe the iterative method to obtain the approximate solution of the fuzzy integral equation (1). Section 6 includes the error analysis of the numerical method. In Sect. 7, we consider some numerical experiments in order to confirm the theoretical results. Finally, some concluding remarks will be presented in the final Section.

2 Preliminaries

Firstly, we consider the basic definition of fuzzy number which is given by Dubois and Prade [19]:

Definition 1 A fuzzy number is a function $u : \mathbb{R} \to [0, 1]$ satisfying the following properties:

(i) u is normal, i.e. $\exists x_0 \in \mathbb{R}$ with $u(x_0) = 1$.
(ii) u is a convex fuzzy set, i.e. $u(\lambda x + (1-\lambda)y) \geq \min\{u(x), u(y)\}$, $\forall x, y \in \mathbb{R}$, $\lambda \in [0, 1]$.
(iii) u is upper semi-continuous on $\mathbb{R}$.
(iv) the $[u]^0 = \overline{\{x \in \mathbb{R} : u(x) > 0\}}$ is compact, here $\overline{A}$ denotes the closure of A.

The space of fuzzy real numbers on $\mathbb{R}$ is denoted by $\mathbb{R}_{\mathcal{F}}$. Note that every real number $a \in \mathbb{R}$ can be interpreted as a fuzzy number $\tilde{a} = \chi_{\{a\}}$ and therefore $\mathbb{R} \subset \mathbb{R}_{\mathcal{F}}$. For $0 < r \leq 1$, we denote the *r-level* set $[u]^r = \{x \in \mathbb{R} : u(x) \geq r\}$, that is a closed interval (see [49]) and $[u]^r = [u^r_-, u^r_+]$, $\forall r \in [0, 1]$. These lead to the usual parametric representation of a fuzzy number. According to [11], for any $0 < r \leq 1$ the fuzzy number u is determined by an ordered pair of functions (u^r_-, u^r_+) where $u^r_-, u^r_+ : [0, 1] \to \mathbb{R}$ satisfies the following three conditions:

(i) u^r_- is a bounded monotonic non decreasing left-continuous function $\forall r \in]0, 1]$ and right-continuous for $r = 0$;
(ii) u^r_+ is a bounded monotonic non increasing left-continuous function $\forall r \in]0, 1]$ and right-continuous for $r = 0$;
(iii) $u^r_- \leq u^r_+$, $\forall r \in [0, 1]$.

For $u, v \in \mathbb{R}_{\mathcal{F}}$, $k \in \mathbb{R}$, the addition and the scalar multiplication operations for fuzzy numbers are defined as follow:

(i) $[u \oplus v]^r = [u]^r + [v]^r = [u^r_- + v^r_-, u^r_+ + v^r_+]$, $\forall r \in [0, 1]$

(ii) $[k \odot u]^r = k \cdot [u]^r = \begin{cases} [ku^r_-, ku^r_+], & k \geq 0, \\ [ku^r_+, ku^r_-], & k < 0. \end{cases}$

The Hausdorff distance between fuzzy numbers is defined by

$$D(u, v) = \sup_{r\in[0,1]} \left\{\max\left(\left|u_-^r - v_-^r\right|, \left|u_+^r - v_+^r\right|\right)\right\}$$

for any $u, v \in \mathbb{R}_{\mathcal{F}}$.

Lemma 1 *The following properties are hold for the Hausdorff distance (see [4, 26]):*

(i) $(\mathbb{R}_{\mathcal{F}}, D)$ *is a complete metric space,*
(ii) $D(u \oplus w, v \oplus w) = D(u, v)\ \forall u, v, w \in \mathbb{R}_{\mathcal{F}}$,
(iii) $D(u \oplus v, w \oplus e) \leq D(u, w) + D(v, e)\ \forall u, v, w, e \in \mathbb{R}_{\mathcal{F}}$,
(iv) $D\left(u \oplus v, \tilde{0}\right) \leq D\left(u, \tilde{0}\right) + D\left(v, \tilde{0}\right)\ \forall u, v \in \mathbb{R}_{\mathcal{F}}$,
(v) $D(k \odot u, k \odot v) = |k|\, D(u, v)\ \forall u, v \in \mathbb{R}_{\mathcal{F}}\ \forall k \in \mathbb{R}$.
(vi) $D(k_1 \odot u, k_2 \odot u) = |k_1 - k_2|\, D\left(u, \tilde{0}\right)\ \forall k_1, k_2 \in \mathbb{R}$ *with* $k_1 \cdot k_2 \geq 0$ *and* $\forall u \in \mathbb{R}_{\mathcal{F}}$.

The property (iv) suggest the definition of a function $\|\cdot\| : \mathbb{R}_{\mathcal{F}} \to \mathbb{R}$ by $\|u\| = D\left(u, \tilde{0}\right)$ that has the properties of usual norms. In [11], the properties of this function are presented as follows:

(i) $\|u\| \geq 0\ \forall u \in \mathbb{R}_{\mathcal{F}}$ and $\|u\| = 0$ iff $u = \tilde{0}$,
(ii) $\|\lambda \odot u\| = |\lambda| \cdot \|u\|$ and $\|u \oplus v\| \leq \|u\| + \|v\|\ \forall u, v \in \mathbb{R}_{\mathcal{F}}\ \forall \lambda \in \mathbb{R}$,
(iii) $|\|u\| - \|v\|| \leq D(u, v)$ and $D(u, v) \leq \|u\| + \|v\|$
$\forall u, v \in \mathbb{R}_{\mathcal{F}}$.

It is easy to verify that $(\mathbb{R}_{\mathcal{F}}, \oplus, \odot, \|\cdot\|)$ is not a normed space because $(\mathbb{R}_{\mathcal{F}}, \oplus)$ is not a group.

Definition 2 For any fuzzy-number-valued function $f : I \subset \mathbb{R} \times \mathbb{R} \to \mathbb{R}_{\mathcal{F}}$ we can define the functions $f_-^r(.,.), f_+^r(.,.) : I \subset \mathbb{R} \times \mathbb{R} \to \mathbb{R},\ r \in [0, 1]$ by $f_-^r(s, t) = (f(s, t))_-^r,\ f_+^r(s, t,) = (f(s, t))_-^r,\ \forall (s, t) \in [a, b] \times [c, d]$. These functions are the left and right r-level functions of f.

Definition 3 (*see* [41]). A function $f : [a, b] \times [c, d] \to \mathbb{R}_{\mathcal{F}}$ is called:
(i) continuous in $(x_0, y_0) \in [a, b] \times [c, d]$ if for any $\varepsilon > 0$ there exists $\delta > 0$ such that for any $(x, y) \in [a, b] \times [c, d]$ with $|x - x_0| < \delta,\ |y - y_0| < \delta$ we have

$$D(f(x, y), f(x_0, y_0)) < \varepsilon.$$

The function f is continuous on $[a, b] \times [c, d]$ if it is continuous in each $(x, y) \in [a, b] \times [c, d]$.
(ii) bounded if there exists $M \geq 0$ such that

$$D\left(f(x, y), \tilde{0}\right) \leq M, \quad \forall (x, y) \in [a, b] \times [c, d].$$

Lemma 2 (see *[4]*). *Let* $f : [a, b] \times [c, d] \to \mathbb{R}_{\mathcal{F}}$ *fuzzy continuous (with respect to metric D), then* $D(f(x, y), \tilde{0}) \leq M, \ \forall \ (x, y) \in [a, b] \times [c, d], \ M > 0$, *that is* f *is fuzzy bounded.*

Definition 4 (*see* [17]). A fuzzy-number-valued function $f : [a, b] \times [c, d] \to \mathbb{R}_{\mathcal{F}}$ is said to be Lipschitz function if there exist $L_1, L_2 \geq 0$ such that

$$D\left(f\left(x_1, y_1\right), f\left(x_2, y_2\right)\right) \leq L_1 \left|x_1 - x_2\right| + L_2 \left|y_1 - y_2\right|. \tag{2}$$

Let $C\left([a, b] \times [c, d], \mathbb{R}_{\mathcal{F}}\right)$, be the space of two dimensional fuzzy continuous functions with the metric

$$D^*(f, g) = \sup_{a \leq s \leq b,\ c \leq t \leq d} D\left(f(s, t), g(s, t)\right),$$

for all $f, g \in C\left([a, b] \times [c, d], \mathbb{R}_{\mathcal{F}}\right)$, which is called the uniform distance between fuzzy-number-valued functions. It is easy to see that $C\left([a, b] \times [c, d], \mathbb{R}_{\mathcal{F}}\right)$ is a complete metric space.

In [41], the notion of Henstock integral for fuzzy-number-valued functions is defined as follows:

Definition 5 Let $f : [a, b] \times [c, d] \to \mathbb{R}_{\mathcal{F}}$ be a bounded mapping. For $\Delta_x : a = x_0 < x_1 < \cdots < x_{m-1} < x_m = b$ a partition of the interval $[a, b]$ and $\Delta_y : c = y_0 < y_1 < \cdots < y_{n-1} < y_n = d$ a partition of the interval $[c, d]$, let us consider the selection points $\xi_i \in \left[x_{i-1}, x_i\right], \ i = \overline{1, m}, \ \eta_j \in \left[y_{j-1}, y_j\right], \ j = \overline{1, n}$, and the functions $\delta : [a, b] \to \mathbb{R}_+$. $\sigma : [c, d] \to \mathbb{R}_+$. The partitions $P_x = \{([x_{i-1}, x_i]; \xi_i), \ i = \overline{1, m}\}$ denoted by $P_x = (\Delta_x, \xi)$ and $P_y = \{([y_{j-1}, y_j]; \eta_j), \ j = \overline{1, n}\}$ denoted by $P_y = (\Delta_y, \eta)$ are said to be δ-fine iff $\left[x_{i-1}, x_i\right] \subseteq \left(\xi_i - \delta\left(\xi_i\right), \xi_i + \delta\left(\xi_i\right)\right), \ \forall i = \overline{1, m}$, and σ-fine iff $\left[y_{j-1}, y_j\right] \subseteq \left(\eta_j - \sigma\left(\eta_j\right), \eta_j + \sigma\left(\eta_j\right)\right), \ \forall j = \overline{1, n}$, respectively. The function f is said to be fuzzy Henstock integrable if there exists $I(f) \in \mathbb{R}_{\mathcal{F}}$ with the property that for any $\epsilon > 0$ there is a function $\delta : [a, b] \to \mathbb{R}_+$ and a function $\sigma : [c, d] \to \mathbb{R}_+$ such that for any partition δ-fine P_x, and for any partition σ-fine P_y, we have $D\left(\sum_{i=1}^{m} \sum_{j=1}^{n} \left(x_i - x_{i-1}\right)\left(y_j - y_{j-1}\right) \odot f\left(\xi_i, \eta_j\right), I(f)\right) < \epsilon$.
The fuzzy number I is named the fuzzy Henstock double integral of f and will be denoted by

$$I(f) = (FH) \int_c^d (FH) \left(\int_a^b f(s, t)\, ds\right) dt.$$

Remark 1 If in the above definition, the functions δ and σ are constant then it obtains the fuzzy-Riemann double integrability. In this case $I(f) \in \mathbb{R}_{\mathcal{F}}$ is called the fuzzy-Riemann double integral of f on $[a, b] \times [c, d]$, being denoted by

$$I(f) = (FR) \int_c^d (FR) \left(\int_a^b f(s, t)\, ds\right) dt.$$

Consequently, the fuzzy-Riemann double integrability is a particular case of the fuzzy-Henstock double integrability, and therefore any valid property for the double integral (FH) will be valid for the double integral (FR), too.

Lemma 3 (see *[41]*). *If* $f \in C([a,b]\times[c,d],\mathbb{R}_{\mathcal{F}})$ *then*

$$(FR)\int_c^d \left((FR)\int_a^b f(s,t)\,ds\right)dt$$

exists and

$$\left[(FR)\int_c^d \left((FR)\int_a^b f(s,t)\,ds\right)dt\right]^r = \left[\int_c^d\int_a^b f_-^r(s,t)\,dsdt, \int_c^d\int_a^b f_+^r(s,t)\,dsdt\right].$$

Lemma 4 (see *[53]*). *If* $f,g:[a,b]\times[c,d]\to\mathbb{R}_{\mathcal{F}}$ *are continuous fuzzy functions then the function* $\varphi:[a,b]\times[c,d]\to\mathbb{R}_+$ *defined by* $\varphi(s,t)=D(f(s,t),g(s,t))$ *is continuous on* $[a,b]\times[c,d]$ *and*

$$D\left((FR)\int_c^d (FR)\left(\int_a^b f(s,t)\,ds\right)dt, (FR)\int_c^d (FR)\left(\int_a^b g(s,t)\,ds\right)dt\right) \le$$
$$\le \int_c^d\int_a^b D(f(s,t),g(s,t))\,dsdt.$$

Lemma 5 (Bica and Popescu) *If* $f,g\in C([a,b]\times[c,d],\mathbb{R}_{\mathcal{F}})$ *and* $\alpha\in C([a,b]\times[c,d],\mathbb{R}_+)$ *then the functions* $\alpha\cdot g:[a,b]\times[c,d]\to\mathbb{R}_{\mathcal{F}}$ *and* $F:[a,b]\times[c,d]\to\mathbb{R}_{\mathcal{F}}$ *given by* $(\alpha\cdot g)(s,t)=\alpha(s,t)\cdot g(s,t)$, $\forall (s,t)\in[a,b]\times[c,d]$ *and* $F(s,t)=(FR)\int_c^d (FR)\int_a^b f(s,t)dsdt$ *are continuous.*

3 Quadrature Formulas

In this section, we present some theorems concerning the quadrature rules obtained in [11], and in this context we focus on the particular case of trapezoidal rule (see Corollary 9).

Theorem 1 *[11] Let* $f:[a,b]\to\mathbb{R}_{\mathcal{F}}$ *be a Lipschitz function with constant L. Then, for any division* $\Delta: a=x_0<x_1<\cdots<x_n=b$, *and any points* $\xi_i\in[x_{i-1},x_i]$, $i=1,\ldots,n$, *we have*

$$D\left((FR)\int_a^b f(t)dt, \sum_{i=1}^n (x_i-x_{i-1})\odot f(\xi_i)\right) \le \frac{L}{2}\sum_{i=1}^n (x_i-x_{i-1})^2.$$

Proposition 1 *[11] Let* $f:[a,b]\to\mathbb{R}_{\mathcal{F}}$ *be a Lipschitz function with constant L. Then*

$$D\left((FR)\int_a^b f(t)dt, [(x-a)\odot f(u)\oplus(b-x)\odot f(v)]\right) \leq L\cdot\left[\frac{(b-a)^2}{4}+\left(x-\frac{a+b}{2}\right)^2\right],$$

for any $x \in [a, b]$, and $u \in [a, x]$, $v \in [x, b]$.

The following corollary is obtained by considering $u = a$, $v = b$ and $x = \frac{a+b}{2}$ from the above proposition.

Corollary 1 *[14] Let $f : [a, b] \to \mathbb{R}_{\mathcal{F}}$ be a Lipschitz function. then the following simple trapezoidal formula holds:*

$$D\left((FR)\int_a^b f(t)dt, \frac{b-a}{2}\odot[f(a)\oplus f(b)]\right) \leq L\cdot\frac{(b-a)^2}{4}.$$

The fuzzy trapezoidal quadrature formula can be extended for uniform partitions.

Remark 2 [14] For uniform patition, $\Delta = a = t_0 < t_1 < \cdots < t_n = b$, with $t_i = a + i.\frac{b-a}{n}$, $\forall\,\overline{0, n}$, the following composite trapezoidal inequality holds:

$$D\left((FR)\int_a^b f(t)dt, \sum_{i=1}^{n}\frac{t_i - t_{i-1}}{2}\odot[f(t_{i-1})\oplus f(t_i)]\right) \leq \frac{L(b-a)^2}{4n}. \qquad (3)$$

4 Two Dimensional Fuzzy Integral Equations

In this section, we consider the two dimensional nonlinear fuzzy Fredholm-Volterra integral Eq. (1), under the following conditions

(i) $f : \Omega = [a, b] \times [c, d] \to \mathbb{R}_{\mathcal{F}}$ is fuzzy number-valued continuous function,
(ii) $H : \Omega^2 \to R^+$ is continuous and there exists $M > 0$, such that

$$M_H = \max_{(x,y),(s,t)\in\Omega} H(s, t, x, y)$$

,
(iii) $\psi : \mathbb{R}_{\mathcal{F}} \to \mathbb{R}_{\mathcal{F}}$ is fuzzy continuous and there exists $\alpha > 0$ such that

$$D(\psi(u(x, y)), \psi(v(x, y))) \leq \alpha.D(u(x, y), v(x, y)),$$

for all $(x, y) \in [a, b] \times [c, d]$, $u, v \in C([a, b] \times [c, d], \mathbb{R}_{\mathcal{F}})$,
(iv) $\gamma = \alpha(b-a)(d-c)M_H < 1$.
(v) there exist $\mu, \mu' \geq 0$ such that

$$|H(x, y, s, t) - H(x', y', s, t)| \leq \mu|x - x'| + \mu'|y - y'|,$$

for any $(x, y), (s, t), (x', y') \in [a, b] \times [c, d]$;

(vi) there exist $\lambda, \lambda' \geq 0$ such that

$$\left|H(x, y, s, t) - H\left(x, y, s', t'\right)\right| \leq \lambda \left|s - s'\right| + \lambda' \left|t - t'\right|,$$

for any $(x, y), (s, t), (s', t') \in [a, b] \times [c, d]$;

(vii) there exist $\beta, \beta' \geq 0$ such that

$$D(f(x, y)), f(x', y'))) \leq \beta \left|x - x'\right| + \beta' \left|y - y'\right|,$$

for any $(x, y), (x', y') \in [a, b] \times [c, d]$;

Lemma 6 *Under the conditions (ii)–(vi), for any $z \in C([a, b] \times [c, d], \mathbb{R}_{\mathcal{F}})$ the function $F_z : [a, b] \times [c, d] \to \mathbb{R}_{\mathcal{F}}$ defined by*

$$F_z(x, y) = (FR) \int_c^d (FR) \int_a^x H(x, y, s, t) \cdot \psi(z(s, t)) ds dt$$

is uniformly continuous on $[a, b] \times [c, d]$.

Proof Firstly, for arbitrary fixed $z \in C([a, b] \times [c, d], \mathbb{R}_{\mathcal{F}})$, we prove that the function $\Phi_z : [a, b] \times [c, d] \to \mathbb{R}_{\mathcal{F}}, \Phi_z(s, t) = H(x, y, s, t) \cdot \psi(z(s, t))$ is continuous in arbitrary point $(s_0, t_0) \in [a, b] \times [c, d]$. For this purpose we consider:

$$\begin{aligned}
D(\Phi_z(s, t), \Phi_z(s_0, t_0)) &= D(H(x, y, s, t) \odot \psi(z(s, t)), H(x, y, s_0, t_0) \odot \psi(z(s_0, t_0))) \\
&\leq D(H(x, y, s, t) \odot \psi(z(s, t)), H(x, y, s_0, t_0) \odot \psi(z(s, t))) + \\
&\quad + D(H(x, y, s_0, t_0) \odot \psi(z(s, t)), H(x, y, s_0, t_0) \odot \psi(z(s_0, t_0))) \\
&\leq |H(x, y, s, t) - H(x, y, s_0, t_0)| D\left(\psi(z(s, t)), \tilde{0}\right) + \\
&\quad + |H(x, y, s_0, t_0)| D(\psi(z(s, t)), \psi(z(s_0, t_0))).
\end{aligned}$$

According to the continuity of z, it can be easily verified that $\psi(z)$ is continuous and therefore it is bounded. So, there is $M_\psi \geq 0$ such that $D\left(\psi(z(s, t)), \tilde{0}\right) \leq M_\psi, \forall (s, t) \in [a, b] \times [c, d]$. By, the boundedness of H, $\psi(z)$ and Lipschitz condition (vi), we obtain:

$$D(\Phi_z(s, t), \Phi_z(s_0, t_0)) \leq (\lambda |s - s_0| + \lambda' |t - t_0| M_\psi + \alpha M_H D(z(s, t), z(s_0, t_0)).$$

Let arbitrary $(s_0, t_0) \in [a, b] \times [c, d]$ and $\epsilon > 0$. Since $z \in C([a, b] \times [c, d], \mathbb{R}_{\mathcal{F}})$ we conclude that there exist $\delta(\epsilon)) > 0$ such that for any $(s, t) \in [a, b] \times [c, d]$ with $|s - s_0| + |t - t_0| < \delta(\epsilon))$ it follows that $|s - s_0| + \lambda' |t - t_0| < \frac{\epsilon}{2 \max\{\lambda, \lambda'\} M_\psi}$ and $D(z(s, t), z(s_0, t_0)) < \frac{\epsilon}{2\alpha M_H}$. Then

$$D(\Phi_z(s, t), \Phi_z(s_0, t_0)) < \max\{\lambda, \lambda'\} M_\psi . \frac{\epsilon}{2 \max\{\lambda, \lambda'\} M_\psi} + \alpha M_H . \frac{\epsilon}{2\alpha M_H} = \epsilon.$$

Now, we prove that F_z is uniformly continuous. To this aim, taking arbitrary $(x, y), (x', y') \in [a, b] \times [c, d]$, we have:

$$\begin{aligned}
D(F_z(x,y), F_z(x',y')) &= D\Big((FR)\int_c^d (FR)\int_a^x H(x,y,s,t)\cdot\psi(z(s,t))dsdt,\\
&\quad ,(FR)\int_c^d (FR)\int_a^{x'} H(x',y',s,t)\cdot\psi(z(s,t))dsdt\Big)\\
&\le D\Big((FR)\int_c^d (FR)\int_a^x H(x,y,s,t)\cdot\psi(z(s,t))dsdt,\\
&\quad ,(FR)\int_c^d (FR)\int_a^x H(x',y',s,t)\cdot\psi(z(s,t))dsdt\Big)+\\
&\quad +D\left(\tilde{0},(FR)\int_c^d (FR)\int_x^{x'} H(x',y',s,t)\cdot\psi(z(s,t))dsdt\right)\\
&\le \int_c^d\int_a^x |H(x,y,s,t)-H(x',y',s,t)|D\left(\psi(z(s,t)),\tilde{0}\right)dsdt + \int_c^d\int_x^{x'} M_H M_\psi dsdt\\
&\le (d-c)(x-a)M_\psi(\mu|x-x'|+\mu'|y-y'|)+(d-c)|x-x'|M_H M_\psi\\
&\le \left(\frac{\mu\gamma M_\psi}{M_H}+\frac{M_\psi}{b-a}\right)|x-x'|+\left(\frac{\mu'\gamma M_\psi}{M_H}\right)|y-y'|.
\end{aligned}$$

Consequently, for arbitrary $\epsilon > 0$, there exists $\delta > 0$,

$$0 < \delta < \min\left\{1, \frac{\epsilon}{\max\left\{\frac{\mu\gamma M_\psi}{M_H}+\frac{M_\psi}{b-a}, \frac{\mu'\gamma M_\psi}{M_H}\right\}}\right\}$$

such that for any $(x, y), (x', y') \in [a, b] \times [c, d]$ with $\left|s - s'\right| + \left|t - t'\right| < \delta$ we have:

$$D(F_z(x,y), F_z(x',y')) \le \left(\max\left\{\frac{\mu\gamma M_\psi}{M_H}+\frac{M_\psi}{b-a}, \frac{\mu'\gamma M_\psi}{M_H}\right\}\right)\delta < \epsilon.$$

Therefore F_z is uniformly continuous for any fixed $z \in C([a, b] \times [c, d], \mathbb{R}_{\mathcal{F}})$.

The following theorem presents sufficient conditions for the existence of a unique solution of Eq. (1).

Theorem 2 *Under the conditions (i)–(vii), then the iterative procedure*

$$z_0(x,y) = f(x,y), \tag{4}$$

$$z_k(x,y) = f(x,y) \oplus (FR)\int_c^d (FR)\int_a^x H(x,y,s,t)\odot\psi(z_{k-1}(s,t))dsdt, \tag{5}$$

for all $(x, y) \in [a, b] \times [c, d]$, $k \in \mathbb{N}^*$, *converges to the solution of* F *of* (1). *In addition, the following error bound holds:*

$$D^*(z, z_k) \leq \frac{\gamma^{k+1}}{1-\gamma} D^*(z_1, z_0), \ \ \forall k \in \mathbb{N}^*, \tag{6}$$

and by choosing $z_0 \in C([a, b] \times [c, d], \mathbb{R}_{\mathcal{F}})$ *the inequality* (6) *becomes:*

$$D^*(z, z_k) \leq \frac{\gamma^{k+1}}{\alpha(1-\gamma)} M_0, \ \ \forall k \in \mathbb{N}^*. \tag{7}$$

where $M_0 = \sup_{(x,y)\in[a,b]\times[c,d]} ||G(f(x, y))||_{\mathcal{F}}$. *Moreover, the sequence of successive approximations* (4)–(5) *is uniformly bounded that is, there exists a constant* $M > 0$ *such that* $D\left(z_k(x, y), \tilde{0}\right) \leq M$, *for all* $k \in \mathbb{N}^*$ *and* $\forall (x, y) \in [a, b] \times [c, d]$. *Furthermore, the solution* z^* *is bounded.*

Proof To prove this theorem, on $C([a, b] \times [c, d], \mathbb{R}_{\mathcal{F}})$ we define the integral operator $A : C([a, b] \times [c, d], \mathbb{R}_{\mathcal{F}}) \to C([a, b] \times [c, d], \mathbb{R}_{\mathcal{F}})$, associated with (1) by:

$$A(z(x, y)) = f(x, y) \oplus (FR)\int_c^d (FR)\int_a^x H(x, y, s, t) \odot G(z(s, t))dsdt.$$

Since $f \in C([a, b] \times [c, d], \mathbb{R}_{\mathcal{F}})$, we deduce from Lemma 6 that $A(z) \in C([a, b] \times [c, d], \mathbb{R}_{\mathcal{F}})$ for any $z \in C([a, b] \times [c, d], \mathbb{R}_{\mathcal{F}})$, so the integral operator is well defined. To investigate the existence and uniqueness of the solution of (1), we find a fixed point for the operator A. In order to prove that A is a contraction mapping, we consider arbitrary $u, v \in C([a, b] \times [c, d], \mathbb{R}_{\mathcal{F}})$; then, we have:

$$\begin{aligned}
&D(A(u(x, y)), A(v(x, y))) \leq \\
&\leq \int_c^d \int_a^x D\Big(H(x, y, s, t) \odot G(u(s, t)), H(x, y, s, t) \odot G(v(s, t))\Big)dsdt \leq \\
&\leq \int_c^d \int_a^x \alpha|H(x, y, s, t|D(u(s, t), v(s, t))\, dsdt \leq \alpha M_H(b-a)(d-c)D^*(u, v) = \\
&= \gamma D^*(u, v).
\end{aligned}$$

Since $\gamma < 1$, it follows that the operator A is a contraction mapping. Therefore, it has a unique fixed point $z^* \in C([a, b] \times [c, d], \mathbb{R}_{\mathcal{F}})$ which is the unique solution of Eq. (1). Hence, the Banach's fixed point theorem convinces the existence and uniqueness of the solution of integral equation (1); moreover, the sequence of successive approximations (4)–(4) converges to the solution z^* for any arbitrary initial term, and the following error bound holds:

$$D\left(z^*, z_k\right) \leq \frac{\gamma^k}{1-\gamma} D^*(z_1, z_0).$$

Choosing $Z_0 = f$, we have:

$$D(z_1(x,y),z_0(x,y)) =$$
$$= D\left(f(x,y) \oplus (FR)\int_c^d (FR)\int_a^x H(x,y,s,t) \odot G(z_0(s,t))dsdt, z_0(x,y)\right)$$
$$\leq D\left((FR)\int_c^d (FR)\int_a^x H(x,y,s,t) \odot G(z_0(s,t))dsdt, (FR)\int_c^d (FR)\int_a^x \tilde{0}dsdt\right)$$
$$\leq \int_c^d \int_a^x |H(x,y,s,t|D\left(G(f(s,t)),\tilde{0}\right)dsdt \leq M_H(b-a)(d-c)D^*(G(f),\tilde{0}).$$

According to the continuity of $G(f)$, there exists $M_0 > 0$ such that $D^*(G(f),\tilde{0}) \leq M_0$, and so we obtain:

$$D(z_1(x,y),z_0(x,y)) \leq \frac{\gamma}{\alpha}M_0 \leq \frac{M_0}{\alpha}.$$

Considering the above inequality and inequality (6) leads to estimate (7).
To establish the fact that z^* is uniformly bounded, we consider arbitrary $(x,y) \in [a,b] \times [c,d]$ and $k \in \mathbb{N}$; then, we obtain:

$$D(z_k(x,y), z_{k-1}(x,y)) \leq \int_c^d \int_a^x |H(x,y,s,t)|D(G(z_{k-1}(s,t)), G(z_{k-2}(s,t)))dsdt$$
$$\leq \alpha M_H(b-a)(d-c)D^*(z_{k-1},z_{k-2}) = \gamma D^*(z_{k-1},z_{k-2}).$$

Successive repetition of this inequality gives:

$$D(z_k(x,y), z_{k-1}(x,y)) \leq \gamma^{k-1}D^*(z_1,z_0), \quad \forall (x,y) \in [a,b] \times [c,d], \quad k \in \mathbb{N}.$$

Hence,

$$D(z_k(x,y), z_0(x,y)) \leq$$
$$\leq D(z_k(x,y), z_{k-1}(x,y)) + D(z_{k-1}(x,y), z_{k-2}(x,y)) + \cdots + D(z_1(x,y), z_0(x,y)) \leq$$
$$\leq \left(\gamma^{k-1} + \gamma^{k-2} + \cdots + 1\right)D^*(z_1,z_0) \leq \left(\frac{1-\gamma^k}{1-\gamma}\right)\left(\frac{M_0}{\alpha}\right).$$

Consequently:

$$D(z_k(x,y), z_0(x,y)) \leq \frac{M_0}{\alpha(1-\gamma)}. \tag{8}$$

Considering (8) we have:

$$D\left(z_k(x,y),\tilde{0}\right) \leq D(z_k(x,y), z_0(x,y)) + D\left(z_0(x,y),\tilde{0}\right) \leq \frac{M_0}{\alpha(1-\gamma)} + M_f =: M,$$

where $M_f > 0$ and $D\left(z_0(x, y), \tilde{0}\right) = D\left(f(x, y), \tilde{0}\right) \leq M_f$, for all $(x, y) \in [a, b] \times [c, d]$.

Theorem 3 *Under the conditions (i)–(vii), the sequence of successive approximations* (4) *is uniform Lipschitz; that is, there exist a constant* L_1, L_2 *such that for all* $(x, y), (x', y') \in [a, b] \times [c, d]$, $k \in \mathbb{N}^*$, *the following inequality holds:*

$$D\left(z_k(x, y), z_k(x', y')\right) \leq L_1|x - x'| + L_2|y - y'|. \tag{9}$$

Proof The proof is similar to the proof of Theorem 20 in [3].

5 The Successive Approximations Method

In this section, we propose the numerical approximation method for the solution of the two dimensional nonlinear fuzzy Fredholm-Volterra integral equation (1). In this way, we consider the following uniform partition $\Delta = (\Delta_x, \Delta_s)$ of the rectangle $[a, b] \times [c, d]$, with:

$$\Delta_x : a = x_0 < x_1 < \cdots < x_n = b,$$

$$\Delta_y : c = y_0 < y_1 < \cdots < y_k = d,$$

where $x_i = a + \frac{i(b-a)}{m}$, $y_j = c + \frac{j(d-c)}{n}$, $0 \leq i \leq m$, $1 \leq j \leq n$, $h_1 = \frac{b-a}{m}$, $h_1 = \frac{d-c}{n}$. We apply the quadrature formula (3) to approximate the multiple integrals of the terms of the sequence of successive approximations (4)–(5),

$$z_0(x_i, y_j) = f(x_i, y_j), \quad i = \overline{1, n},\ j = \overline{1, n},\ k \in \mathbb{N}^*,$$

$$z_k(x_i, y_j) = f(x_i, y_j) \oplus (FR)\int_c^d (FR)\int_a^{x_i} H(x_i, y_j, s, t) \odot \psi(z_{k-1}(s, t))dsdt,\ k \in \mathbb{N}^*,$$

obtaining the following iterative process:

$$\begin{aligned} z_k\left(x_i, y_j\right) &= f\left(x_i, y_j\right) \oplus \\ &\oplus \frac{h_1 h_2}{4} \odot \sum_{p=1}^{i}\sum_{q=1}^{n}\Big[\Psi_{k-1}\left(x_i, y_j, s_{p-1}, t_{q-1}\right) \oplus \Psi_{k-1}\left(x_i, y_j, s_{p-1}, t_q\right) \oplus \\ &\oplus \Psi_{k-1}\left(x_i, y_j, s_p, t_{q-1}\right) \oplus \Psi_{k-1}\left(x_i, y_j, s_p, t_q\right)\Big] \oplus R_{k,i,j}, \\ &\quad i = \overline{1, n},\ j = \overline{1, n},\ k \in \mathbb{N}^*, \end{aligned} \tag{10}$$

where

$$\Psi_k(x, y, s, t) = H(x, y, s, t) \odot G(z_k(s, t)), \quad \forall k \in \mathbb{N}^*.$$

According to (3) and (9), we get:

$$D\left(R_{k,i,j}, \tilde{0}\right) \leq \frac{(b-a)(d-c)}{4}\left(L_1.\frac{b-a}{m} + L_2.\frac{d-c}{n}\right). \tag{11}$$

Considering the above presented expressions it obtains the following iterative algorithm:
Step 0: Give the data $a, b, c, d, \epsilon', m, n$ and functions f, G, H.
Step 1 (initialization step): For $i = \overline{1,n},\ j = \overline{1,n}$ we set:

$$\overline{z_0}(x_i, y_j) = z_0(x_i, y_j) = f(x_i, y_j).$$

Step 2 (the first iterative step): For $k = 1$ and for all $i = \overline{1,n},\ j = \overline{1,n}$, compute

$$\begin{aligned}\overline{z_1}(x_i, y_j) &= f(x_i, y_j) \oplus \\ &\oplus \frac{h_1 h_2}{4} \odot \sum_{p=1}^{i}\sum_{q=1}^{n}\Big[\overline{\Psi_0}(x_i, y_j, s_{p-1}, t_{q-1}) \oplus \overline{\Psi_0}(x_i, y_j, s_{p-1}, t_q) \oplus \\ &\oplus \overline{\Psi_0}(x_i, y_j, s_p, t_{q-1}) \oplus \overline{\Psi_0}(x_i, y_j, s_p, t_q)\Big].\end{aligned} \tag{12}$$

Steps 3 (the general iterative step): By induction for $k \in \mathbb{N}^*$, $k \geq 2$,, using the values computed at the previous step, we obtain for $i = \overline{1,n},\ j = \overline{1,n}$, the values:

$$\begin{aligned}\overline{z_k}(x_i, y_j) &= f(x_i, y_j) \oplus \\ &\oplus \frac{h_1 h_2}{4} \odot \sum_{p=1}^{i}\sum_{q=1}^{n}\Big[\overline{\Psi_{k-1}}(x_i, y_j, s_{p-1}, t_{q-1}) \oplus \overline{\Psi_{k-1}}(x_i, y_j, s_{p-1}, t_q) \oplus \\ &\oplus \overline{\Psi_{k-1}}(x_i, y_j, s_p, t_{q-1}) \oplus \overline{\Psi_{k-1}}(x_i, y_j, s_p, t_q)\Big].\end{aligned} \tag{13}$$

Steps 4 (the stopping criterion step): If the following condition succeeds:

$$D\left(\overline{z_k}(x_i, y_j), \overline{z_{k-1}}(x_i, y_j)\right) < \varepsilon, \quad \forall i = \overline{1,m},\ j = \overline{1,n}. \tag{14}$$

then we stop to this "k" and restore its value ($k = \ldots$), and the values $z_k(x_i, y_j),\ i = \overline{1,m},\ j = \overline{1,n}$, computed at this last iterative step. This condition is active after Step 2.
Steps 5 (final step): Print "k" and print $z_k(x_i, y_j),\ i = \overline{1,m},\ j = \overline{1,n}$. STOP.

6 Error Analysis

In this section, we give the error estimate of the distance between the approximate solution and the exact solution of Eq. (1).

Theorem 4 *Suppose that the conditions (i)–(vii) hold. Then for the exact solution z of (1) and the approximations $\overline{z_k}$ given by (10), the following error estimate holds for all $k \in \mathbb{N}^*$*

$$D^*(z, \overline{z_k}) \leq \frac{\gamma^{k+1}}{\alpha(1-\gamma)} M_0 + \frac{(b-a)(d-c)}{4(1-\gamma)} \left(L_1 . \frac{b-a}{m} + L_2 . \frac{d-c}{n} \right), \quad \forall k \in \mathbb{N}^*.$$

where $\gamma = \alpha M_H (b-a)(d-c)$ *and* $M_0 = \sup_{(x,y) \in [a,b] \times [c,d]} ||G(f(x,y))||_{\mathcal{F}}$.

Proof According to (7) we have:

$$D^*(z, z_k) \leq \frac{\gamma^{k+1}}{\alpha(1-\gamma)} . M_0, \quad \forall k \in \mathbb{N}^*. \tag{15}$$

Since

$$D^*(z, \overline{z_k}) \leq D^*(z, z_k) + D^*(z_k, \overline{z_k}). \tag{16}$$

Now, we need to approximate $D^*(z_k, \overline{z_k})$ for all $k \in \mathbb{N}^*$.
From (10), (11) and (13) for $k = 1$ we obtain:

$$D(z_1(x_i, y_j), \overline{z_1}(x_i, y_j)) \leq D(R_{1,i,j}, \tilde{0}) \leq \frac{(b-a)(d-c)}{4} \left(L_1 . \frac{b-a}{m} + L_2 . \frac{d-c}{n} \right).$$

Using (10) and (13) we get

$$\begin{aligned} D(z_k(x_i, y_j), \overline{z_k}(x_i, y_j)) &\leq D(R_{k,i,j}, \tilde{0}) + \\ &+ \frac{h_1 h_2}{4} \cdot \sum_{p=1}^{i} \sum_{q=1}^{n} \Big[D\left(\Psi_{k-1}\left(x_i, y_j, s_{p-1}, t_{q-1}\right), \overline{\Psi_{k-1}}\left(x_i, y_j, s_{p-1}, t_{q-1}\right)\right) + \\ &+ D\left(\Psi_{k-1}\left(x_i, y_j, s_{p-1}, t_q\right), \overline{\Psi_{k-1}}\left(x_i, y_j, s_{p-1}, t_q\right)\right) + \\ &+ D\left(\Psi_{k-1}\left(x_i, y_j, s_p, t_{q-1}\right), \overline{\Psi_{k-1}}\left(x_i, y_j, s_p, t_{q-1}\right)\right) + \\ &+ D\left(\Psi_{k-1}\left(x_i, y_j, s_p, t_q\right), \overline{\Psi_{k-1}}\left(x_i, y_j, s_p, t_q\right)\right) \Big]. \end{aligned}$$

Considering $\Psi_k(x, y, s, t) = H(x, y, s, t) \odot G(z_k(s, t))$ and $\overline{\Psi_k}(x, y, s, t) = H(x, y, s, t) \odot G(\overline{z_k}(s, t))$, we get:

$$D(z_k(x_i,y_j),\overline{z_k}(x_i,y_j)) \le D(R_{k,i,j},\tilde{0})+$$
$$+\frac{h_1h_2}{4}\odot\sum_{p=1}^{i}\sum_{q=1}^{n}\Big[|H\left(x_i,y_j,s_{p-1},t_{q-1}\right)|D\left(G\left(z_k\left(s_{p-1},t_{q-1}\right)\right),G\left(\overline{z_k}\left(s_{p-1},t_{q-1}\right)\right)\right)+$$
$$+H\left(x_i,y_j,s_{p-1},t_q\right)|D\left(G\left(z_k\left(s_{p-1},t_q\right)\right),G\left(\overline{z_k}\left(s_{p-1},t_q\right)\right)\right)+$$
$$+H\left(x_i,y_j,s_p,t_{q-1}\right)|D\left(G\left(z_k\left(s_p,t_{q-1}\right)\right),G\left(\overline{z_k}\left(s_p,t_{q-1}\right)\right)\right)+$$
$$+H\left(x_i,y_j,s_p,t_q\right)|D\left(G\left(z_k\left(s_p,t_q\right)\right),G\left(\overline{z_k}\left(s_p,t_q\right)\right)\right)\Big].$$

Taking into account the conditions (ii) and (iii), we derive:

$$D(z_k(x_i,y_j),\overline{z_k}(x_i,y_j)) \le \frac{(b-a)(d-c)}{4}\left(L_1.\frac{b-a}{m}+L_2.\frac{d-c}{n}\right)+$$
$$+\frac{\gamma}{4mn}\cdot\sum_{p=1}^{i}\sum_{q=1}^{n}\Big[D\left(z_k\left(s_{p-1},t_{q-1}\right),\overline{z_k}\left(s_{p-1},t_{q-1}\right)\right)+D\left(z_k\left(s_{p-1},t_q\right),\overline{z_k}\left(s_{p-1},t_q\right)\right)+$$
$$+D\left(z_k\left(s_p,t_{q-1}\right),\overline{z_k}\left(s_p,t_{q-1}\right)\right)+D\left(z_k\left(s_p,t_q\right),\overline{z_k}\left(s_p,t_q\right)\right)\Big], \tag{17}$$

Now, from (17) for $k=2$ we conclude that:

$$D(z_2(x_i,y_j),\overline{z_2}(x_i,y_j)) \le \frac{(b-a)(d-c)}{4}\left(L_1.\frac{b-a}{m}+L_2.\frac{d-c}{n}\right)+$$
$$+\frac{\gamma}{4mn}\cdot\sum_{p=1}^{m}\sum_{q=1}^{n}\frac{4(b-a)(d-c)}{4}\left(L_1.\frac{b-a}{m}+L_2.\frac{d-c}{n}\right)$$
$$\le(1+\gamma).\left[\frac{(b-a)(d-c)}{4}\left(L_1.\frac{b-a}{m}+L_2.\frac{d-c}{n}\right)\right],\quad i=\overline{1,m},\ j=\overline{1,n}.$$

By induction for $k\in\mathbb{N}^*,\ m\ge 3$, we have:

$$D(z_k(x_i,y_j),\overline{z_k}(x_i,y_j)) \le$$
$$\le\left(1+\gamma+\cdots+\gamma^{k-1}\right).\left[\frac{(b-a)(d-c)}{4}\left(L_1.\frac{b-a}{m}+L_2.\frac{d-c}{n}\right)\right]$$
$$=\left(\frac{1-\gamma^{k-1}}{1-\gamma}\right).\left[\frac{(b-a)(d-c)}{4}\left(L_1.\frac{b-a}{m}+L_2.\frac{d-c}{n}\right)\right]$$
$$\le\left[\frac{(b-a)(d-c)}{4(1-\gamma)}\left(L_1.\frac{b-a}{m}+L_2.\frac{d-c}{n}\right)\right],\quad i=\overline{1,m},\ j=\overline{1,n}.$$

Therefore, in view of $\gamma<1$ we obtain:

$$D^*(z_k,\overline{z_k}) \le \frac{(b-a)(d-c)}{4(1-\gamma)}\left(L_1.\frac{b-a}{m}+L_2.\frac{d-c}{n}\right). \tag{18}$$

The assertion follows from (15), (16) and (18).

Remark 3 Since $\gamma < 1$, it is easy to show that

$$\lim_{m,n,k\to\infty} D^*(z, \overline{z_m}) = 0$$

This result confirms that the proposed method is convergent.

7 Numerical Examples

In this section, we provide an example in order to demonstrate the validity and accuracy of the proposed method. In order to explain the maximum absolute error between exact and approximate solution of the method at the mesh points we introduce the following notations:

$$e_-^r = \max |z_-^r(x_i, y_j) - \overline{z_-^r}(x_i, y_j)| \quad i = \overline{1, m},\ j = \overline{1, n},$$
$$e_+^r = \max |z_+^r(x_i, y_j) - \overline{z_+^r}(x_i, y_j)| \quad i = \overline{1, m},\ j = \overline{1, n},$$

where $\overline{z}$ is the approximation of z and $x_i = \frac{i}{m},\ i = \overline{1, m},\quad y_j = \frac{j}{n},\ j = \overline{1, n}$. Table 1 includes the maximum errors for different values of *m,n,k*.

Example 1 Consider the following nonlinear two dimensional fuzzy Fredholm-Volterra integral equation:

$$z(x, y) = f(x, y) \oplus$$

$$(FR)\int_0^1 (FR)\int_0^x H(x, y, s, t) \odot (z(s, t))^2 ds dt,$$

where $(x, y)t \in [0, 1]^2$ and

$$H(x, y, s, t) = \frac{te^{x-2s}}{10 + xy}, \qquad x, y, s, t \in [0, 1],$$

$$f_-^r(x, y) = rye^x - \frac{r^2 xe^x}{4(10 + xy)}, \quad x, y, r \in [0, 1],$$

$$f_+^r(x, y) = (2 - r)ye^x - \frac{(2 - r)^2 xe^x}{4(10 + xy)}, \quad x, y, r \in [0, 1],$$

the exact solution is

$$\left(z_-^r(x, y), z_+^r(x, y)\right) = \left(rye^x, (2 - r)ye^x\right).$$

Table 1 Maximum absolute errors for Example 1 on the certain level sets

	m = n = 10,	**k = 5**	**m = n = 16,**	**k = 5**	**m = n = 30,**	**k = 5**
r	e_-^r	e_+^r	e_-^r	e_+^r	e_-^r	e_+^r
0.25	7.143E−5	1.467E−2	4.741E−5	1.332E−2	3.640E−5	1.270E−2
0.50	4.223E−4	9.258E−3	3.242E−4	8.291E−3	2.792E−4	7.847E−3
0.75	1.270E−3	5.420E−3	1.044E−3	4.764E−3	9.409E−4	4.463E−3
1	2.851E−3	2.851E−3	2.440E−3	2.440E−3	2.252E−3	2.252E−3

The absolute errors of the approximate solution and the exact solution are shown in Table 1.

8 Conclusion

In this paper, we develop a numerical iterative method for the solution of two-dimensional mixed fuzzy Fredholm-Volterra integral equations of the second kind. This method uses the Picard iterations technique and numerical integration for approximating the iterated integrals based on a quadrature rule involved in each iterative step. Consequently, an iterative scheme that converges to the exact solution of the integral equation is provided. Moreover, we take into account the conditions that ensure the convergence of the sequence of successive approximations to the solution at each point. To prove the existence of a unique solution for the given equation, we have applied the Banach fixed point principle. Also, we have applied the composite trapezoidal rule for the numerical integration of the iterates. One of the advantages of the proposed method compared to other numerical methods such as Galerkin, Nyström methods is that it does not need to solve a system of algebraic equations, which often is ill-conditioned and, thus, difficult to solve. We have considered a numerical example in order to test the obtained theoretical results and to show the error between the approximate solution and the exact solution in some mesh points.

References

1. Abbasbandy, S., Allahviranloo, T.: The Adomian decomposition method applied to the fuzzy system of Fredholm integral equations of the second kind. Int. J. Uncertain Fuzziness Knowl. Based Syst. **14**, 101–110 (2006)
2. Abbasbandy, S., Babolian, E., Alavi, M.: Numerical method for solving linear Fredholm fuzzy integral equations of the second kind. Chaos Solitons Fractals **31**(1), 138–146 (2007)
3. Akhavan Zakeri, K., Ziari, S., Fariborzi Araghi, M.A., Perfilieva, I.: Efficient numerical solution to a bivariate nonlinear fuzzy Fredholm integral equation. IEEE Trans. Fuzzy Syst. (2019). https://doi.org/10.1109/TFUZZ.2019.2957100

4. Anastassiou, G.A.: Fuzzy Math.: Approx. Theory. Springer, Berlin (2010)
5. Attari, H., Yazdani, Y.: A computational method for for fuzzy Volterra-Fredholm integral equations. Fuzzy Inf. Eng. **2**, 147–156 (2011)
6. Baghmisheh, M., Ezzati, R.: Numerical solution of nonlinear fuzzy Fredholm integral equations of the second kind using hybrid of block-pulse functions and Taylor series. Adv. Diff. Equ. (2015). https://doi.org/10.1186/s13662-015-0389-7
7. Baghmisheh, M., Ezzati, R.: Error estimation and numerical solution of nonlinear fuzzy Fredholm integral equations of the second kind using triangular functions. J. Intell. Fuzzy Syst. **30**(2), 639–649 (2016)
8. Balachandran, K., Parkash, P., Kanagarajan, K.: Existence of solutions of nonlinear fuzzy Volterra-Fredholm integral equations. Ind. J. Pure Appl. Math. **3**, 329–343 (2002)
9. Balachandran, K., Kanagarajan, K.: Existence of solutions of general nonlinear fuzzy Volterra-Fredholm integral equations. J. Appl. Math. Stoch. Anal. **3**, 333–343 (2005)
10. Barkhordari Ahmadi, M., Khezerloo, M.: Fuzzy bivariate Chebyshev method for solving fuzzy Volterra-Fredholm integral equations. Int. J. Ind. Math. **3**(2), 67–77 (2011)
11. Bede, B., Gal, S.G.: Quadrature rules for integrals of fuzzy-number-valued functions. Fuzzy Sets Syst. **145**, 359–380 (2004)
12. Behzadi, S.S., Allahviranloo, T., Abbasbandy, S.: The use of fuzzy expansion method for solving fuzzy linear Volterra-Fredholm integral equations. J. Intell. Fuzzy Syst, **26**(4), 1817–1822 (2014)
13. Bica, A.M.: Error estimation in the approximation of the solution of nonlinear fuzzy Fredholm integral equations. Inf. Sci. **178**, 1279–1292 (2008)
14. Bica, A.M., Popescu, C.: Approximating the solution of nonlinear Hammerstein fuzzy integral equations. Fuzzy Sets Syst. **245**, 1–17 (2014)
15. Bica, A.M., Popescu, C.: Iterative numerical method for nonlinear fuzzy Volterra integral equations. J. Intell. Fuzzy Syst. **32**(3), 1639–1648 (2017)
16. Bica, A.M., Popescu, C.: Fuzzy trapezoidal cubature rule and application to two-dimensional fuzzy Fredholm integral equations. Soft Comput. **21**(5), 1229–1243 (2017)
17. Bica, A.M., Ziari, S.: Iterative numerical method for solving fuzzy Volterra linear integral equations in two dimensions. Soft Comput. **21**(5), 1097–1108 (2017)
18. Bica, A.M., Ziari, S.: Open fuzzy cubature rule with application to nonlinear fuzzy Volterra integral equations in two dimensions. Fuzzy Sets Syst. **358**, 108–131 (2018)
19. Dubois, D., Prade, H. (1987). Fuzzy numbers: an overview. In: Analysis of Fuzzy Information, vol. 1, pp. 3–39 CRC Press, BocaRaton
20. Ezzati, R., Ziari, S.: Numerical solution and error estimation of fuzzy Fredholm integral equation using fuzzy Bernstein polynomials. Aust. J. Basic. Appl. Sci. **5**(9), 2072–2082 (2011)
21. Ezzati, R., Ziari, S.: Numerical solution of nonlinear fuzzy Fredholm integral equations using iterative method. Appl Math. Comput. **225**, 33–42 (2013a)
22. Ezzati, R., Ziari, S.: Numerical solution of two-dimensional fuzzy Fredholm integral equations of the second kind using fuzzy bivariate Bernstein polynomials. Int. J. Fuzzy Syst. **15**(1), 84–89 (2013b)
23. Ezzati, R., Sadatrasoul, S.M.: Application of bivariate fuzzy Bernstein polynomials to solve two-dimensional fuzzy integral equations. Soft Comput. **21**(14), 3879–3889 (2017)
24. Fariborzi Araghi, M.A., Parandin, N.: Numerical solution of fuzzy Fredholm integral equations by the Lagrange interpolation based on the extension principle. Soft Comput. **15**, 2449–2456 (2011)
25. Friedman, M., Kandel, A., Ming, M.: Solutions to fuzzy integral equations with arbitrary kernels. Int. J. Approx. Reason **20**, 249–262 (1999b)
26. Gal, S.G.: Approximation theory in fuzzy setting. In: Anastassiou, G.A. (ed.) Handbook of Analytic-Computational Methods in Applied Mathematics, pp. 617–666 (2000). Chapman & Hall/CRC Press, Boca Raton (Chapter 13)
27. Kaleva, O.: Fuzzy differential equations. Fuzzy Sets Syst. **24**, 301–317 (1987)
28. Karamseraji, S., Ezzati, R., Ziari, S.: Fuzzy bivariate triangular functions with application to nonlinear fuzzy Fredholm-Volterra integral equations in two dimensions. Soft Comput. **24**, 9091–9103 (2020)

29. Mokhtarnejad, F., Ezzati, R.: The numerical solution of nonlinear Hammerstein fuzzy integral equations by using fuzzy wavelet like operator. J. Intell. Fuzzy Syst. **28**(4), 1617–1626 (2015)
30. Molabahrami, A., Shidfar, A., Ghyasi, A.: An analytical method for solving linear Fredholm fuzzy integral equations of the second kind. Comput. Math. Appl. **61**, 2754–2761 (2011)
31. Mordeson, J., Newman, W.: Fuzzy integral equations. Inf. Sci. **87**, 215–229 (1995)
32. Nieto, J.J., Rodriguez-Lopez, R.: Bounded solutions of fuzzy differential and integral equations. Chaos Solitons & Fractals **27**, 1376–1386 (2006)
33. Nouriani, H., Ezzati, R.: Numerical solution of two-dimensional linear fuzzy Fredholm integral equations by the fuzzy Lagrange interpolation. Adv. Fuzzy Syst. **2018**, 8 (2018)
34. Nouriani, H., Ezzati, R.: Application of fuzzy bicubic splines interpolation for solving two-dimensional linear fuzzy Fredholm integral equations. Int. J. Ind. Math. **11**, 131–142 (2019)
35. Parandin, N., Fariborzi Araghi, M.A.: The numerical solution of linear fuzzy Fredholm integral equations of the second kind by using finite and divided differences methods. Soft Comput. **15**, 729–741 (2010)
36. Park, J.Y., Han, H.K.: Existence and uniqueness theorem for a solution of fuzzy Volterra integral equations. Fuzzy Sets Syst. **105**, 481–488 (1999a)
37. Park, J.Y., Jeong, J.U.: A note on fuzzy integral equations. Fuzzy Sets Syst. **108**, 193–200 (1999b)
38. Park, J.Y., Jeong, J.U.: On the existence and uniqueness of solutions of fuzzy Volterra-Fredholm integral equations. Fuzzy Sets Syst. **115**, 425–431 (2000)
39. Park, J.Y., Lee, S.Y., Jeong, J.U.: The approximate solutions of fuzzy functional integral equations. Fuzzy Sets Syst. **110**, 79–90 (2000)
40. Rivaz, A., Yousefi, F., Salehinejad, H.: Using block pulse functions for solving two-dimensional fuzzy Fredholm integral equations of the second kind. Int. J. Appl. Math. **25**(4), 571–582 (2012)
41. Sadatrasouland, S.M., Ezzati, R.: Quadrature rules and iterative method for numerical solution of two-dimensional fuzzy integral equations. Abs. Appl. Anal. (2014) Article ID 413570
42. Sadatrasoul, S.M., Ezzati, R.: Iterative method for numerical solution of two-dimensional nonlinear fuzzy integral equations. Fuzzy Sets Syst. **280**, 91–106 (2015)
43. Sadatrasouland, S.M., Ezzati, R.: Numerical solution of two-dimensional nonlinear Hammerstein fuzzy integral equations based on optimal fuzzy quadrature formula. J. Comput. Appl. Math. **292**(15), 430–446 (2016)
44. Samadpour Khalifeh Mahaleh, V., Ezzati, R.: Numerical solution of linear fuzzy Fredholm integral equations of second kind using iterative method and midpoint quadrature formula. J. Intell. Fuzzy Syst. **33**, 1293–1302 (2017)
45. Samadpour Khalifeh Mahaleh, V., Ezzati, R.: Numerical solution of two dimensional nonlinear fuzzy Fredholm integral equations of second kind using hybrid of Block-Pulse functions and Bernstein polynomials. Filomat **32**, 4923–4935 (2018)
46. Seikkala, S.: On the fuzzy initial value problem. Fuzzy Sets Syst. **24**, 319–330 (1987)
47. Seifi, A., Lotfi, T., Allahviranloo, T.: A new efficient method using Fibonacci polynomials for solving of first-order fuzzy Fredholm-Volterra integro-differential equations. Soft Comput. **23**, 9777–9791 (2019)
48. Vali, M.A., Agheli, M.J., Gohari Nezhad, G.: Homotopy analysis method to solve two-dimensional fuzzy Fredholm integral equation. Gen. Math. Notes **22**(1), 31–43 (2014)
49. Wu, C., Song, S., Wang, H.: On the basic solutions to the generalized fuzzy integral equation. Fuzzy Sets Syst. **95**, 255–260 (1998)
50. Ziari, S., Bica, A.M.: New error estimate in the iterative numerical method for nonlinear fuzzy Hammerstein-Fredholm integral equations. Fuzzy Sets Syst. **295**, 136–152 (2016)
51. Ziari, S., Ezzati, R., Abbasbandy, S.: Numerical solution of linear fuzzy Fredholm integral equations of the second kind using fuzzy Haar wavelet. In: Greco, S., Bouchon-Meunier, B., Coletti, G., Fedrizzi, M., Matarazzo, B., Yager, R.R. (eds.) Advances in Computational Intelligence. IPMU 2012. Communications in Computer and Information Science, vol 299. Springer, Berlin, Heidelberg (2012). https://doi.org/10.1007/978-3-642-31718-7_9
52. Ziari, S., Ezzati, R. (2016). Fuzzy block-pulse functions and its application to solve linear fuzzy Fredholm integral equations of the second kind. In: Carvalho, J., Lesot, M.J., Kaymak,

U., Vieira, S., Bouchon-Meunier, B., Yager, R. (eds.) Information Processing and Management of Uncertainty in Knowledge-Based Systems. IPMU 2016. Communications in Computer and Information Science, vol. 611. Springer, Cham. https://doi.org/10.1007/978-3-319-40581-0_67

53. Ziari, S.: Iterative method for solving two-dimensional nonlinear fuzzy integral equations using fuzzy bivariate block-pulse functions with error estimation. Iran J. Fuzzy Syst. **15**(1), 55–76 (2018)
54. Ziari, S.: Towards the accuracy of iterative numerical methods for fuzzy Hammerstein-Fredholm integral equations. Fuzzy Sets Syst. **375**, 161–178 (2019)
55. Ziari, S., Perfilieva, I., Abbasbandy, S.: Block-pulse functions in the method of successive approximations for nonlinear fuzzy Fredholm integral equations. Differ. Equ. Dyn. Syst. (2019). https://doi.org/10.1007/s12591-019-00482-y
56. Ziari, S., Bica, A.M.: An iterative numerical method to solve nonlinear fuzzy Volterra-Hammerstein integral equations. J. Intell. Fuzzy Syst. **37**(5), 6717–6729 (2019)

The manufacturer's authorised representative in the EU is Springer Nature Customer Service Centre GmbH, Europaplatz 3, 69115 Heidelberg, Germany. If you have any concerns regarding our products, please contact ProductSafety@springernature.com

Printed and bound by CPI Group (UK) Ltd, Croydon, CR0 4YY
30/09/2026
02225329-0005